Process Analytical Chemistry

Process Analytical Chemistry

Edited by

F. McLENNAN
ZENECA Fine Chemicals Manufacturing Organisation
Huddersfield
UK

and

B. KOWALSKI
CPAC
University of Washington
Seattle
Washington
USA

BLACKIE ACADEMIC & PROFESSIONAL
An Imprint of Chapman & Hall

London · Glasgow · Weinheim · New York · Tokyo · Melbourne · Madras

Published by
Blackie Academic and Professional, an imprint of Chapman & Hall,
Wester Cleddens Road, Bishopbriggs, Glasgow G64 2NZ

Chapman & Hall, 2–6 Boundary Row, London SE1 8HN, UK

Blackie Academic & Professional, Wester Cleddens Road, Bishopbriggs, Glasgow G64 2NZ, UK

Chapman & Hall GmbH, Pappelallee 3, 69469 Weinheim, Germany

Chapman & Hall USA, 115 Fifth Avenue, Fourth Floor, New York NY 10003, USA

Chapman & Hall Japan, ITP-Japan, Kyowa Building, 3F, 2-2-1 Hirakawacho, Chiyoda-ku, Tokyo 102, Japan

DA Book (Aust.) Pty Ltd, 648 Whitehorse Road, Mitcham 3132, Victoria, Australia

Chapman & Hall India, R. Seshadri, 32 Second Main Road, CIT East, Madras 600 035, India

First edition 1995

©1995 Chapman & Hall

Typeset in 10/12 pt Times by Pure Tech Corporation, Pondicherry, India
Printed in Great Britain by St Edmundsbury Press, Bury St Edmunds, Suffolk

ISBN 0 7514 0038 6

A catalogue record for this book is available from the British Library
Library of Congress Catalog Card Number: 95–75554

∞ Printed on acid-free text paper, manufactured in accordance with ANSI/NISO Z39.48-1992 (Permanence of Paper)

Preface

Traditional methods for controlling chemical manufacturing processes have relied exclusively on the measurement of temperature, pressure and flow rate. Only when more information was essential for the safe operation of a plant would the addition of other types of process analysers be considered. Measurement of oxygen in the manufacture of ethylene is a case in point, since oxygen can lead to runaway reactions and the loss of lives and equipment.

More recently, the manufacture of new polymers, materials and other complex products has demanded more timely composition data in order to ensure that the highest possible quality product be made at the lowest possible cost. Better process control with the use of detailed, real-time chemical measurements has become the key to lowering quality costs, i.e. costs associated with reprocessing, destroying or selling off-spec material. Quality costs in chemical and materials manufacturing are estimated to be ten per cent of sales!

Sophisticated on-line and in-line chemical analyses are also required when it is necessary to determine not only product composition, but also product performance during manufacturing. For example, octane numbers for gasoline, and several other performance parameters for all fuels, are today determined on-line during blending from near infrared spectral data analysed by multivariate calibration methods. Another application involves spectral data acquired during polymerization processes to predict quality parameters such as hardness, elongation or dyeability of the polymer product.

Finally, recent environmental regulations require data on aspects such as impurities, solvents and wastewater, to ensure that chemical manufacturing is safe for workers, for communities near chemical plants, and for the environment. These demands for real-time quantitative chemical information on a growing list of manufacturing processes present new challenges to analytical chemists, instrument engineers and plant supervisors.

In response to these needs, the Center for Process Analytical Chemistry was established in 1984 at the University of Washington to work with industry to identify, prioritize and address generic needs in the newly emerging area of Process Analytical Chemistry. Since then a journal (*Process Control and Quality*) has been introduced, several International Forums on Process Analytical Chemistry (IFPAC) have been held, and in 1993 the *Application Reviews* issue of *Analytical Chemistry* contained the first review on the field, authored by chemists from Dow Chemical Company.

Although a few books are available on process analysers, these focus primarily on commercially available technology. This is the first book on *Process Analytical Chemistry* to cover the present and future of this new field. Our international team of contributors has been brought together from academia, equipment suppliers and different sectors of the chemical industry, to produce a volume which covers a list of topics currently under development as well as in real-life applications. Written for a broad range of scientists and engineers educated in the physical sciences and working in the chemical and allied industries, the book can also be used as the basis for a course at advanced undergraduate or graduate level. Some familiarity with standard laboratory chemical analysis is assumed. The editors hope that the book will provide the basis for more academic involvement in the field. The future of analytical chemistry calls for a partnership between analytical laboratories filled with the most sophisticated instrumentation and in-field chemical sensors/analysers capable of long-term, maintenance-free operation even in the most hostile environments.

Beneficiaries of this book include students and practitioners of analytical chemistry, process engineering, plant supervision and control/intelligence. The book opens the opportunity for analytical chemists to work closely with chemical engineers to design, build and operate safer and more efficient manufacturing processes for the present and future. The editors see this possibility as the key to manufacturing excellence as we move into the 21st century.

F. McLennan
B.R. Kowalski

Contributors

K.N. Andrew Department of Environmental Sciences, University of Plymouth, Drake Circus, Plymouth, Devon PL4 8AA, UK

R.M. Belchamber Process Analysis & Automation, Falcon House, Fernhill Road, Farnborough, Hampshire GU14 9RX, UK

L.E.A. Berlouis Department of Pure and Applied Chemistry, University of Strathclyde, 295 Cathedral Street, Glasgow G1 1XL, UK

K.S. Booksh Department of Chemistry and Biochemistry, The University of South Carolina, Columbia, SC 29208, USA

J.J. Breen US Environmental Protection Agency, Office of Pollution Prevention & Toxics, Economics, Exposure and Technology Division, Industrial Chemistry Branch, Washington DC 20460, USA

L.W. Burgess Center for Process Analytical Chemistry, University of Washington, Seattle, WA 98195, USA

W.P. Carey Center for Process Analytical Chemistry, University of Washington, Seattle, WA 98195, USA

K.J. Clevett Clevett Associates Inc., 26 Upper Drive, PO Box 7047, Watchung, NJ 07060, USA

G.L. Combs Applied Automation, Inc./Hartmann & Braun, PO Box 9999, Bartlesville, OK 74005-9999, USA

R.L. Cook Applied Automation, Inc./Hartmann & Braun, PO Box 9999, Bartlesville, OK 74005-9999, USA

N.C. Crabb Process Technology Department, ZENECA Fine Chemicals Manufacturing Organisation, North of England Works, Huddersfield, West Yorkshire HD2 1FF, UK

J.J. Gunnell Exxon Chemical Olefins Inc., Fife Ethylene Plant, Mossmorran, Cowdenbeath, Fife KY4 8EP, UK

M.L. Hitchman   Department of Pure and Applied Chemistry, University of Strathclyde, 295 Cathedral Street, Glasgow G1 1XL, UK

P.W.B. King   ICI Engineering Technology, The Heath, Runcorn, Cheshire WA7 4QF, UK

B.R. Kowalski   Center for Process Analytical Chemistry, University of Washington, Mail Stop BG10, Seattle, Washington 98195, USA

E.A. McGrath   Center for Process Analytical Chemistry, University of Washington, Seattle, WA 98195, USA

P. MacLaurin   Process Technology Department, ZENECA Fine Chemicals Manufacturing Organisation, North of England Works, Huddersfield, West Yorkshire HD2 1FF, UK

F. McLennan   ZENECA Fine Chemicals Manufacturing Organisation, North of England Works, Leeds Road, Huddersfield, West Yorkshire HD2 1FF, UK

B.M. Wise   Eigenvector Technologies, 4154 Laurel Drive, West Richland, Washington 99352, USA

P.J. Worsfold   Department of Environmental Sciences, University of Plymouth, Drake Circus, Plymouth, Devon PL4 8AA, UK

Contents

4 Process chromatography

G.L. COMBS and R.L. COOK

9   Environmental monitoring for a sustainable future

From pollution control to pollution prevention   313

E.A. McGRATH, K.S. BOOKSH and J.J. BREEN

10   Non-invasive techniques   329

R.M. BELCHAMBER

11 Future developments of chemical sensor and analyzer systems   **353**
W.P. CAREY and L.W. BURGESS

1 Process analytical chemistry in perspective

F. McLENNAN

1.1 Introduction

Process Analytical Chemistry (PAC) is the application of analytical science to the monitoring and control of industrial chemical process [1,2]. This information may be used to both control and optimise the performance of a chemical process in terms of capacity, quality, cost, consistency and waste reduction.

PAC is not new. It has been applied in the petroleum and petrochemical industries since the 1950s but is presently going through a reincarnation and is a rapidly developing field in all areas of chemical production – petroleum, fine chemicals, commodity chemicals, petrochemicals, biotechnology, food, pharmaceuticals, etc. being fuelled by technological advances in analytical chemistry together with changing needs within the chemical industry.

In a traditional chemical manufacturing plant, samples are taken from reaction areas and transported to the analytical laboratory which is typically centralised. Here the samples are analysed by highly qualified technical staff using state-of-the-art equipment producing results typically in a few hours to a few days. Such analysis is generally used retrospectively to measure process efficiency, to identify materials which need to be reworked or discarded or in a multistage batch synthesis to assess the charge for the next stage. Where these results are critical to the continuation of the process, the process is usually designed to accommodate this time delay giving rise to longer cycle times and reduced plant utilisation.

Process control in this environment is effected by an experimental correlation of physical parameters during the process such as flow rates, times, temperatures, pressures with chemical composition, quality and yield of the derived material followed by subsequent control of these physical parameters.

Implementation of PAC dramatically changes this scene. PAC analysers are situated either in or immediately next to the manufacturing process. They are designed to withstand the rigours of a manufacturing environment and to give a high degree of reliability. They are operated either automatically or by non-technical staff such as process operatives and produce real or near-real-time data which can be used for process control and optimisation.

The move towards PAC has been fuelled by two developments. Firstly, increasing international competitiveness within the chemical industry has lead to the widespread adoption of 'right first time' and 'just in time' approaches to manufacturing and quality. This has placed the emphasis on building quality into all stages of the process, increased manufacturing flexibility, reduced inventory and improved control of processes. Secondly, during the past decade advances in analytical chemistry and in particular the development of the microcomputer and improved algorithms for data handling, have enabled almost instantaneous generation of information.

Moving from a traditional analysis approach to a PAC approach is not easy, not only does it require significant technical developments but it also requires a 'cultural' change. This change needs to be embraced not only by the analyst community, but also by manufacturing, R&D and engineering, etc. This change process requires a 'champion' or better still a number of champions at both the managerial and technical levels in order to be successful.

Figure 1.1 outlines the key differences between the traditional and PAC approaches to process control.

While this chapter and most of this book will concentrate on the role of in-plant analysis, this does not mean that there is not a place for the specialised analytical laboratory and it is the author's belief that an integrated approach to process analysis is essential to meet all the needs of a modern chemical plant [3, 4]. It also does not mean that manufacturing processes are the only processes to benefit from moving analytical chemistry from its centralist role to the in-situ or distributed role. The human body, the air over a city and a mountain lake all represent complex chemical processes, the study of which would benefit from on-line, real-time and even non-invasive analysis.

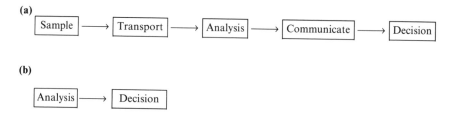

Figure 1.1 (a) Traditional approach to process control. Analysis employs technical staff, high-tech equipment and typically takes several hours. (b) Process analytical chemistry approach to process control. Analysis is either automatic or employs non-technical staff, utilises rugged and reliable equipment and takes seconds or minutes.

1.2 Terminology

For the purpose of this volume, we propose to use the following definitions (Figure 1.2).

Off-line analysis: This involves manual removal of the sample, transport to the measurement instrument which is located in a specialised central laboratory using highly qualified technical staff. This is typified by relatively low sample frequency, complex sample preparation, flexible and complex analysis. The advantages of this approach arise from the economy of sharing expensive instruments and skilled staff.

At-line analysis: Many of the deficiencies of off-line analysis—time delay, administration costs, prioritisation of work—may be addressed by carrying out the analysis at-line. This still involves manual sampling but in this case the measurement is carried out on a dedicated analyser by the process operative. At-line analysis is usually accompanied by significant method development work to simplify the sample preparation and to modify the measurement technique to permit the use of robust, reliable instrumentation. It is a mistake to simply transfer the laboratory analysis to the plant floor – time and effort spent in the evaluation of what information is required to control the process invariably leads to the development of a more robust solution.

On-line analysis: We use this definition to describe all examples of fully automated analyser systems. Other authors have subdivided this further into on-line, in-line and non-invasive analysis but we will consider all these as one group.

Table 1.1 highlights the pros and cons of each of these approaches.

Table 1.1 The pros and cons of each approach

	Advantages	Disadvantages
Off-line	Expert analysts available. Flexible operation. Controlled environment. Sophisticated instrumentation. Low unit costs/test.	Slow. Lack of ownership of data. Conflicts of priorities. Addition admin costs.
At-line	Dedicated instrument. Faster sampling process. Simpler instrumentation. Ownership of data by production personnel. Control of priorities.	Low equipment utilisation. Equipment needs to be robust to cope with production environment.
On-line	Fast. Automatic feedback possible. Dedicated analyser.	Minimum downtime required. Long/expensive method development. 24 h troubleshooting/maintenance resource required. Electrical classification required.

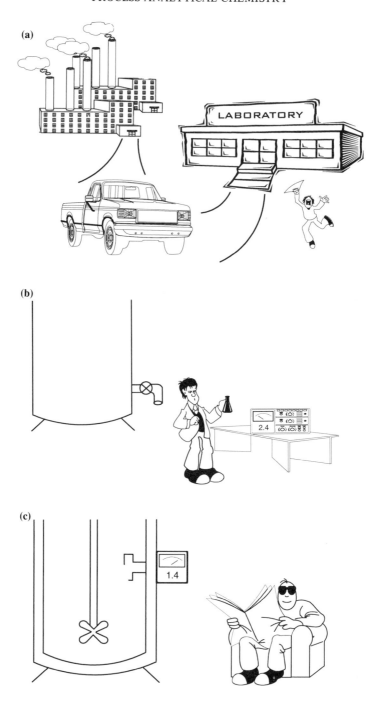

Figure 1.2 (a) Off-line, (b) at-line, and (c) on-line analysis.

1.3 An integrated approach

PAC is a truly multidisciplinary science. The aim of each project should be to define the optimum process control, measurement and analysis philosophy for the process. This key stage requires contributions from the analytical chemist, measurement engineer, control engineer, process development chemist, chemical engineer and production manager. In the past we have had on-line analysis teams and off-line analysis teams, often competing with each other and nobody looking at the important area of at-line analysis.

While the move from off-line to in-plant analysis can be retrofitted to a process and bring substantial benefits, experience has shown that the biggest impact comes when PAC is integrated with the process development and plant design (e.g. rapid on-line or at-line analysis may allow efficient manufacture without intermediate storages). In the former case, much of the in-built inefficiency (i.e. to cope with time consuming off-line analysis) cannot be recouped. An effective way of getting in at a sufficiently early stage, is to apply the PAC philosophy (especially on-line) to process research and development. Substantial gains in development time, process cycle times, materials efficiency and energy efficiency are achievable even where such work does not lead to process-scale PAC projects.

On a project by project basis, PAC should be problem rather than technique driven. It is nonetheless essential that high level research and development is undertaken to enhance the range of techniques available. This is especially true for on-line analysis where the application of even well established laboratory techniques may offer huge challenges. In addition to instrument development, the application of commercially available systems to specific process problems can also be complex. A progressive PAC organisation needs high calibre staff and support from senior management to ensure that project implementation against today's problems is undertaken in parallel with longer term research to solve tomorrow's problems. Such research should make full use of existing company research infrastructure; the PAC specialist cannot be an expert in all analytical techniques. It is therefore important that 'traditional' researchers embrace the PAC philosophy to some extent.

Great progress in PAC has clearly been made over the past 5–10 years. Most of the R&D has been undertaken by the world's major chemical companies; it was not easy to follow developments since much of the work is deemed proprietary and confidential. Over the years, this has been supported by relatively small 'niche' instrument companies, close to particular sectors of the chemical industry. More recently, the major suppliers of analytical instrumentation have become interested in PAC. The strategic alliance between Perkin Elmer and Dow Chemical, to develop and commercialise Dow's PAC expertise, is a good example.

The standing of PAC as an academic discipline has grown enormously in recent times. To a considerable extent, this has been catalysed by the creation of the Centre for Process Analytical Chemistry (CPAC) at the University of Washington in 1984. This industry/academic collaboration continues to thrive with around 50 industrial funding partners. Similarly, the Measurement and Control Engineering Centre at the University of Tennessee is an industry/university collaboration. Other academic institutions are active in PAC to some extent and in some cases small specialist groups have been established particularly in The Netherlands, and the UK.

1.4 Technology and instrumentation

All laboratory techniques can be applied to process analysis. The only limitation is the cost and time involved in making the technology safe and robust. The remaining chapters of this volume deal with detailed descriptions of the key technologies involved in PAC. What is presented here is a brief outline of trends and future technologies as perceived by the author.

PAC has its origins in the Petrochemicals and Petroleum industry with large-scale continuous processes where extreme pressures on unit production costs required the development of on-line analysers for control and feedback. This led to the development of physical property analysers (see chapter 3) and the implementation of on-line gas chromatographs and mass spectrophotometers.

Since these pioneering days, significant developments have greatly improved analyser robustness, reliability and performance.

In the past 10 years, we have seen a minor revolution in the technology and instrumentation available to the process analytical community. This has come about by the development of new laboratory analytical techniques, improvements in electronics and the development of chemometrics software.

During the past decade, we have probably seen the most significant development in the area of spectroscopy. Infrared (IR) spectroscopy both in the mid-IR (4000–400 cm^{-1}) and near-IR (0.8–2.5 µm) ranges which offer high-speed full-spectrum scanning, have been considerably enhanced through the use of chemometrics together with the development of fibre optic and ATR probes. Mass spectrometry continues to improve and recent applications have included multistream, multicomponent analysis particularly for environmental monitoring and control. Other areas where we would expect major developments in the future are Raman spectroscopy which has enormous potential as a fast, multicomponent, non-invasive technique. Process nuclear magnetic resonance

(NMR) instruments are becoming available and this area is likely to grow.

Gas chromatography (GC) was one of the first techniques developed for the measurement of individual components in mixtures. The relatively poor chromatographic efficiency effected by packed GC columns led to the development techniques such as column switching and back flushing to improve separation efficiency. More recently, the introduction of capillary columns has provided the required separation power with the added benefit of significantly reduced method development time. Other enhancements have come from improved electronics, component engineering, detector sensitivity and stability and temperature control and programming.

High-performance liquid chromatography (HPLC) continues to be the Cinderella of process analytical techniques. Numerous applications have been reported of at-line HPLC but although there seems to be enormous potential for the technique, particularly within the Pharmaceutical and Fine Chemicals sectors, its applications to on-line analysis appears to be limited by robustness and reliability. Since, historically, the development of process analysers has lagged behind laboratory analysers by 5–10 years, we would expect to see applications of supercritical fluid chromatography and capillary zone electrophoresis appear before the end of this decade.

Flow injection analysis still offers much untapped potential as a means of automating 'wet chemical' methods but recent improvements in autotitrimetry have led to the implementation of this technology both at-line and on-line within the author's manufacturing plants.

The market for process analysis instrumentation is growing rapidly worldwide. A recent paper by Clevett [5] has indicated that the demand for process analytical instrumentation throughout the EEC will grow by $\approx 40\%$ throughout the period 1992–1997. Much of this will be in the water and gas analyser area fuelled by both new and stricter enforcement of environmental regulations. The complex process analyser market is seen to grow by about 8% per annum over this period as shown in Figure 1.3.

Despite these advances in technology and instrumentation, many applications are bedevilled by the complexity of presenting a representative sample of the process stream to the analyser.

The difficulty of this part of the process must never be underestimated be it for off-line, at-line or on-line analysis. The technical problem in sampling anything other than a homogenous gas or liquid are significant. Further developments of sampling systems to handle heterogenous media is urgently required. It has been estimated that 80–90% of all problems with process analysis systems involve sample handling.

This explains why techniques which do not require sampling such as fibre optic spectroscopy and non-invasive techniques are receiving considerable

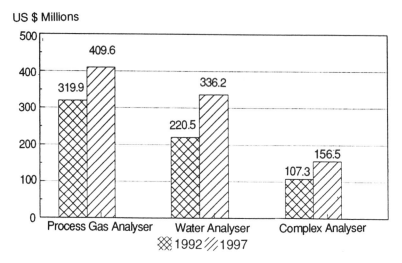

Figure 1.3 The European market for process analytical instruments.

attention. However, this is not a panacea. There needs to be a balance between automation of primary methods often involving extractive sampling and deconvolution of non-specific signals from in-line instruments.

It is also important to note that in-line analysis may also suffer from sampling problems if the probe is incorrectly positioned or the mixing characteristics of the process is not understood.

1.5 PAC maintenance and support

All too often in discussions with colleagues in the chemical industry have we heard of in-plant analysis projects failing to deliver after a period of time. Often this occurs despite clear business objectives having been identified and met and despite excellent pieces of work to identify the appropriate technical solution. The reason for this failure is invariably not a failure to implement the project successfully but a failure to operate, maintain or keep the system up to date. Often this is due to a failure to identify a change in the production requirement but more generally it is due to a lack of adequate operational support.

The activities required for the success of a process analysis solution are outlined in Figure 1.4. Steps 2–6 are the project phase usually led by a technical function and assuming an appropriate level of technical competence are usually carried out successfully. In many organisations, the involvement of the technical functions ends here and the system is 'handed over' to production. This is usually a recipe for disaster. There is an ongoing need for a high level of technical involvement.

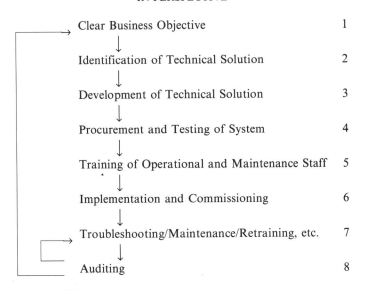

Figure 1.4 The activities required for the success of a process analysis solution.

Analyser systems invariably do not work forever. Typical downtimes in a laboratory vary from 0.1 to 5% depending on the complexity of the analyser system and the ability of the technicians to identify and solve the problem. In the much less controlled environment of a chemical production plant, despite attempts during the development phase to develop robust systems, downtime will rarely be less. There is a need to have a well thought out calibration, preventative maintenance and quality assurance systems in place together with access to appropriate troubleshooting and breakdown maintenance resources.

New staff (both operational and maintenance) are invariably recruited within production and they require the same degree of training as was given prior to project implementation. Finally, the overall performance of the system requires periodic auditing to ensure compliance.

This can lead to identification of changes to the maintenance and training regime as well as to identify technical problems. Equally importantly it can lead to a change in the measurement requirement which may at the extreme necessitate a complete recycling through many and occasionally all of the steps in Figure 1.4.

1.6 Safety

Installing instrumentation into a process environment raises many complex and diverse issues around safety, which can only be superficially covered here. The nature of the hazards are numerous but the most important and potentially the most devastating problem, is that of the instrument acting as an ignition source in a flammable atmosphere and causing a fire or an explosion.

To produce an explosion or fire requires three constituents; a flammable gas or dust, air and an ignition source. The first two are commonly present in process plant, so it is essential to prevent the third occurring. Not all mixtures of flammable gases and air will generate an explosive mixture, too little or too much of the gas will prevent combustion. These extreme conditions are referred to as the lower and upper explosive limits (LEL and UEL). Almost all finely divided powders will burn in air, these dust explosions have the added complication that a small blast can disturb more dust which will further fuel the flame. The study of the probability of flammable atmosphere has given rise to the area classification system in which process plant is categorised as per the potential hazard. This classification requires an in-depth knowledge of the process and all the risks necessitating a multidisciplinary team of experts. In Europe a plant would be divided into three zones as outlined in Table 1.2. The USA uses a slightly different system based on two divisions, of which division 1 is similar to the European zone 0 and division 2 relates to zones 1 and 2. All areas not classified under a zone are referred to as non-hazardous. For simplicity the following text refers only to the European system, but there are parallels in America and elsewhere in the world.

Table 1.2 Explanation of the zones used in Europe

Zone 0	An explosive gas–air mixture is continuously present
Zone 1	An explosive gas–air mixture is likely to occur in normal operation
Zone 2	An explosive gas–air mixture is unlikely to occur and will only exist for a short duration

Considerable time and expense can be saved if it is possible to locate an instrument in a non-hazardous location. If this is not possible then the equipment must be chosen to be suitable for installation in the relevant zone. Such equipment will have been proven to be safe by an independent body and have been awarded an IEC code [6] dictating the limits of their safe installation.

1.7 Information sources

1.7.1 Literature

Engineering and analytical chemistry are central to sound PAC. Practitioners need to maintain a good general appreciation of these broad areas through regular review of the literature. In addition, many practitioners will have special interests relating to particular techniques and need to access the more specialist literature in those areas. The literature relating specifically to PAC is limited but growing. This text provides an introductory overview while other accounts of PAC can be obtained from books by Clevett [7], Huskins [8] and Mix [9] together with a major review of PAC in *Analytical Chemistry* [10]. For specific applications, literature from instrument manufacturers can be invaluable although such literature, for obvious reasons, tends to stress the positive aspects of the instrumentation. A specialist journal, *Process Control and Quality* (published by Elsevier), was launched in 1991 but the vast majority of papers are distributed throughout many science, technology and engineering publications. Practitioners therefore need to be thorough in literature searching and aware of the diversity of sources.

To illustrate this, a number of databases were searched for PAC papers. The search statement was as follows:

(process (w) analytical (w) chemistry or Process (w) (analysis or monitoring or measurement) or (online or inline or atline)
(w) analysis or online (w) (analyser or analyzer) and PY = year

The top ten databases (based on number of papers) are shown in Table 1.3. Papers per year over the past ten years are shown to illustrate any trends. The exercise was validated by selecting at least 20 titles and abstracts from each database and checking relevance to PAC. Although there were some false hits (e.g. on-line derivatisation prior to chromatographic analysis or business process analysis), most papers were relevant (75–100%) depending on database.

Table 1.3 The top ten databases

Database	Number of papers					Number of papers				
	1984	1985	1986	1987	1988	1989	1990	1991	1992	1993
INSPEC	91	95	87	155	109	133	159	148	143	103
NTIS	24	28	38	38	24	35	26	45	29	37
COMPENDEX PLUS	150	175	134	121	92	123	126	141	164	164
ENERGY SCITECT	60	62	61	95	45	75	52	75	61	89
PASCAL	38	32	22	28	20	26	45	45	67	75
ANALYTICAL ABS	1	2	9	13	9	16	27	28	51	32
CHEMENG & BIOTECH ABS	321	539	573	614	385	318	213	121	137	126
CHEMICAL ABS	38	47	57	49	56	78	82	89	101	88
SCISEARCH	37	33	26	31	40	40	51	117	136	134
PTS PROMPT	35	28	52	89	74	98	101	150	137	121

A number of interesting observations can be made. Firstly, the multidisciplinary nature of PAC and the need for wide searching is confirmed. The practitioner with a background in analytical chemistry, for example, clearly needs to look beyond ANALYTICAL ABSTRACTS!

Databases with a strong engineering flavour (INSPEC, COMPENDEX PLUS, ENERGY SCITEC, CHEMENG & BIOTECH ABS) have generally yielded high numbers of papers over the past ten years. No real trends emerge except the surprising decline in papers from the search of CHEMENG & BIOTECH ABS.

Databases with an emphasis on chemistry or science and technology generally (CHEMICAL ABS, ANALYTICAL ABS, NTIS, SCISEARCH) show substantial increases in the number of PAC-related papers over the past 10 years. This shift towards chemistry is the result of increased analyser complexity. Process measurement and analysis has traditionally been an extension to the role of control engineers. Increasing demands for quality, complex chemical entities and improved environmental performance, however, have resulted in the need for complex analysers often configured for specific proprietary applications. Delivery of such systems is dependent on a marriage of engineering and chemistry – this is becoming evident in the literature.

A final observation is the rise in papers from PTS PROMPT. This database has a commercial focus targeting business literature worldwide. It is interesting and pleasing that PAC awareness outside of the technical community is increasing.

1.7.2 Conferences

Conferences offer invaluable opportunities for learning of new developments, evaluating new instrumentation, informal benchmarking and sharing frustrations! As a discipline, PAC is reasonably well catered for in this area. There are two major international PAC conferences targeting delegates from a range of disciplines. The venue for the biannual Anatech conference alternates between Europe and the USA (details from Elsevier, Amsterdam). The annual International Forum on PAC (IFPAC) is held in the USA (details from InfoScience Services, Northbrook, IL, USA).

In addition to these international conferences, many local events are available including those associated with membership of academic/industry PAC consortia. Specialist conferences in key technology areas with wide PAC applicability (e.g. near-IR, chemometrics) should also be of interest.

References

1. J.B. Callis, D.L. Illman and B.R. Kowalski, *Anal. Chem.* **59** (1987) 624A.
2. M.T. Rube and D.J. Eustace, *Anal. Chem.* **62** (1990) 65A.
3. N.C. Crabb and F. McLennan, *Process Control Qual.* **3** (1992) 229.
4. P. Van Vurren, *Process Control Qual.* **3** (1992) 101.

5. K.J. Clevett, *Indust. Perspect. Control Instrumentation* **26**(3) (1994) 29.
6. BS5345 Part 1. British Standards Institution, London, UK, 1976.
7. K.J. Clevett, *Process Analyzer Technology.* John Wiley & Sons, New York, USA, 1986.
8. D.J. Huskins, *Quality Measuring Instruments in On-line Process Analysis.* Halstead Press, New York, USA, 1982.
9. P.E. Mix, *The Design and Application of Process Analyser Systems.* Wiley Interscience, New York, USA, 1984.
10. Process Analytical Chemistry, *Anal. Chem.* **65**(12) (1993) 199R.

2 Sampling systems

J.J. GUNNELL

2.1 Introduction

In most technologies, the tasks which are routinely tackled by experts in the field can seem all but impossible to a worker in a somewhat related discipline. Such a person can readily imagine the many problems of applying a technology but is not familiar with the solutions to the problems and it can be informative to look at a technology from the viewpoint of just such an educated outsider. Consider for example the thoughts that might go through the mind of an analyst whose normal environment is a traditional laboratory in which measurements are made using bench-top analysers. A typical challenge would be to provide an analysis of the trace components in a stream from an ethane/ethylene splitter tower. The physical conditions are that the process stream is at a pressure of 8–10 bar, the temperature is −100°C and the sample is a fluid which is at its bubble point. Furthermore, the sample is to be taken from a point near to the top of the tower which is about 100 m above the ground and fresh data are required every 5 min in order to control the operation of the tower. The analyst with a laboratory background might readily identify gas chromatography as a suitable measurement technology but then he/she starts to list the difficulties.

1. The stream is at its bubble point. Am I sampling a gas or a liquid? Wouldn't the composition be different if I took a sample which was a gas to if I took one which was a liquid? How can I make sure that the sample which I take fairly represents what's going on in the process?
2. The physical conditions of the sample are that it's at a moderate pressure but a very low temperature. I can't introduce a sample like that into an analyser; it will wreck it. What do I have to do to make the sample compatible with the analyser?
3. They need a new analysis every 5 min. The chromatography will take about four minutes so that doesn't leave me long to extract a sample and get it to the analyser. How can I make sure that the results are timely?
4. The sample is flammable. Some of the components might be poisonous too. How can I carry out the analysis in a safe way?

Also, I'll have to make sure that nothing I do has an adverse effect from the environmental point of view.

5. They need this data right round the clock in order to control the process. That means that however I solve this problem, one thing is for sure: the equipment I design is going to have to be very reliable.

6. It sounds like I'll be extracting quite a lot of sample from the process. That could cost quite a bit if I waste it. Also, the equipment I'm designing could be pretty expensive. Then there's the cost of maintaining it too. I'm going to have to think of ways of making my design cost-effective.

In this way, our laboratory analyst has characterised the job of the designer of process analyser systems. The analysis might be straightforward but the preparation of the sample might require considerable expertise and the main issues to consider are:

- Representativeness
- Compatibility
- Timeliness
- Safety and environmental concerns
- Reliability
- Cost

Given that suitable analysers are available which can make the required measurement, the objective is to design a sample system which can present samples to the analyser in an appropriate form. The requirements of industries which use process analysers are widely different and so this chapter tends not to give detailed information about the components which comprise a system. For example, the requirements for a filter in, say, food processing are rather different to those for a filter in the petrochemical industry. Instead of detailing the components which are suitable for various industrial applications, this chapter outlines the considerations which should be taken into account when selecting components. There are such a variety of detailed requirements and constraints on any individual process measurement that the design of sample systems cannot be prescriptive but must instead be tackled on a case by case basis. The principles of good design are, however, fairly few and the aim of this chapter is to introduce them in a general context. In order to illustrate some principles or to demonstrate how calculations are performed, the analysis of a stream from an ethylene splitter is sometimes used as a convenient example. However, the underlying ideas can be transferred to completely different industries and applications.

Some basic assumptions are made about process analytical chemistry at a manufacturing site and although individual situations might not fit this model, it represents the mainstream of design in many industrial applications. The elements of this philosophy are that:

- There are many process analysers distributed throughout the plant and, when available, process analysers are used in preference to laboratory methods.
- The analysers make measurements round the clock and the data generated are used for automatic control and monitoring of the processes and for certification of products.
- In order to facilitate installation and operation, to minimise costs and to ensure safety, analysers are grouped together in purpose-designed analyser houses which provide a suitable environment for the operation of the analysers and the activities of maintenance staff.
- Data are transmitted back to central locations where they are used by people or computers in order to operate the plant. It is possible to exercise some operations on the analyser system (such as switching stream or making validation checks) from the central locations.

Figure 2.1 shows this philosophy in a schematic way. Adopting this approach has an important impact on the design of any individual process analyser system: because analysers are grouped together it is usually necessary to bring the sample to them. Although the locations of

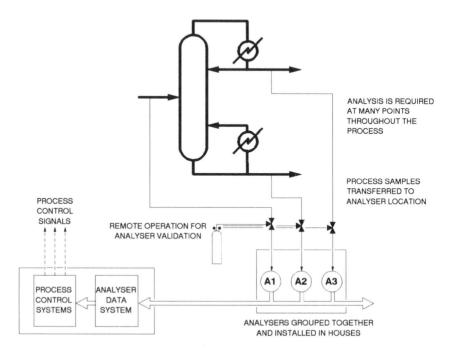

Figure 2.1 Process analysers in an operating plant.

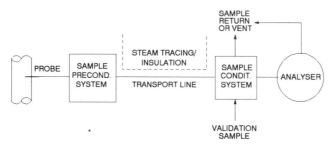

Figure 2.2 General sample system schematic.

analyser houses are selected to minimise the distances from the process to the analysers, necessarily it will often happen that an individual installation will have a substantial separation between the place where the sample is taken and the place where the analysis is performed. The piping and components which bridge the gap are, of course, the sample system and the correct design of this system, is an essential ingredient in delivering correct analytical data to the users of the data.

Sample systems can be regarded as a sequence of individual units, each of which has specific parts to play in the general task of dealing with process samples. The overall scheme is shown in Figure 2.2 and the primary roles for each of the components are as follows:

1. Sample point   To extract a sample from the process.
2. Preconditioning system   To prepare the sample for transportation.
3. Sample transport system To transfer the sample to the analysis location.
4. Conditioning system   To prepare the sample for the analyser.
5. Validation system   To enable validation or calibration of the analyser.
6. Sample disposal system   To deal with any remaining sample.

2.2 Sample points

2.2.1 Representative and timely sampling

Representativeness from the perspective of the analyser system designer means that changes in the measured variables are proportional to changes of those variables in the process itself. Timeliness means that the delay between a process variable changing and the data from the analyser reflecting that change is acceptable to the user of the data. Clearly, the task of ensuring representativeness and timeliness begins at the point where a sample is extracted from the process. The location of the sample point should be at a place where the process stream itself represents the overall condition of the process which is to be monitored and there are two considerations to take into account.

Firstly, is there anything in the general arrangement of the process equipment which could affect representativeness or timeliness? Things to ensure are that the sample extraction point is at a place where there is a good flow of the process fluid; that intervening equipment between the process of interest and a proposed location of a sample point does not modify the process stream or add an unacceptable delay to the response time of the system; and that the process stream is not otherwise changed, for example, by being downstream of a recycle line. Figure 2.3 shows some examples of how not to do it. The correct time to avoid pitfalls such as these is right at the beginning of the design of the system. A detailed examination of the process flow schematics must be made and a location for the sample point which satisfies the requirements of representativeness of the process must be specified. This activity should be the joint responsibility of the process designer, the user of the analytical data and of the analyser system designer.

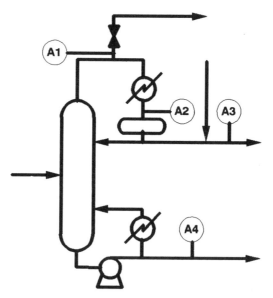

Figure 2.3 Badly located sample points. **A1** When V1 is closed there is no process flow past A1. Sample is not representative of the process. **A2** The knock-out drum leads to a large process lag. The analyser will not reflect the dynamics of the process. **A3** The analyser is downstream of a recycle line. The result reflects the stream going forward but not the tower overhead stream. **A4** If the pump speed varies, the process lag will vary giving timeliness problems and under some conditions there may not be a flow to the analyser.

Secondly, there may be localised or smaller scale effects which can modify the composition of a process sample due to separation or poor mixing of process fluids in a pipe. Such effects include picking a place where two process streams which are combined together have not had the opportunity to become fully mixed, or else the components in a stream have had an opportunity to separate. This can be countered by moving the sample point to a location where good mixing will be guaranteed, such as downstream of a strainer or a pipe bend. To ensure that a good physical location is selected, it is necessary to study isometric drawings or to 'walk the line' in order to check for potential problems.

Selection of a suitable sample point illustrates two attributes which contribute to the design of good analytical systems: the designer is not merely a purveyor of data, but needs to understand the way in which the process works and in which the data will be used; and good design cannot be carried out on the drawing board alone, but requires field work too in order to confirm that there are no practical problems.

The equipment which penetrates the plant's process envelope is called a sample probe. Although it is essential to design the probe in such a way that it will extract a sample which fairly represents the aspects of the process for which a measurement is required, this is not quite the same as saying that it should *fully* represent the *entire* process stream. For example, the required measurement might be of a trace component in a gas stream, but the stream might (as many process streams do), carry solid particles or droplets of liquid and it might not be necessary to extract these materials along with a sample of the gas; indeed it could lead to extra difficulties later on in preparing the sample for the analyser if such materials were extracted.

Not withstanding the cautions just given about picking a sample point which is downstream of a place where the process stream is modified, there can be considerable advantages to the analyser system designer in looking for just such a position. If the process stream contains high levels of condensates, particles or reactive components, but if there are also process units which eliminate them, then picking a sample point downstream of those units will simplify the design of the sample system. Whether such a step can be justified is, of course, a subject for negotiation between the data user and the analyser system designer and depends on considerations of representativeness and timeliness. The crucial factor is always the use which will be made of the analytical data.

Similarly, although localised effects should be considered to make sure that there are no problems of representativeness, localised effects can also be used to the advantage of the analyser system designer. The orientation of the pipeline, the physical location of the sample probe in it, and the detailed design of the sample probe itself all offer opportunities to improve the quality of the sample which is extracted. The most

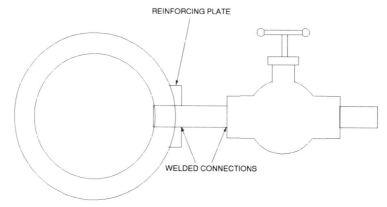

Figure 2.4 Simple sample tap.

elementary connection to a process line (or vessel) is a simple sample tap, as shown in Figure 2.4. A reinforcing boss is welded to the pipeline, a hole is drilled, the edges are cleaned and a stub of pipe is welded in place. An isolation valve is welded (or sometimes connected by flanges or screwed couplings) to the stub. Naturally all materials and construction must match the physical design requirements of the process envelope itself. With sample taps such as this, it is important to pick the right orientation for the sample point.

If a pipeline is horizontal, then the worst place for a sample tap is on the bottom surface. In many processes, particularly in petrochemical or oil refining industries, apart from the main process fluids there can also be dirt, rust, pipe-scale, heavy oils or polymers present. A sample tap at the bottom of a pipe will be susceptible to collecting any solid material or sludge which passes through the pipe and will suffer the greatest risk of blockage. For a gas stream, the top of a pipe is a good idea: that might allow particles of solids of droplets of liquid to fall back down into the main process line; this effect can be enhanced by using a comparatively wide bore take-off pipe so that the flow velocity is low. For liquid streams, however, the top of the pipeline is a bad choice: vapours or gases will be found here so the best orientation in order to extract a liquid-only sample would be on the side of the process pipe. The localised separation of phases due to gravity can be minimised by putting the sample point in a vertical section of process piping. Figure 2.5 illustrates these effects.

A simple sample tap is not always the best choice. In a pipeline where the flow velocity is low enough to give laminar flow, then (theoretically) the flow at the surface of the pipe is static, with the velocity increasing to a maximum at the centre of the pipe. This means that sampling at the edge of the pipe is poor from the point of view of timeliness. Alternatively,

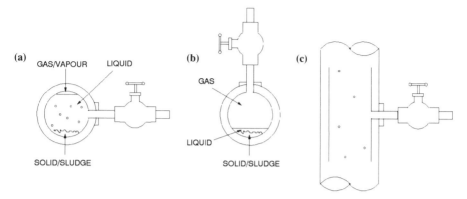

Figure 2.5 Effect of tap orientation on representativeness. (a) Orientation minimises risk of blockage by solids or sludge and minimises amount of gas or vapour which is sampled. (b) Orientation allows solids, liquids or condensates to fall back into the process pipe. (c) In vertical pipes effect of localised separation of phases is minimised.

physical constraints might prevent access from the desired direction. Finally, extra advantages may be possible if the sample is extracted using a probe which extends some way into the pipeline or vessel. Figure 2.6 shows the general design. Welded, flanged or screwed connections are used, depending on engineering constraints. Samples probes of this type can offer the chance to do some initial separation of particles from the process stream. If the entry to the pipe points downstream, if the process velocity is high enough and if the velocity of the sample into the probe is low enough, then some filtering of particles of solids can occur because under these conditions they cannot change direction and flow back into the probe: arrangements like this are sometimes called 'inertia filters' because it is the relatively high inertia of particles which carries them past the sample point.

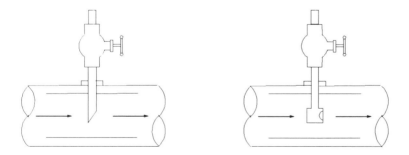

Figure 2.6 Insertion probes. Inertial filtering takes place if the velocity of the process is much higher than the velocity of the sample entering the probe; particles are unable to change direction so the amount of solid which is sampled is small.

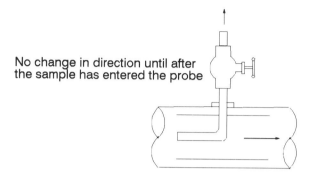

Figure 2.7   Isokinetic probe.

There is another side to inertia filters. If it is important that the sample which is extracted actually does fully represent a two-phase process stream, then that implies that there should not be any such inertial filtration effect. Probes which avoid this are called 'isokinetic probes'. Their design is such that the flow velocity into the probe matches the flow velocity past the probe, and there is no direction change until after the sample has entered the probe. Figure 2.7 illustrates this approach.

The sample probes considered so far have all taken a sample from a small part of the process, essentially from the region at the tip of the probe and this has been justified on the basis that process stream is well mixed and a sample from that small region is a good representation of the whole process. In some applications it might be known that this is not a valid assumption and in that case sampling from several points across the process stream is necessary in order to get an average sample which represents the total stream. This is sometimes done for steam streams or flue gases. The usual approach is to employ an insertion probe with several ports drilled into it. However, if it is known that the composition of the process fluid varies across the pipeline, and perhaps the flow velocity across the pipeline varies too, then that implies that to obtain a representative average sample, the amount of sample which is extracted at each point across the pipeline should not be equal. Making the ports have different sizes can account for such effects. However, it must be recognised that the basis of this approach comes from a belief about flow and composition distributions across the sampling probe and in practice it is difficult to confirm such beliefs either by measurement or

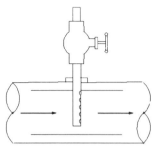

Figure 2.8 Multiport sample probe. Representativeness is only achieved if the amount of sample withdrawn at each entry is proportional to the flow velocity at each entry.

experimentation. Furthermore, it will be extremely difficult to tell if partial or complete blockages at some ports have affected the composition of the sample which is withdrawn. Figure 2.8 illustrates the design.

2.2.2 Reliability

There are two aspects to reliability. The first is to prevent failures by careful design and location of the equipment. The second is to anticipate failure and have a maintenance plan in mind for when such failures happen. Correct orientation of the sample tap in the pipeline can minimise the risk of blockage. However, a problem with all designs of sample tap concerns what to do if they do become blocked whilst in service. If there is no operational problem from the point of view of the process in breaking into the process envelope, then the simplest approach is to design the sample probe using screwed or flanged fittings so that the equipment can be easily dismantled. Unfortunately, in many installations it is not acceptable to open up the process envelope in this way and so extra facilities must be designed into the sample probe.

If it is expected that a blockage will be fairly light, then arranging a backflush to the probe might be enough. This could be operated either automatically on a regular, timed basis or else manually. For heavier blockages a hydraulic pump can be used to apply a greater pressure to clear the probe. Of course, the medium which is used to flush the probe must be acceptable from the point of view of the process; it would not do to use compressed air to clear a blockage in a hydrocarbon line, for example!

For both simple sample taps and insertion probes where heavier or more compact fouling is anticipated, it is possible to make a design which allows the opportunity to clear blockages by rodding or drilling. This can be achieved by the arrangement shown in Figure 2.9(a). The valve is of a type which, when open, allows a straight-through path so that a rod or

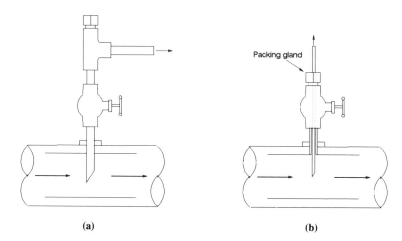

Figure 2.9 Clearance of probe blockages. (a) Cap can be removed to allow drilling or rodding to clear blockages. (b) Packing gland can be loosened to allow probe to be withdrawn. For high-pressure systems a retainer is required to avoid danger of 'launching' the probe.

drill can be inserted through it: gate or ball valves are both suitable. Once again, process considerations dictate whether such an approach is viable; the process envelope will be breached by the clearing operation and so high-pressure processes or processes which contain hazardous materials will not be amenable to this approach.

For circumstances where blockage at a probe is anticipated but it is not acceptable to breach the process envelope in order to maintain the probe, then it is necessary to design a probe which can be withdrawn without interruption to the process whilst the process is still in service. Figure 2.9(b) shows such a design. The process envelope is maintained by glands but they may be slackened sufficiently to allow the probe to be withdrawn. The process isolation valve can then be closed and the assembly dismantled to allow cleaning. Safety concerns must be taken into account. If the process is at high pressure then retainers are required to prevent the probe being blown completely out of the assembly. If the process fluid is hazardous, then care may also be required in venting or removing the process fluid from the assembly before dismantling.

While reliability can be improved by good probe design, and enhanced by making provision for clearing blockages, another important requirement is to make maintenance as easy as possible by providing good access and lighting. Cutting corners to save installation cost can seem like a good idea during the project stages but can turn out to be expensive later on when an analyser which controls the process plant fails because the sample probe blocks and it is impossible to get to it to make a repair.

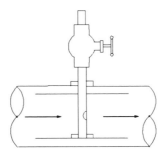

Figure 2.10 Supported probe to avoid vibrational stress. Extended probe is supported by a welded cup. Alternative approaches such as drilling through and welding are also possible.

Blockages are one source of unreliability. There are also mechanical failures which could lead to sample probe failure. Clearly the sample probe assembly must be resistant to chemical attack, corrosion or electrochemical erosion caused by the process fluids. Such effects are avoided by checking for the compatibility of the materials of the sampling system with the process fluids: this is routinely done by the designers of the process equipment.

Insertion probes can also suffer from mechanical failure. In particular, an unsupported insertion probe will have a natural vibration frequency. It is possible for the flow of process fluid to excite vibration in the probe, leading to stress fracture at the point of attachment of the probe to the pipe. This can be a particular risk in fast flowing process streams and in this case it may be necessary to provide additional supports or braces for the probe, as shown in Figure 2.10.

A final point to keep in mind concerns the maintenance of the process lines themselves. In some processes pipelines are periodically cleared by pigging: that is, driving a purpose-designed plug (or pig) through them to scour out material building up on the walls. If this is going to happen, then don't design an insertion-type probe!

2.3 Preconditioning system

Once a sample has been extracted from the process vessel or pipeline it must then be transferred to the analyser. Sometimes the distance from the sample point to the analyser is very short, or the sample might already be in a suitable form for transport without further treatment. However, frequently the distance is long, perhaps 100 m, and the sample might need to be modified in order to improve the reliability or the safety of the transport system. If this is the case, then a preconditioning system, situated close to the sample point, will be required.

The design objective of the preconditioning system is to eliminate any problems that could affect the transport of the sample to the analyser. Questions to be considered are:

(1) Will any of the constituents of the sample, such as solid particles, droplets or condensate, lead to blockage or fouling of the sample transport line?
(2) Is the physical nature of the sample suitable for transport?

It has already been shown how intelligent design of the sample probe can minimise the amount of unwanted material that is extracted. However, it is frequently necessary to provide extra protection for the sample transport line in the form of a filter to remove solids, or a knockout pot to trap condensates.

Filtration is often carried out in several stages through a sample system, the objective at each being to provide protection for the next unit in the sequence. Thus, an inertial filter at the sample probe provides a coarse level of filtration to protect both the probe and the short length of piping going to the preconditioning panel from blockage. A filter at the preconditioning system is there to provide protection for the sample transport line, but it need not provide a finer level of filtration for the analyser itself; that will be done at a later stage in the sequence. Selection of the correct type of filter at the preconditioning panel is dictated only by the need to protect the sample transport line from blockage. In many cases all that is required is a Y-strainer with a coarse filter mesh. However, the selection of this component obviously depends on the details of the process stream and the specific industry requirements.

Pressure adjustment may also be required at the preconditioning system. For processes that operate at high pressures it might be desirable to lower the pressure at the preconditioning panel by using a pressure regulator. This could mean that the mechanical specification for sample transport piping could be relaxed, but there is one caution: failure of the regulator must be considered and so if this approach is taken, then adequate pressure relief must be designed into the system. An alternative reason for installing a pressure regulator at the preconditioning system is to stabilise the flow through the rest of the system against fluctuations in the process pressure.

Ideally, all of the motive force to transfer a sample through the sample handling system comes from the process. However, if there is insufficient pressure available then a pump is required at the preconditioning panel to raise the pressure (or a suction device could be installed at the sample conditioning panel). The detailed choice of the pump will of course be industry and application specific.

Temperature adjustment might be required at the preconditioning panel. Although it might be necessary to change the temperature of very hot or very cold samples for safety reasons, it is more common to modify the temperature in order to avoid change-of-phase effects such as unwanted condensation in gas mixtures or vaporisation in liquid streams.

Cooling can be achieved by finned tubing where ambient air is the cooling medium; or jacketed heat exchangers where a cooling medium such as water is used; or by refrigerated chillers when it is necessary to get down to subambient temperatures. Which approach to use depends on the amount of heat which must be removed from the sample.

Heating can be achieved using finned tubing (if the sample is at subambient temperatures), heat exchangers or, for higher temperatures, electrical or stream-based systems. Once again, the precise details will be application and industry specific. If heating is used to avoid condensation, then a downstream catch-pot should be included in the design to removal any liquid that gets through, and it is preferable to fit an automatic drainage system to the pot in order to avoid it over filling and causing large amounts of liquid to be entrained.

Maintenance, hence system reliability, can be facilitated by including indicators to show that the components of the preconditioning system are operating as designed. Thus, if pressure adjustment takes place, a pressure gauge should be installed, if the temperature is changed then a temperature indicator is required. A pressure gauge downstream of a filter can also be useful to check that the filter has not become blocked. Finally, although flow indicators are often installed near to the analyser, it can be useful to have one also available at the preconditioning system. However, strictly speaking, indicators are best located at the place where adjustment takes place. If the flow is set or controlled at the sample conditioning panel, that is where the flow indicator should be. If the role of the preconditioning panel is to minimise pressure fluctuations in order to help to stabilise the flow, then a pressure gauge is required at the preconditioning panel.

The last facility that is often installed at the preconditioning panel is a laboratory sample point. This can be used in order to validate the on-line analyser by comparison to a laboratory analysis, or else to take a sample which can be analysed for components which are not measured by the on-line analyser, to assist in process troubleshooting activities. Thought must be given to timeliness, safety and environmental issues. Samples can be taken by diverting the sample flow through a vessel (often called a 'sample bomb'). However, there are two effects to consider. Firstly, the previous contents of the bomb could be swept into the system and subsequently measured by the analyser. This must be avoided either by purging the vessel or by appropriate vessel preparation (such as evacuating it). Secondly, if an extra volume is put into the sample flow, this will slow down the response time of the system by increasing the volume that must be purged and by introducing the

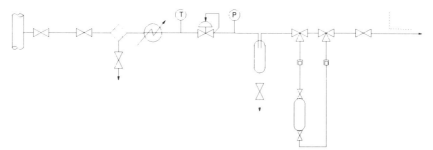

Figure 2.11 Preconditioning system. 1. The filter provides protection for the sample transport line and the regulator. A manually operated valve enables the pressure to be released prior to removal of the filter element for maintenance. 2. Temperature is adjusted by a heat exchanger. The indicator enables correct operation of the exchanger to be checked. 3. Pressure is set and controlled by a regulator in order to match the pressure requirements of downstream equipment and to achieve flow stability. 4. A knock-out vessel with a blow-down valve eliminates small amounts of liquid from vapour streams. 5. The flow can be diverted into a bomb in order to take a sample for off-line analysis.

opportunity for diffusion effects in a vessel which is not swept by plug-flow of the sample.

Figure 2.11 illustrates some general designs of sample preconditioning systems.

2.4 Sample transport

Having extracted a sample from the process and preconditioned it, the next task is to get it near to the analyser. Timeliness and representativeness are the key issues.

If the analyser is situated some distance from the sample point, then a good design must pay attention to timeliness. The general considerations are straightforward:

(1)   What data update frequency is required? What is the response time of the analyser? How long does that leave to transfer the sample from the tapping point to the analyser?

(2)   What is the effective volume of the complete sample conditioning system? What is the volumetric flow of sample required to achieve a timely data update? Would the flow to the analyser alone bring down a fresh sample quickly enough? If a higher flow rate is required (and it frequently is), then a fast loop system will be required. Figure 2.12 shows schematically how a fast loop can be used to achieve a timely response. The disposal of this extra amount of sample is considered in section 2.7.

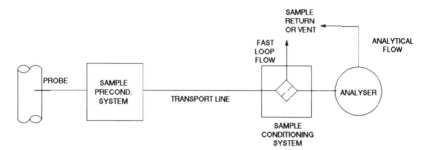

Figure 2.12 Sample fast loop system. Only the analytical flow is filtered. It is convenient to take the fast loop flow from the unfiltered side of the filter. This reduces the load on the filter and the fast loop flow can be used to provide some cleaning of the filter element.

(3) Having established the required flow rate, is there a sufficient driving force to achieve that flow? If not, then an extra driving force is required.

Illustrating the general theme by the specific example described in the introduction in section 2.1, the considerations are as follows:

(1) *The data update frequency is 5 min. The analyser takes 4 min to complete an analysis, so 1 min is left to transfer the sample to the analyser.*

(2) *The effective volume of the sample conditioning system is the volume of the sample transport line plus the volume of the extra components in the system.*

For sample transport tubing, it is not uncommon to use 6 mm outside diameter tubing with a wall thickness of 1 mm. For this tubing size and a run of 100 m, the volume of the tubing is about 1250 ml ($\pi \times radius^2 \times length$).

Suppose an estimate for the volume of the rest of the components is 500 ml (for a filter and assorted valves, regulators and indicators). An extra allowance must be made to account for the time required to sweep sample through these components, due to diffusion effects. As a rule of thumb, a reasonable estimate is to multiply the actual total volume of all components by three, giving about 1500 ml for this example.

Thus, the effective volume of the sample system is about 2750 ml and this much sample must be transferred in 1 min. Constraints such as a desire to limit the amount of sample which is piped into an analyser house often limit the flow to the analyser. Typically, the flow required by gas chromatographs (and many other analysers) is about 50 ml min. Consequently, in our example, a sample fast loop carrying about 2700 ml min will be required.

(3) *The driving force required to achieve this flow can be estimated using the equations of fluid dynamics. Proprietary software programs are available which allow the sizes, required sample transport times and fluid properties to be 'plugged in' and the required pressure differential to be given as an output. For incompressible fluid flow in pipes, the equation giving the required pressure drop is:*

$$\Delta p = 6.47 Fl\rho V^2/D^5$$

where Δp is the pressure drop (Pa);
       F is the friction factor;
       I is the length of pipe (m);
       ρ is the density of fluid (kg/m^3);
       V is the volumetric flow rate (m^3/s); and
       D is the diameter of pipe (m);

   and

$$F = \tau/\rho v^2$$

where τ is the shear stress at surface (N/m^2); and
       v is the velocity of fluid (m/s).

The equations of flow for compressible fluids are more complex because of the effective change in density in the fluid which occurs as it flows through the system. Modelling the flow is made particularly complex by the pressure steps which occur each time the flow passes through restrictions such as valves or filters and the exact location of these will have an effect on the total driving force required. However, since gases have viscosities which are typically 100 times less than the viscosities of liquids, it turns out that very often the main effect which limits gas flow is the approach to a critical flow velocity at a valve or fitting.

In our example, the driving force required to move 2700 ml min though 100 m of 6 mm outside diameter tubing with 1 mm thick walls is about 0.4 barg: this gives a sense of scale of the pressure differential which is required for a fairly long sample transport line with a fairly small bore pipe.

The required pressure drop should be compared to the available pressure drop which is the difference between the pressure at the sample point and the pressure at the sample disposal point. If this sufficiently outweighs the required driving force, then no extra equipment is required. However, if insufficient pressure is available, then a pump (or suction device) will be required. The minimum process pressure in the example is 8 barg and so provided the pressure at the disposal point is less than about 7 barg, there should not be a need for a pump.

It is important that data are timely and the above considerations have illustrated how to check for a good sample transport design. However, as well as being fast enough for the data user, it is normally important that the variation in the age of data is acceptable too. This is particularly so if the data are used in process control applications when the dynamic tuning of a control scheme (or loop) is critical. In designs in which the driving force for sample transport is derived from the process itself, particularly if the sample disposal point is back into the process rather than into a general sample collection header, it is essential to check that the process pressure at the take off and return points are suitably stable. For example, locating the sample point and the disposal point on either side of a control valve is clearly unsuitable; as the valve opens and closes to control the process, the pressure drop will vary and consequently the time it take the sample to be transported to the analyser will also vary.

Similarly, all operating process regimes must be considered. If there are alternative process flow paths, the effect on driving force during and after the changes to such alternatives must be checked to make sure that no unwanted disturbances occur.

Pressure regulators or flow controllers might be required in order to achieve acceptable stability in the data update time.

Considerations of timeliness give rise to the general design of the sample transport line. It is also necessary to consider representativeness of the sample, that is to say: can any unexpected or unwanted changes happen in the sample transport line?

Changes of phase are one effect to consider. In general, it is necessary to consider whether unwanted condensation can occur in gaseous samples or vaporisation can occur in liquid samples. Both effects are approached by consideration of pressures and temperatures.

For gases, condensation can be avoided by lowering the pressure or increasing the temperature in order to move away from the dew point of the sample. Electrical or steam heat tracing of sample transport lines is commonly used to prevent condensation during transport.

For liquids, vaporisation can be avoided by maintaining a high pressure and by lowering the temperature of the fluid in order to move away from the bubble point. Using finned tubing to enable air cooling, and not insulating the transport line could be all that is required; consideration of the fluid properties, process conditions and ambient temperatures will indicate whether extra cooling is necessary.

Some substances can give special representativeness problems because of the way in which they may interact with the walls of the sample transport line. The measurement of moisture at low concentrations (say, at less than 50 ppm vol or at dew points of less than $-40°C$) can be affected by the absorption and desorption of water molecules. Thus, if a sample transport system were to start off wet (that is to say, it had previously been filled by a gas which was saturated with water), and a very dry process sample was then run through the tube, moisture would be picked up from the walls and the sample delivered to the other end of the tube would be much wetter than the sample taken from the process. It can take hours before the system reaches an equilibrium which properly represents the process stream.

2.5 Conditioning system

Sample conditioning panels ensure that the process sample is compatible with the requirements of the analyser. This means making sure that the phase, pressure, temperature and flow rate to the analyser are suitable; that the sample is filtered to the appropriate degree; that condensates are removed from gas streams or gases are removed from liquid streams; and any other treatment such as drying of gases or removal of reactive components is carried out. Facilities are also usually provided to enable validation or calibration of the analyser and also to take a sample for laboratory analysis.

Clearly, the requirements and constraints are set by the analyser, but the possible treatments depend on the process stream. The great variety in process streams and process analysers means, of course, that there are a tremendous number of potential designs for sample conditioning systems (and this is just due the physical nature of processes and analysers; add an extra dimension of variation when the individual preferences of designers are included). Within an overall design, considerable choice is also available for the individual components which are used. There are, for example, many different classes of filter and within in each class, different designs are available from different manufacturers. There is not room in this section to cover all of these options. Instead, some general principles are highlighted and a few of the more common designs are described.

Given that a sample stream can require several treatments in order to prepare it for the analyser, the challenge for the designer is to select the right components and get them in the correct order. Some samples are straightforward to prepare. Gases which do not contain components which might condense under the conditions required by the analyser or liquid samples which contain neither entrained gas nor components which might vaporise need only filtration and adjustment of temperature, pressure and flow.

Figure 2.13 gives an outline of a suitable sequence of components for such easy-to-handle samples. Note that reliability is assured not only by trying to make a design which is reliable in the base case, but also by

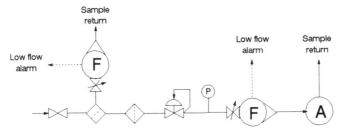

Figure 2.13 Simple sample system: gases and liquids. 1. A fast loop flow ensures timeliness of data. This flow need not be filtered. Taking the fast loop at the filter enables the filter to be swept clean, prolonging the period between maintenance. The filter should be fine enough to provide protection for the pressure regulator. 2. The pressure regulator sets a suitable pressure for the analyser and other downstream components and also ensures flow stability. 3. The flow meters are equipped with alarms to indicate low flows.

anticipating potential failures and including monitors which provide alarms in the event of a failure. Thus, a major risk is that the flow of sample fails. It may not be apparent from the analytical result that this has happened, particularly if the analyser is in an application where its result is zero for much of the time. Sometimes software alarms in the analyser or the control system which uses the analyser's data are used to detect sample failure by making sure that the value isn't 'frozen'. However, it makes sense to provide all sample systems with a low flow alarm which alerts both the data user and the analyser maintenance team. In fact, having seen that timeliness is important to the tuning of control loops, the alarm should be designed for low flow, not no flow, and an alarm should be raised if the flow moves outside of an acceptable range.

Gas samples which do contain condensable components need conditioning according to whether or not the condensables are to be retained for analysis. Figures 2.14 and 2.15 show appropriate sequences for each case.

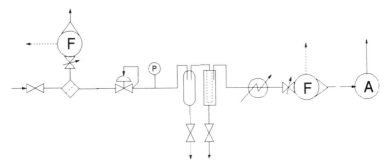

Figure 2.14 Gas: condensable retained. 1. The sample is filtered while hot, at the fast loop filter. 2. The pressure is reduced if necessary. 3. A knock-out pot and coalescing filter remove any liquid droplets which are present. 4. The sample is heated to several degrees above its dew point. 5. The pressure is adjusted and the flow rate is set and measured.

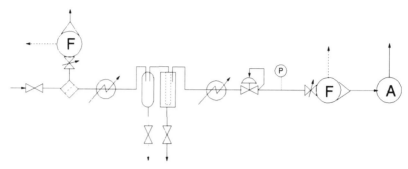

Figure 2.15 Gas: condensable rejected. 1. The sample is filtered while hot, at the fast loop filter. 2. The sample is then cooled to the required dew point. This step is not required if the sample is already below the required dew point. 3. The liquids which condense are eliminated by a knock-out pot and coalescing filter. 4. The sample is then heated to several degrees above its dew point. 5. The pressure is adjusted and the flow rate is set and measured.

The sequence of components for liquid samples which contain gases or have a potential to vaporise are shown in Figure 2.16.

The best way to validate an analyser system is by injecting a test sample and checking the analyser's result. Statistical methods are frequently used to determine whether recalibration or maintenance activity is required. It is often necessary to carry out a validation check quickly. For example, if an analytical result indicates an unexpected process change it can be useful to delay responding to the change for a couple of minutes while the analyser is validated. To enable this to happen, it must be possible to initiate in the test sample to be run from the process control panel; this requires remote stream switching for the test sample. The facility of being able to initiate test runs from a central location is also efficient for the maintenance teams: they can easily check a number of analysers according to a routine schedule and if any problems are indicated, they can go out to the analyser prepared

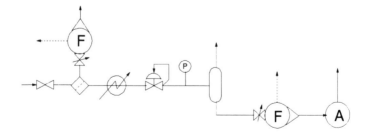

Figure 2.16 Liquid: degassing. 1. The sample is filtered at the fast loop filter 2. The sample is then cooled and the pressure is reduced. This will promote the release of gas from the liquid. 3. Gas is eliminated by a de-bubbler. 4. The flow rate is set and adjusted.

with the correct maintenance equipment. Remote introduction of a test sample is a special case of stream switching and it is dealt with in section 2.6.

Sometimes analysers are validated by taking a sample and analysing it in a laboratory, then comparing the result with the analyser's reading at the time the sample was taken. This is a less effective strategy: if there is a discrepancy between the two results, it is not known which is in error and the on-line analyser (and laboratory) would still need to be tested using standard samples. Nevertheless, it is still appropriate to provide facilities which allow a sample to be withdrawn from the analyser for measurement elsewhere. Firstly, such a back-up may be necessary in the event of a prolonged failure of the on-line analyser. Secondly, trouble-shooting of unexpected process problems can often be facilitated by laboratory analysis for components which are not measured by the on-line analyser. Laboratory sampling points can be installed either at the preconditioning panel as described in section, 2.3 or at the conditioning panel. In either event, the design must be such that the processes of taking a sample does not affect the results given by the on-line analyser, either in terms, of representativeness or timeliness. Taking the sample from the fast loop, using a system of three-way valves as shown in Figure 2.11 is a good approach.

Sometimes, because of the complexity of an analysis or the range of measurements which must be made, it is necessary to use more than one analyser on a process stream. If this is the case, the design should ensure independence between the analysers so that it is possible to take one out of service to run validation checks or to carry out maintenance without affecting the other(s). Normally, this would mean that the analysers cannot be installed in series with each other. It is permissible for them to share the sample probe, preconditioning and transport systems. Some parts of the conditioning system can be common too, but Figure 2.17 shows how the necessary independence can be achieved.

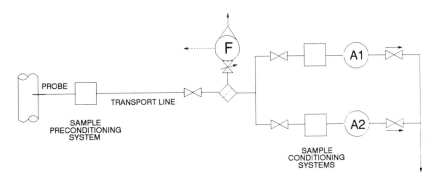

Figure 2.17 Two analysers, one stream.

2.6 Stream selection

In order to improve the cost effectiveness of an analyser system, a single analyser is sometimes connected to more than one process stream. This is worthwhile if the installed cost of the analyser is high (because the analyser itself is expensive or because the housing and facilities it requires are costly) but the data user must only need to measure one stream at a time and must be able to tolerate delays while the analyser is switched from stream to stream.

In multistream designs, each stream will have its own probe, preconditioning and transport system but the streams will come together at the conditioning panel. Compatibility with the analyser will be achieved in the ways described in section 2.5. However, special attention is required to ensure that representativeness is not lost due to cross-contamination of the streams. Timeliness must also be considered: after switching from one stream to another, how long does it take for the analyser to correctly reflect the new stream?

In order to avoid cross-contamination, it is necessary that the process streams share as few components of the sample conditioning system as possible: it is impractical for the same filter to be used for two streams because it is certain that each time the stream is switched, traces of the previous stream linger on and contaminate the new sample. Even components which have fairly small volumes, little surface area and are well swept by the sample (flow meters for example), will require some time before the condition of the new stream is accurately represented after stream switching and this must be taken into account when considering the timeless of the system.

The manner of stream selection requires careful design in order to guard against component failure causing cross-contamination. Manufacturers take great care in the design of valves but even so, seals wear and leaks occur. In a system in which only a single valve isolates one stream from another, if the valve leaks then there will be cross-contamination. Furthermore, it will be extremely difficult to tell if such a failure has occurred.

The most reliable stream isolation is to completely disconnect the piping from streams which are not being analysed. This can be achieved manually if stream switching was extremely infrequent and when it was required, there was no particular urgency in getting the change-over made quickly. However, if this is not suitable, then the system must be designed for remote or automatic stream selection.

There is one special case of stream switching which should apply to almost all analysers: the need to introduce a validation sample. This is an aspect of reliability which is connected to the confidence which a user of analytical data has in the analyser system. The need for a validation check can come from two directions. Firstly, the team who are responsible for maintenance of the analyser will need from time to time to check its performance by introducing a test sample. This is often carried out according to a maintenance schedule and the record of results provides a good overview of the performance of the analyser. Secondly, it can happen that for one reason or another the user of the data (for example, the plant operator), might request that the analyser be checked. In both cases it is necessary to introduce a test sample and there are great advantages in speed and convenience in being able to do this from a remote location. It is still necessary to eliminate the possibility of stream to stream contamination, but where one of the streams is a validation sample, it is not necessary to maintain a constant flow up to the stream selection valves: this would just waste the validation sample. Figure 2.18 shows the schematics for a three-stream system with validation sample selection.

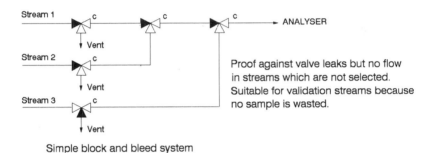

Figure 2.18   Multistream designs with validation sample selection.

2.7 Sample disposal

Collection and disposal of all of the sample which has been extracted from the process is the final step in designing a sample handling system. Venting to atmosphere is only acceptable if there are no safety implications, if the sample does not present a source of environmental pollution and if it is inexpensive.

If these conditions are not met, then the best option is to return the sample to the process. For this to be possible there has to be an adequate pressure differential. Sometimes suitable sample and return points can be found which allow the sample to flow without the need for any extra source of motive power. Care must be taken that the pressure differential is stable enough so that there are no issues of timeliness (having the sample and return points on either side of a control valve would not meet this requirement). Pressure regulators can give the necessary stability. It is also appropriate to provide a non-return valve so that there is no danger of a sample not flowing back from the return point. Finally, it must be recognised that a path between two areas of the process is thus being

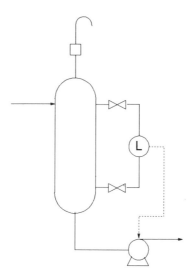

Figure 2.19 Liquid sample return.

provided. It must be confirmed that the process can tolerate this and in particular, that such a routing does not lead to any safety problems.

An alternative sample disposal routing is into a flare header. Such a system can be used to collect samples from a number of locations and route them to a single disposal point. Normally, these systems operate at quite low and fairly stable pressures. However, this method of disposal leads to the value of the sample being lost.

If there is insufficient pressure to get the sample away to the process or flare, then pumping is required. The choice of the pump depends on the nature of the sample and so it is not dealt with here. A system which is sometimes used for liquid samples is that rather than have a pump continuously running, the sample is collected into a vessel and a level controller runs the pump intermittently to drive the sample back to a suitable process point. Figure 2.19 shows a design for such a system.

Bibliography

K. J. Clevett, *Process Analyzer Technology*. John Wiley & Sons, New York, USA, 1986.
D. C. Cornish, G. Jepson and M. J. Smurthwaite, *Sampling Systems for Process Analysers*. Butterworths, London, 1981.
E.A. Houser, '*Principles of Sample Handling System Design for Process Analysis*' Proc. ISA Symp., Pittsburgh, PA, USA, 1977.
J.H. Perry, *Chemical Engineering Handbook*: McGraw-Hill Chemical Engineering Service. ISBN 0-07-766 482-X.
Flow of Fluids in Valves, Fittings and Pipe. Crane Company, Publ 410, 1969.

3 Physical property analyzers
K. J. CLEVETT

3.1 Introduction

3.1.1 Background

In this chapter, we shall deal with a group of instruments categorized as *physical property analyzers*, designed (as their name implies) to measure a physical property of an industrial process stream. For the purposes of the ensuing discussion, the following parameters are classified as physical properties, but it is important to realize that these parameters are also related to the composition of the process fluid.

Viscosity	Boiling point (distillation)
Flash point	Pour point
Cloud point	Freeze point
Cold filter plug point	Vapor pressure
Density/specific gravity	Octane number
Refractive index	Thermal conductivity

Thus, carrying out an on-line measurement of the above is usually done in effect to infer the process stream composition, as the following will serve to illustrate:

- an increase in liquid viscosity means an increase in concentration of higher boiling constituents in the sample; and
- an increase in flash point means an increase in concentration of lower boiling constituents in the sample.

Traditionally, the majority of the physical property measurements listed above have been used in the petroleum, petrochemical and associated industries. The following are exceptions:

- Refractive index analyzers are used extensively in the food, beverage and brewing industries.
- Density and specific gravity analyzers have been used in gas, liquid and slurry service on a multi-industry basis.
- Viscosity measurement is used extensively in several industries, both for batch and continuous manufacturing processes.

The first physical property analyzers, such as viscometers, boiling point, flash point and density analyzers, were installed in petroleum refineries in the late 1950s and so there is widespread experience in their use. The first analyzers were not very reliable. Therefore operator confidence in the data generated was low and their use for the purposes of process control was not considered. In fact, many installations were of an experimental type only.

Over the years, as with all process analyzer technology, we have seen many advances in analytical techniques, materials of construction, electronics, etc., that have improved the reliability and performance of analyzer systems. Today, the process analyzer is used primarily to provide real-time data for industrial process control and optimization and is recognized as a integral part of any process control system.

In many petroleum refineries, one can still see the traditional physical property analyzers used extensively, but many have been replaced or upgraded to provide more reliable operation and microprocessor-based control units are now the rule rather than the exception. As in the instrument world in general, the microprocessor has caused a major revolution in process analyzer design and operation, providing a powerful tool with self-diagnostics and report writing capability that can be used as an aid to maintenance and day-to-day operation in the analyzer area.

Another development we have seen in recent years in the physical property analyzer area is the encroachment into this area by other, relatively new on-line analytical techniques. This subject is discussed in more detail later, in relation to the specific parameters measured.

For more detailed information on physical property analyzers and many other important topics in the on-line process analyzer field, the reader is referred to Ref. 1.

3.1.2 Calibration and validation

The subject of calibration and validation has a particularly significant bearing on physical property analyzers. First, let us distinguish between these two terms:

- *Calibration:* Generally refers to initial calibration, where the analyzer is checked over its full measurement range. This is usually done during factory checkout, prior to site commissioning, or both.
- *Validation:* Generally carried out during normal operation, on demand or on a periodic basis. Usually only one point within the analyzer range is checked.

Many physical property analyzers measure parameters that have no meaning except in relation to a manmade method and scale that has been devised by organizations such as the American Society for Testing

Materials (ASTM). In the petroleum industry, the traditional ASTM laboratory test methods were first devised in the period from about 1914 to 1933. Examples are flash point, pour point, cloud point, vapor pressure and octane number, and many have remained virtually the same, except that some tests are now carried out with automatic testing apparatus rather than manually.

In the case of calibration/validation of a process gas chromatograph (see chapter 4), usually standard reference samples of accurately known composition are used, and there are relatively few problems with this approach. In the case of physical property analyzers, the traditional approach used by petroleum refineries is the 'grab sample' method, whereby the analyzer reading is compared with the result obtained on a process sample using the standard ASTM test method.

To illustrate, let us consider the measurement of octane number on a typical gasoline sample. The standard ASTM test method uses a knock engine. Engine knock is an easily observable phenomenon which is related to the performance of an engine under road conditions. This is a typical example of a test based on a manmade scale, where n-heptane was arbitrarily given an octane number of 0 and iso-octane (2,2,4-trimethyl pentane) was given an octane number of 100. Then a procedure and a test engine were developed to measure the octane number of gasoline samples.

The point here is that generally physical property analyzers must be calibrated/validated against the standard laboratory test, which in many cases has a lower precision than the process analyzer. This is against the usual principles of calibration, where the test should have a higher precision than the instrument being calibrated. This has been one of the major problems in analyzer calibration/validation.

3.2 Viscosity measurement

3.2.1 Background and definitions

Viscosity is a measure of the internal friction of a fluid (liquid or gas). Since a fluid cannot support a shearing stress, however small, various layers of the fluid will move relative to each other, and therefore a velocity gradient will exist at right angles to the direction of flow.

Newton's definition of viscosity can be explained by reference to Figure 3.1, which shows two parallel planes AB and CD of area A cm^2, separated by a distance of Y cm, moving in a fluid. The upper layer of fluid has a velocity greater than the lower layer and will tend to drag the lower layer forward. At steady flow conditions there is no acceleration and therefore an equal and opposite drag will be exerted on the upper

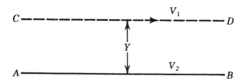

Figure 3.1 Newton's definition of viscosity. (Reproduced from Ref. 1, by permission of John Wiley & Sons Inc., New York, USA.)

layer by the slower-moving lower layer. Assuming that the velocities of the upper and lower layers are V_1 and V_2, respectively, the shear rate (D) is equal to ($V_1 - V_2$). The force per unit area of the layer required to act on the surface A to maintain this shear rate is known as the shear stress (τ), where $\tau = F/A$.

Newton postulated that shear stress is proportional to shear rate, and defined the ratio τ/D as the coefficient of viscosity or more simply the viscosity (η) of the fluid. He assumed this to be constant for each fluid (Newton's Law of Constancy), although this proved NOT to be the case.

In the CGS system, the unit of viscosity is the *poise (P)*, where $t\,P = 1$ dyne sec.cm^{-2} and is the tangential force per square centimeter required to maintain a velocity difference between layers 1 cm apart. In the SI system the unit of measurement is the Newton sec.m^{-2} or Pa.sec. Industry generally uses the *centipoise (cP)*, 100th of the poise. Newton's expression T/D defines the *absolute viscosity*. The ratio of absolute viscosity and density is known as the *kinematic viscosity* (v), where:

$$v = \eta/\rho$$

In the CGS system the unit of kinematic viscosity is the stokes (St), and the industrial measurement unit the *centistokes (cSt)*. The reciprocal of the absolute viscosity is known as the *fluidity*.

The term *consistency* is often used in industry to describe the flow properties of a fluid. The consistency refers to a point viscosity measurement at a given set of conditions (e.g. temperature, flowrate). If this single measurement is sufficient to define the shape of the shear stress/shear rate curve, then the material is Newtonian and its viscosity and consistency are the same. Therefore the term consistency is more related to what is known as *non-Newtonian* behavior, since the viscosity characteristics of these materials cannot be defined by a single viscosity measurement. In the process industries, consistency is most often used to describe for example the solid content of a slurry or suspension such as pulp stock (paper industry) or batter mixtures (food industry), where it is a measure of the degree of firmness of the fluid.

Many fluids obey Newton's Law of Constancy and for these substances the rate of shear at which their viscosities are measured is unimportant. Several substances however, particularly heavy hydrocarbon streams, show deviations from Newton's Law and exhibit ratios of shear stress/shear rate that are NOT constant. For non-Newtonian fluids, the rheological behavior can vary and allows a classification of these fluids as follows:

(1) *Plastic flow behavior:* Usually occurs with crowded dispersed substances, the dispersed part of which is wetted. Examples include printing inks, paints, vaseline, mayonnaise and toothpaste.

(2) *Pseudo-plastic flow behavior:* Occurs with long molecular chains in solvents, sometimes known as 'shear thinning'.

(3) *Dilatant flow behavior:* Occurs mostly with tightly packed, dispersed substances that are unwetted or only slightly wetted. Examples are starch in water and oil-based paint sediment.

(4) *Thixotropic flow behavior:* This is a time-dependent effect that is completely reversible. The viscosity is observed to decrease during shearing at a constant rate and to increase when the substance is at rest. The rheogram exhibits a hysteresis loop.

(5) *Rheopectic flow behavior:* This is the reverse of thixotropy. Such substances are rare, whereas thixotropic substances are common.

(6) *Visco-elastic flow behavior:* This type of flow behavior is said to exist in substances that exhibit both elastic and viscous properties.

We have seen from the above that for certain substances viscosity varies with shear stress. It should also be noted that viscosity varies with temperature, density and composition of mixtures. The most well-known parameter relating viscosity with temperature is the *viscosity index*, which is an important specification for lubricating oils used in the auto industry. With most liquids, the viscosity–temperature curve is exponential, and a small temperature change can cause a relatively large change in viscosity. It is therefore essential that viscosity measurements be made at constant temperature, or some form of temperature compensation employed. Because the exponential shape of the curve makes compensation difficult except over a relatively narrow range, often the simplest way to eliminate temperature effects is to carry out the measurement at a constant temperature.

3.2.2 Classical methods of measurement

Some knowledge of the classical techniques used for viscosity measurement will give the reader an insight into the development of laboratory methods and how these methods have been adapted for on-line viscosity measurement. All of the techniques described below are used in one form or another in industry today.

3.2.2.1 Poiseuille's method. This method, developed in 1842, involves measurement, under laminar flow conditions, of the volume of liquid at a constant head, flowing through a narrow-bore tube in a given time. The viscosity of the liquid can be calculated from the following equation:

$$\eta = \frac{(P_1 - P_2) \times R^4}{8LV}$$

where P_1 is the upstream pressure, P_2 is the downstream pressure, R is the radius of the tube, L is the length of the tube, and V is the volume passing in unit time.

Capillary type viscometers based on this principle are used extensively in industry for on-line viscosity measurement.

3.2.2.2 The U-tube method. This method is essentially a more refined version of Poiseuille's method described above. The glass U-tube viscometer is filled with the sample under test and immersed in a constant temperature bath. Measurement is made of the time taken for the liquid level to fall between two marks on the viscometer. The viscometer has a calibration factor that takes account of the viscosity and density of a standard liquid (e.g. water), and therefore provides a direct measurement of sample viscosity in kinematic units (i.e. viscosity/density). The U-tube method is recognized as a standard for routine laboratory testing, particularly in the petroleum industry. There are several different types of U-tube viscometers available commercially, such as the Ostwald and modified Ostwald. Often these viscometers are calibrated in 'time' units (e.g. Saybolt-Seconds (USA), Redwood-Seconds (UK) and Engler-Degrees (Germany)).

Efflux cup viscometers are adaptations of the U-tube method, where the efflux time of a specified volume of the test fluid is measured through a fixed orifice at the base of the cup, and is representative of the viscosity of the fluid. Examples include the Saybolt, Ford and Zahn cup viscometers. The first is a standard for testing petroleum products, whereas the other two are used primarily in the paint and varnish industries.

3.2.2.3 The rotating cylinder method. In this method, the sample under test fills the annular space between two concentric cylinders, the outer one of which is fixed and the inner one is rotated evenly about its vertical axis. The torque (τ) required to rotate the inner cylinder at a constant angular velocity (ω) is related to the sample viscosity (η) by the expression:

$$\omega = \frac{\tau}{4\pi L \eta} \left(\frac{1}{R_1^2} - \frac{1}{R_2^2} \right)$$

where R_1 and R_2 are the radii of the inner and outer cylinder, respectively.

This method is very useful for studying the effect of variations in shear rate on the viscosity of fluids (rheological investigations). Two companies, Brookfield Engineering (USA) and Contraves (Switzerland) are particularly well known in this field, and manufacture a range of laboratory and on-line viscometers operating on this principle.

3.2.2.4 The falling sphere (Stokes' method). This method is used for measuring the viscosity of very viscous fluids. An adaptation of the method is used for on-line viscosity measurement. For a spherical body travelling at low velocity in a fluid, in which the outer boundary is far removed from the sphere, Stokes calculated the retarding force on the sphere to be equal to $6\pi\eta rv$, where r is the radius of the sphere and v its velocity. When the sphere reaches its limiting velocity (no acceleration), the resultant downward force is equal to the retarding force due to fluid viscosity. The viscosity of the fluid may be calculated from the expression:

$$\eta = \frac{2r^2 g}{9v} (D - d)$$

where D is the density of the sphere, d is the density of the fluid, v is the velocity of the sphere, and r is the radius of the sphere.

In the standard laboratory apparatus, measurement is made of the time taken by the sphere to fall between two marks, which is related to the fluid viscosity. An adaptation of the Stokes method is utilized in the range of on-line viscometers manufactured by Norcross Corporation (USA) and used extensively in the printing industry.

3.2.3 On-line process viscometers

Commercially available process viscometers may be conveniently classified into the following types, according to their operating principle:

- Capillary type (Poiseuille's method)
- Vibrational type
- Rotational type
- Falling piston type

The vibrational type is the only one not specifically related to the classical methods described above. Process viscometers commercially available today are listed in Table 3.1, giving measurement principle and operating viscosity range.

Table 3.1 Commercially available on-line process viscometers

Manufacturer	Model number	Operating principle	Viscosity range (cP)
Automation Products	Dynatrol Series CL-10	Vibration	1–100 000
Brookfield Engineering Labs	VTA-120	Rotating Disk	0–2000
	TT-100	Rotating Disk	10–50 000
	TT-220	Rotating Disk	
C.W. Brabender	Convimeter Series	Rotating Cylinder	1–100 000
Cambridge Applied Systems	SIH/SIL-200	Electromagnetic	1000–50 000
	CAS	Electromagnetic	0.5–0.2000
ABB Process Analytics	Ultraviscoson 1800	Vibration	0.1–50 000 x g/cm^3
Contraves AG	DC-40 Series	Rotating Disc	1–1 000 000
Eurcontrol USA	VISC 21.5	Rotating Disc	0–1500
Rotork Analysis Ltd	1077	Capillary DP	0–10 to 0–2500
	422/491	Capillary DP	0–10 to 0–2500
Nametre Company	C4P Series	Vibration	0.1–100 000
		Vibration	
Norcross Corporation	M Series	Falling Piston	0.1–100 000

3.2.3.1 Capillary viscometers. The first on-line viscometers of this type were introduced by Hallikainen Instruments (USA) in the late 1950s, and have been used extensively in the petroleum and petrochemical industries. The latest instruments are manufactured by Precision Scientific Inc. (USA) and Rotork Analysis Ltd. (UK).

A typical example is the Precision Scientific Model 44860 viscometer, shown in Figure 3.2, which is designed for applications such as fuel oil blending, lube oil production and residual fuel oil dilution. The analyzer is provided with an integral sampling system and a fast-acting temperature control system to protect the analyzer from major changes in process stream temperature. The booster pump/back pressure regulator system protects the analyzer from changes in supply pressure and return pressure. The metering pump, which has about two thirds the capacity of the booster pump and is driven by the same motor, provides a constant flowrate of sample through the capillary. The pumps, heat exchangers, capillary and differential pressure sensor are mounted in a constant temperature bath as shown. All components submerged in the bath are suspended from the bath cover, providing easy access for maintenance. The analyzer is supplied with eight capillary range tubes allowing operation at spans from 2 to 4000 cP at bath operating temperature.

3.2.3.2 Vibrational viscometers. Analyzers of this type measure the viscosity of the process fluid by measuring the extent of dampening of vibratory motion of the sensor caused by the 'viscous drag' of the process fluid. A typical example, shown in Figure 3.3, is the oscillating sphere viscometer manufactured by Nametre Company (USA). The sensing

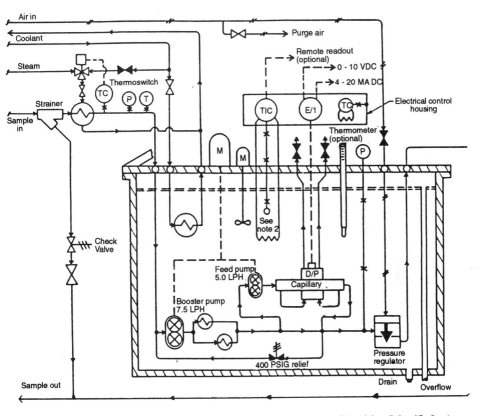

Figure 3.2 Precision Scientific Model 44860 viscometer. (Courtesy of Precision Scientific Inc.)

element is a 1.26 in. stainless-steel sphere which is immersed in the process fluid. Viscosity is determined by measuring the power required to maintain torsional vibration of the sphere at constant amplitude. The output signal from the transducer is proportional to the product of viscosity × density, and the dynamic range of the instrument is 0.1–100 000 cP with six linear decades selectable by pushbutton. Provision is allowed for entering density values so that the viscosity can be displayed in absolute units (cPO), kinematic units (cSt) or even Saybolt-Second units.

3.2.3.3 Rotational viscometers. A typical example is the Model VISC 21.5 viscosity transmitter manufactured by Eurcontrol Kalle and shown schematically in Figure 3.4. This is available with either pneumatic or electronic transmission. The analysis unit consists of a flow-through sample chamber, drive motor and assembly, and the transducer assembly. The measurement system consists of a rotating disk, which is driven at constant speed, and a stationary disk. The torque exerted on the stationary disk is proportional to the fluid viscosity and is transmitted

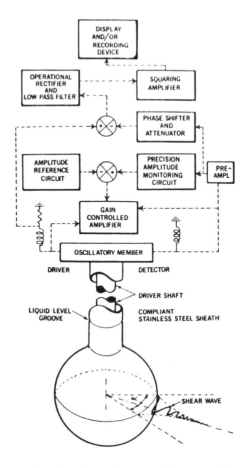

Figure 3.3 Nametre vibrating sphere viscometer. (Courtesy of Nametre Company.)

through a measuring shaft to the position sensing system. Maximum sample flowrate is dependent upon sample chamber size. Sample pressures up to 1200 psig and temperatures up to 310°C can be accommodated. The transmitters have ranges and spans as shown in Table 3.2.

Table 3.2 Ranges and spans of the transmitters

	Range (cP)	Minimum span (cP)
Pneumatic transmitter	1500	25
Electronic transmitter	700	25

3.2.3.4 Falling piston viscometers. The wide range of in-line falling piston viscometers manufactured exclusively by Norcross Corporation (USA) has been used extensively in a variety of industries for about 30

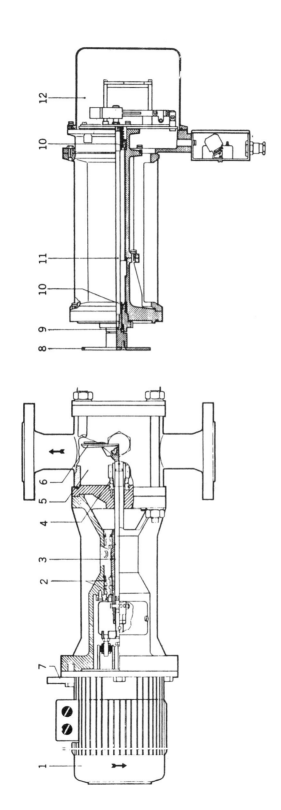

Figure 3.4 Eurocontrol rotating disc viscometer. 1, electric motor; 2, angular contact ball bearings; 3, drive shaft; 4, mechanical seal; 5, measuring vessel; 6, rotating disc; 7, support bracket; 8, support bracket; 9, O-ring seal; 10, measuring disc (stationary); 9, O-ring seal; 10, angular contact ball bearing; 11, measuring shaft; 12, transducer cover. (Courtesy of Eurcontrol Kalle.)

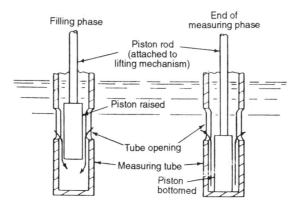

Figure 3.5 Norcross falling piston viscometer. (Courtesy of Norcross Corporation.)

years. The measurement principle is an adaptation of the Stokes' method described earlier. The product line includes instruments for open tanks or vessels operating at atmospheric pressure, closed vessels operating above or below atmospheric pressure, and for in-line and sidestream applications.

The basic sensing element, shown schematically in Figure 3.5, is a piston and tube assembly, which is immersed in the process fluid. The piston is periodically raised by a lifting mechanism, an action that draws sample in through the tube openings. The liquid then passes down through the clearance between the piston and cylinder bore into the space formed below the piston when it is lifted. The piston is then allowed to fall under gravity and the sample is expelled through the same path by which it entered the tube. The time taken for the piston to fall from its start position to a predetermined level is a measure of the liquid viscosity.

The measurement ranges, temperature and pressure limitations for the Norcross product line are as shown in Table 3.3.

Table 3.3 Limitations for the Norcross product line

Model Number	Viscosity range (cP)	Temperature (°F)	Pressure (Psig)
M16	0.1–10 000	400	150
M24	0.1–9000	500	600
M28	0.1–100 000	1000	7000
M30	0.1–9000	400	1500
M31	0.1–9000	500	1500
M32	0.1–9000	500	1500

3.2.4 Recent developments

One development of note in the area of on-line viscosity measurement was the recent introduction of a range of instruments by Cambridge

Applied Systems Inc. (USA) which operate on an electromagnetic principle and are designed for in-process control applications. In the 'slide-ring' sensor, alternating magnetic forces are applied to a ring mounted on a cylinder that contains two magnetic coils. As the ring is pulled magnetically by a coil, the time the ring takes to travel to the coil is related to the viscosity of the sample. The magnetic field is then reversed and a measurement is taken in the opposite direction. The two viscosity values are then averaged and displayed as absolute viscosity. An RTD probe is integrated into the sensor to provide instantaneous temperature information. The back-and-forth movement of the slide ring continuously samples fresh fluid and provides a clean-in-place capability when used with solvents during clean-up.

Each slide-ring has a 20:1 dynamic range. For example, slide-ring R2 has a viscosity range of 1000–20 000 cP, whereas slide-ring R5 has a range of 2500–50 000 cP. A range change involves just changing the slide-ring and recalibrating the electronics. The sensor is designed as an immersion unit for direct mounting in pipes or vessels and is available in several configurations to suit the particular application. The electronics unit, which can be supplied as an industrial NEMA 4 or explosion-proof wall-mount enclosure, includes a 12 VDC power supply, circuit-breaker switch, digital displays of viscosity and temperature, with temperature compensation circuity as a option.

3.3 Boiling point measurement

3.3.1 Background and definitions

Distillation forms the backbone of many industrial processes and is particularly important in the petroleum, petrochemical and chemical industries. On-line boiling point (distillation) analyzers were among the first process analyzers to be used in these industries. Distillation may be defined as a process for the separation of a homogeneous liquid mixture, in which the liquid is partially vaporized and the vapor thus formed is subsequently condensed. The condensed vapor is enriched in the lower-boiling components of the mixture and the residual liquid in the higher boiling components. There are several types of distillation processes in use today and these may be classified as follows.

- According to the number of components to be separated:

    Binary (two components)
    Ternary (three components)
    Multi-component
    Complex

- According to the type of separation required:

  Equilibrium flash vaporization (EFV)
  Fractionation

They may also be classed as batch, continuous and as pressure, vacuum or steam distillation. Various combinations of the above may be used, depending upon the process requirements.

3.3.1.1 Fractionation processes. In most industrial processes a mixture is required to be separated into products that must meet fairly tight specifications, based for example on purity or boiling point. Separation is carried out in a vertical column known as a *fractionating column* or *fractionator*. Vapor from one stage passes to the stage above and liquid to the stage below. In turn, each stage receives liquid from the stage above and vapor from the stage below. The concentration of lower boiling components is increased in the vapor phase in the direction of vapor flow (up the column) and decreased in the liquid phase in the direction of liquid flow (down the column). The reverse is true for higher boiling components.

There is therefore a temperature gradient in the column, with a minimum as the final vapor is withdrawn from the top of the column, and a maximum as the final liquid is withdrawn from the bottom of the column. A pressure gradient is induced across the column to provide the necessary counterflow of liquid and vapor. Heat energy is supplied to the process at the last stage from which liquid is withdrawn (*reboiler*) and by the feed to the column. The final vapor is condensed and part of the condensate produced is returned to the column (*reflux*). Such fractionators are widely used in industry, the number of stages in the column depending upon the degree of separation required. A typical example of a fractionator and its associated controls is shown in Figure 3.6. From the above description it can therefore be seen that the fractionator separates a mixture into products characterized by their boiling point (or distillation) range.

3.3.1.2 Complex system fractionation. Some of the more important complex systems are petroleum-based materials, such as crude petroleum, natural gas and their associated products. Crude distillation is the primary process in any oil refinery, since it provides feedstocks for all downstream refinery units, including petrochemical plants. Crude distillation is carried out in complex fractionating columns known as *atmospheric* and *vacuum pipestills*. An example of an atmospheric pipestill is shown schematically in Figure 3.7. Each fraction withdrawn from the column can be characterized by an initial and final boiling point (distillation range) and there is a close relationship between the boiling

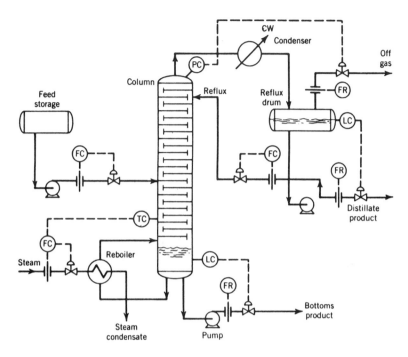

Figure 3.6 Typical fractionator and associated controls. (Reproduced from Ref. 1, by permission of John Wiley & Sons Inc., New York, USA.)

ranges of the various products. Precise control of this complex fractionator is a very important factor in overall refinery operations, since it is the initial step in the process and provides intermediate products and feedstock for all downstream operations.

Another major problem is that in many refineries the pipestill's operation may change frequently to use a different crude source and produce a different range of products. During the changeover period, frequent testing is required to ensure that on-grade production is achieved as soon as possible. The traditional method is to use laboratory testing but this suffers from inherent delays in sampling, testing and reporting the test results. With today's emphasis on maximum process efficiency and optimization, on-line process analyzers are used extensively to provide real-time quality data for use in advanced pipestill control systems.

3.3.2 Laboratory methods of measurement

Process boiling point analyzers must be correlated against the standard laboratory methods and therefore an understanding of the ASTM

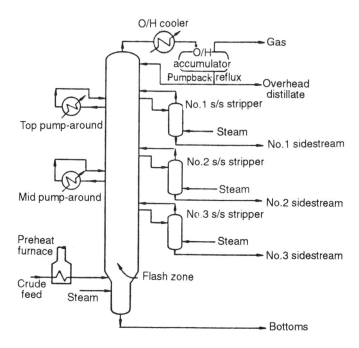

Figure 3.7   Typical atmospheric pipestill. (Reproduced from Ref. 1, by permission of John Wiley & Sons Inc., New York, USA.)

methods and their precision are essential to the analyzer engineer. The standard ASTM distillation test methods have been used by the petroleum industry for about 80 years.

ASTM method D86 covers the atmospheric distillation of a wide range of petroleum products, and equipment specifications are laid down in ASTM E133. The test is a single-plate batch distillation with empirical heat input and distillation rate. A 100 ml sample of the process stream is distilled in a flask at atmospheric pressure and the vapor condensed and collected in a graduated cylinder. Observations of vapor temperature and volume of condensate are made over the full boiling range of the sample. The test can take up to 15 min to complete and has a precision (repeatability) that can vary from 1 to 5°F, depending upon the rate of change of temperature and percent recovered. The D86 test is usually carried out now using automatic distillation apparatus such as the Precision Scientific Inc. ADA IV.

Another test of importance is ASTM D1160, which covers the distillation of petroleum products under reduced pressure, and is applicable to high boiling products that are partially or completely vaporized at 750°F and at pressures down to 1 mm Hg. There are several other standard distillation test methods applicable to specific petroleum products.

In addition to the conventional distillation test methods listed above, the ASTM has recognized the technique of gas chromatography simulated distillation (GC SimDis) for boiling point estimations. The temperature-programmed GC has been used in the laboratory for many years for this purpose. Two ASTM test methods are appropriate here:

ASTM D2887   Boiling point distribution of petroleum products (boiling range 100–1000°F)

ASTM D3710   Boiling point distribution of gasoline and gasoline fractions (FBP 500°F)

3.3.3 Process boiling point analyzers

Process boiling point analyzers may be classified into the following types:

- The so-called 'continuous' analyzers
- Batch type distillation analyzers
- GC simulated distillation analyzers
- Supercritical fluid chromatographs

These instruments can measure initial boiling point, final boiling point, intermediate boiling points, percent recovered at a given temperature, and multipoint distillation. Some are capable of providing a complete distillation curve for a process sample. There is thus considerable overlap in these arbitrary categories.

It should be noted that the so-called 'continuous' analyzers cannot, by virtue of their operating principle, measure IBP or FBP of a process stream. They are limited to a single point between 5 and 95%. Although they have continuous sample flow, neither are they truly continuous since the distillation process has a certain finite response time. The advantage of this type over the batch type distillation analyzer is that it will respond to a process change within a few minutes even though the final value may not be reached for 10–15 min. The batch type analyzer is cyclic in operation and has a finite cycle time of 10–15 min. Therefore, it is possible under certain conditions that a process change could remain undetected for this time period.

Commercially available process boiling point analyzers are listed in Table 3.4.

3.3.3.1 Continuous boiling point analyzers. A typical example of this type is the Precision Scientific Model 44520, shown schematically in Figure 3.8. This is the latest microprocessor-based instrument in the PSI product line, uses a film evaporation technique to separate overheads and bottoms and is capable of monitoring any boiling point between 5 and 95% within the temperature range 100–600°F.

Table 3.4 Commercially available on-line process boiling point analyzers

Manufacturer	Model number	Type	Range (°F)
Traditional			
Carlo Erba (Italy)	PD-155	5–95% Point	100–650
Fluid Data (USA)	1463	5–97% Point	150–620
	1463M	5–97% Point	150–620
Rotork Analysis (UK)	Distillar 2000	IBP to FBP	100–650
Precision Scientific (USA)	44520	5–95% Point	100–650
	44522	5–95% Point	100–650
	BPA-II	5–95% Point	100–650
	41468	Vac. Dist.	650–1000
GC SimDis			
ABB Process Analytics (USA)	Vista GC		Up to 900
	Vista SFC		High Boiling Hcbns
Applied Automation (USA)	Advance GC		Up to 900
Fluid Data (USA)	MS-IV-BP		Up to 900
Foxboro Analytical (USA)	Model 931C		Up to 900
Rosemount/Beckman (USA)	Model 6750		Up to 900
UOP Monirex (USA)	BPt. Monitor		C_3 to C_{22} Hcbns

Two modes of operation are accommodated. In Mode 1, percentage recovered at a given temperature, power to the heater is controlled to maintain the desired set point temperature. Bottoms flowrate is calculated according to the standpipe volume, subtracted from the total sample flowrate and the difference coverted to percent recovered, which is transmitted as the measured variable. In Mode 2, temperature at a given percentage recovered, bottoms pump speed is controlled and heater voltage regulated to maintain constant standpipe volume. The microprocessor controls all unit operations, monitors all parameters, provides internal self-diagnostic routines and performs barometric correction of temperature measurement (according to ASTM D86 test procedure).

3.3.3.2 Batch type distillation analyzers. Analyzers of this type duplicate the ASTM test method. A typical example is the Rotork Analysis Ltd. (Hone) distillar, shown schematically in Figure 3.9. This is the latest microprocessor-based instrument of a range of distillation analyzers originally manufactured by Hone Instruments Ltd in the UK since the early 1960s. The analyzer unit is basically a miniature ditillation unit with sample burette, flask, heater, condenser, calibrated receiver and uses a thermocouple to sense vapor temperature. The burette takes 100 ml of process sample and the liquid level in the receiver is visible throughout the analysis.

For a single-point distillation, the detector is mounted on the receiver at a preselected point (e.g. 90 ml = 90% point). When the condensate liquid level reaches the detector, several actions are initiated:

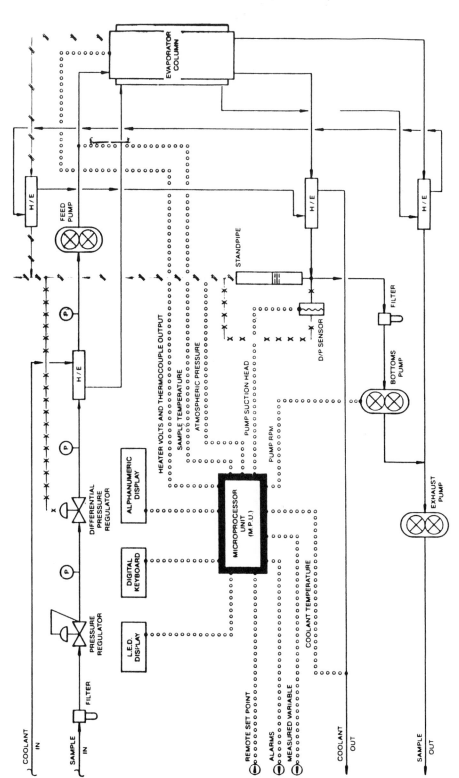

Figure 3.8 Precision Scientific Model 44520 boiling point analyzer. (Courtesy of Precision Scientific Inc.)

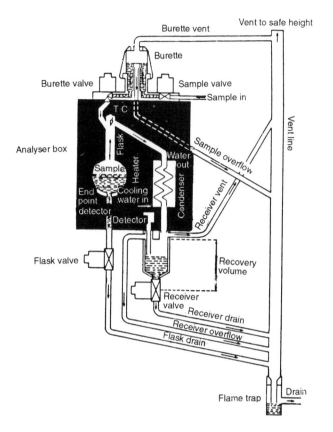

Figure 3.9 Rotork Analysis (Hone) distillar boiling point analyzer. (Courtesy of Rotork Analysis Ltd.)

- The heater is switched off.
- The programmer is started to carry out the next cycle of operation.
- The new boiling point temperature is recorded remotely via a peak picker and mV/I transducer.

The programmer operates the pneumatic valves in the correct sequence to flush and drain the flask and receiver ready for the next analysis. The cycle time of the analyzer is 5–15 min, depending upon the boiling range of the sample.

The distillar is also capable of performing multipoint analysis, a percent recovered analysis at a given temperature and a complete distillation curve for a process sample. Final boiling point can also be measured. This is achieved by a resistance thermometer mounted in the base of the flask, which detects the large temperature change that occurs when the last drop of liquid is vaporized.

3.3.3.3 Simulated distillation by GC. GC SimDis has been used in the laboratory for many years and as mentioned earlier is recognized as a standard analytical techique for boiling point measurement by the ASTM. In recent years there has been much interest shown in isothermal and temperature-programmed GC for on-line boiling point measurement. The major applications have been limited to relatively low boiling hydrocarbon streams such as gasoline, solvents and naphtha. The use of the latest microprocessor-based GC systems coupled with the use of capillary column technology has had considerable success and this technique is beginning to move into the domain previously dominated by the traditional analyzers decribed earlier. This is expected to continue in the future as more field operating experience is gained. Most of the major process GC manufacturers are active in this area.

3.3.3.4 Supercritical fluid chromatography (SFC). SFC, which is a well-proven laboratory analytical technique, has started to migrate to the on-line process analysis area. This is a powerful analytical technique that has the attributes of both gas and liquid chromatography and uses supercritical carbon dioxide as the carrier. The high-resolution liquid chromatographic columns can be used in conjunction with the high sensitivity detectors used in gas chromatography. The first on-line SFC was introduced by ABB Process Analytics (USA) at the 1986 ISA show. Initial investigations showed excellent resolution of higher boiling hydrocarbon species by molecular weight, indicating promise in characterizing process streams such as gas oils, fuel oils, and lube oils by boiling range, applications that have not been possible until now. Although there were some teething problems during field evaluation, which was conducted in conjunction with several major oil companies, the manufacturer now reports that these have been solved and the equipment is available for on-line applications. It is anticipated that SFC will eventually become a competitor to the traditional boiling point analyzers and we may well see other companies entering this field in the future.

More general process analysis applications of GC and SFC is given in chapter 4.

3.4 Flash point measurement

3.4.1 Background and definitions

Flash point is an important characteristic of petroleum distillates and is defined as the lowest temperature, corrected to a pressure of 760 mm Hg, at which the application of an ignition source causes the vapor to ignite under the specified conditions of the test. At the flash point temperature, the sample liquid produces sufficient vapor to form an ignitable mixture with air.

Flash point measures the response of the sample to heat and flame under controlled laboratory conditions. It is only one of a number of properties that must be considered in assessing the overall flammability hazard of a particular material. The presence of a highly volatile and flammable material in a relatively non-volatile or non-flammable material will be indicated by a decrease in flash point temperature (e.g. contamination of kerosene by gasoline) Therefore, determination of the flash point temperature of middle distillate products such as kerosene, gas oil, white spirit and diesel fuel provides a criterion of product purity.

3.4.2 Laboratory methods of measurement

The ASTM standard test methods for flash point measurement are among the oldest and most widely used of all tests on petroleum products and as such provide a major workload in refinery laboratories. These tests have changed little over the years, except that today they are often carried out using automatic apparatus rather than manually. A brief description follows of the three ASTM methods that concern us here.

3.4.2.1 Tag closed-cup method ASTM D56. This method covers the testing of all mobile liquids with a flash point below 200°F and a viscosity less that 5.5 cSt at 104°F or less than 9.5 cSt at 77°F. A typical manual Tag closed-cup test apparatus is shown in Figure 3.10. The test sample, at a temperature at least 20°F below the expected flash point, is placed in the test cup. With the lid closed, heat is supplied so that the sample temperature increases by 2°F/min for samples with flash points below 140°F, or by 5°F/min for samples with flash points above 140°F. When the sample is about 10°F below the expected flash point, the lid of the test cup is removed and a small test flame inserted momentarily in the vapor space. Application of the flame is repeated after each 2°F rise in temperature. The source of heat is removed and testing discontinued when a distinct flash is observed in the interior of the flash cup. The temperature at which this occurs is recorded as the flash point. This value is corrected to 760 mm Hg barometric pressure. The repeatability of the test is ±2°F for flash points below 140°F and ±5°F for flash points between 140 and 199°F.

3.4.2.2 Pensky–Martens closed-cup method ASTM D93. This method covers the testing of fuel oils, lube oils and other viscous materials with flash points up to 700°F. The manual test apparatus and procedure is similar to that for the previous test, except that a stirrer assembly is included to ensure complete mixing of the sample, and the test cup is

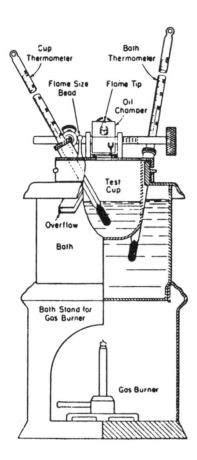

Figure 3.10 Tag closed cup test apparatus. (Reproduced from Ref. 1, by permission John Wiley & Sons Inc., New York, USA.)

heated directly rather than being immersed in a heated bath. The rate of increase of sample temperature during testing is 9–11°F/min. The test flame is applied every 2°F for samples with flash points less than 230°F and every 5°F for samples with flash points above 230°F. Repeatability is quoted as ±4°F for samples with flash points between 95 and 220°F and ±10°F for flash points above 220°F.

3.4.2.3 Cleveland open-cup method ASTM D92.   This method covers the determination of flash points and fire points of all petroleum products except fuel oils, with an open cup flash point below 175°F. The sample is placed in an open test cup and heat applied so that the temperature increases by 9–11°F/min. Starting at least 50°F below the expected flash point, a test flame is passed across the center of the cup. This is repeated thereafter every 5°F rise in temperature. The lowest temperature at which a flash appears at any point on the liquid surface is recorded as the open cup flash point. To determine the *fire point*, the liquid is heated further and the test flame applied every 5°F until the liquid ignites and continues to burn for at least 5 s. The temperature at which this occurs is recorded as the fire point. Repeatability is quoted as ± 15°F for both flash and fire points.

3.4.3 Process flash point analyzers

3.4.3.1 Discontinuous analyzers. This type of analyzer basically duplicates the ASTM test methods. Although the sample flow through the analyzer is continuous, the instrument operates on a cyclic basis. A typical example is the Precision Scientific general flash point analyzer Model 44650, shown schematically in Figure 3.11, which is designed for flash point measurement of hydrocarbon streams with viscosities up to 2000 cSt at 100°F and flash point temperatures from 140 to 600°F. Its main applications are in hydrocarbon middle distillate production, heavy fuel oil blending and solvent refining.

Process sample is pumped into the analyzer by one head of the duplex pump, heated and then mixed with preheated air, prior to entering the flash chamber, where it floods the base of the cup and overflows to drain. An annular orifice provides for uniform distribution of hot vapor throughout the flash chamber. The vapor rises past an electrode that provides an HV spark every second during the heating cycle. Vapor passes out of the flash chamber into a separator, together with the liquid flow. Liquid and any condensed vapor flow by gravity to the second head of the pump assembly where it is returned to the fast loop circulating system.

At the start of the cycle, the electrical heater supplies heat to the sample increasing its temperature at a constant rate until the heated vapor ignites when exposed to the HV spark at the electrode. The force of ignition is sensed by a pressure switch, which initiates the cooling cycle. During cooling, which is normally less than 30 s in duration, the sample heater is switched off and a solenoid valve is actuated which diverts a stream of cool purge air into the flash chamber. A thermocouple located in the liquid at the base of the flash chamber measures the flash point temperature. Repeatability is quoted as ± 3°F.

The sample temperature, as measured by the thermocouple, exhibits a sawtooth pattern, the maximum reached during the heating cycle being the flash point temperature of the sample, the minimum being the temperature reached during the cooling cycle. The analyzer therefore has a finite cycle time which is dependent upon the sample flash point, and

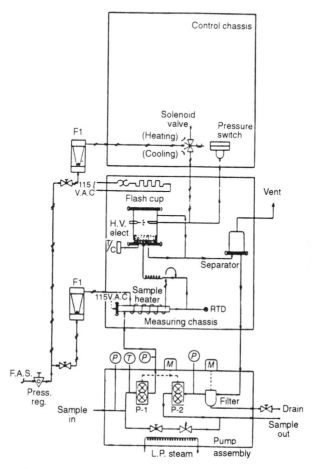

Figure 3.11 Precision Scientific general flash point analyzer. (Courtesy of Precision Scientific Inc.)

peak picking circuitry is required to provide a continuous analog output of the flash point temperature.

3.4.3.2 Continuous analyzers. The only process flash point analyzer of this type is the Rotork Analysis Ltd. (Hone) flashar, shown schematically in Figure 3.12, designed to measure the flash point of petroleum products up to a maximum of 480°F. When hydrocarbon vapors are mixed with air and heated, a vapor/air concentration is reached at which the components will react together spontaneously when exposed to a flame or spark. The flashar relies on the fact that the vapor concentration above a hydrocarbon liquid is essentially constant at the flash point temperature and is independent of actual flash point temperature.

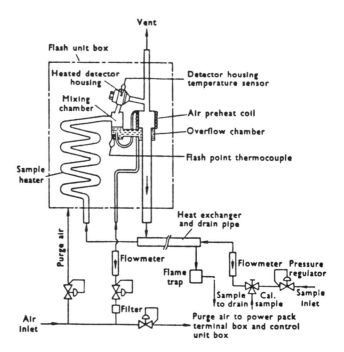

Figure 3.12 Rotork Analysis (Hone) flashar flash point analyzer. (Courtesy of Rotork Analysis Ltd.)

The analyzer uses a catalytic detector to monitor the hydrocarbon vapor concentration that corresponds with the flash point. The vapor reacts with the catalyst to give a reproducible catalyst temperature. The output of the detector is the input to a control system that seeks to maintain the sample temperature at a value that will sustain the required vapor concentration above the liquid. The sample temperature maintained is the flash point of the sample.

3.5 Pour point, cloud point and associated measurements

3.5.1 Background and definitions

Pour point and cloud point are two important characteristics of petroleum products. The *pour point* is the lowest temperature at which the material will pour (or flow) under a prescribed set of conditions. The *cloud point* is the temperature at which a waxy cloud appears in the liquid when cooled under a prescribed set of conditions.

They are important specifications of fuels for jet engines, marine and auto diesel engines and domestic heating equipment. They define the lowest temperature at which the product can be handled without causing pumping difficulties and wax plugging of filters. The ASTM standard laboratory tests, to be described later, are lengthy, tedious, of low precision and subject to human error and judgement, and have a lower precision than the corresponding laboratory test method.

Frequent and accurate measurement of these two parameters is required by petroleum refineries particularly when products are blended in-line to a pour or cloud point specification. Continuous on-line measurement enables finished product quality to be controlled precisely and allows more economic use of available feedstocks, thereby aiding optimization of refinery operations. In addition, on-grade production can be increased, with consequent savings in valuable feedstocks and blendstocks. Applications include monitoring of finished products from in-line blending processes, monitoring of feedstocks for jet fuels, kerosene, marine diesel fuels, auto diesel fuels and in lube oil production. A good example is the blending of heavy naphtha with catalytic gas oil to produce home heating oil or diesel fuel. The repeatability of the process analyzer permits fine tuning of the blending system to control the product to within 2°F of the specification. This saves a significant amount of heavy naphtha, which can then be used for jet fuel or kerosene production, or for feedstock to a catalytic reformer to produce high-octane gasoline blendstock. All of these products produce more revenue for the refinery.

3.5.2 Laboratory methods of measurement

3.5.2.1 Pour point (ASTM D97). The ASTM standard test method was adopted in 1914 as Method GO11 and has remained essentially the same except that today it is frequently carried out using automatic apparatus. It may be used to determine the pour point of any petroleum oil, with certain reservations regarding reproducibility and repeatability. For waxy oils such as fuel oils, the low temperature flow characteristics of the material depend to a certain extent upon handling, storage and thermal history. Therefore these characteristics may not be truly indicated by pour point and it may be difficult to obtain reproducible results.

For the maual test a cylindrical, flat-bottomed test jar of clear glass is used. The jar is fitted with an ASTM thermometer and mounted in a jacket. The complete assembly is immersed in a cooling bath. For pour points less than 50°F, two or more cooling baths are required. The sample is first heated to 115°F and then cooled to a temperature based on the expected pour point. The test jar is placed in the first cooling bath maintained at 30–35°F. At every 5°F drop in temperature, the test jar is removed, tilted just enough to determine whether there is any movement of the liquid, and then replaced.

If there is still movement of the liquid at 50°F, the jar is transferred to the second cooling bath maintained at 0–5°F and testing continued. If the sample has not ceased to flow at 20°F, the test jar is transferred to the third cooling bath maintained at −30 to −25°F. For measurement of very low pour points further cooling baths are required, each bath being maintained at 30°F below the preceding bath. If the sample does not flow when held in a horizontal position for 5 s, the temperature is noted, 5°F added and the result recorded as the pour point. The repeatability of the test is ±5°F and its reproducibility ±10°F.

3.5.2.2 Cloud point (ASTM D2500). This test originally formed part of ASTM D97, but was adopted as a separate test ASTM D2500 in 1967. This method covers only petroleum products that are transparent and have cloud points below 120°F. The apparatus used is similar to that for ASTM D97. The sample is brought to a temperature at least 25°F above the expected cloud point and moisture removed by a suitable method until the liquid is perfectly clear. The test jar is then filled to the mark and immersed in the first cooling bath. At every 2°F drop in temperature the jar is removed, inspected for cloud (haze) and replaced. The procedure thereafter is the same as for pour point except that inspections are made every 2°F. When inspection reveals a distinct cloud or haze in the bottom of the test jar, the temperature reading is noted and recorded as the cloud point. For distillate oils, the repeatability and reproducibility of the test are ±4 and ±8°F, respectively. For other oils both are ±10°F.

3.5.2.3 Freeze point (ASTM D2386). The apparatus for this test is a sample tube designed to hold 25 ml of the sample under test, which is immersed in a Dewar-type vaccum flask filled with a refrigerant such as acetone/solid carbon dioxide. The sample tube is fitted with a stirrer and a standard ASTM thermometer of range −80 to +20°C. As the sample is cooled while stirring continuously, the temperature is noted at which hydrocarbon crystals first appear. The sample tube is then removed from the coolant vessel and its temperature is allowed to rise, again stirring continuously. The temperature, to the nearest 0.5°C, at which the hydrocarbon crystals disappear is recorded as the freeze point. The repeatability and reproducibility of the test are ±0.7 and ±2.6°C, respectively.

3.5.2.4 Additional tests for residual fuel oils. As mentioned earlier, it is difficult to obtain reproducible results using the standard pour point test on residual fuel oils and heavy lubricating oils. In the past, pour point has been used as a guide for the pumpability of heavy oils, but it has been found that viscosity is a better criterion for determining the minimum temperature at which these products may be handled. The Institute of Petroleum (UK) carried out considerable work in this area with several laboratories and recommended that the viscosity of residual fuel oils should be used as a measure of *pumpability*. The standard test for pumpability uses a rotational viscometer (Ferranti Model VH is recommended) and is covered by ASTM D3245. The test is designed to give the storage point·and handling point for a given fuel oil, that is, the minimum temperatures required to store and handle the oil in normal fired boiler installations. The test takes about an hour to complete, but it does give better data on product characteristics than the pour point test. Another test for residual fuel oils is the measurement of *maximum fluidity temperature* and is covered by ASTM D1659.

3.5.2.5 Additional tests for distillate fuels. Neither pour point nor cloud point really give a good indication of how well a distillate fuel will perform in vehicle service. The temperature at which the operability limit is reached lies somewhere between the two. The pour point test is not sensitive enough, whereas the cloud point test is too sensitive, particularly when flow improvers are used. Considerable research, including field trials, was done in the mid-1960s in Europe which resulted in the *cold filter plug point* (CFPP) test method. This test has subsequently been shown to give an excellent indication of low-temperature performance of diesel and heating oils, particularly when they are treated with flow improvement additives. The procedure is covered in standard AFNOR M70/042 and in IP 309. Other similar test methods are DIN 51428 and SIS 155122.

A prescribed volume of sample is placed in the test jar and cooled under specified conditions. At every 1°F drop in temperature, a suction of 200 mm water is applied to the pipette to aspirate the sample through a metal filter. The CFPP of the sample is the temperature in °C at which it is no longer possible to draw 20 ml of sample through the filter in 60 s. Three cooling baths are required to cover the full temperature range and the test apparatus must be moved from one bath to the next as the test proceeds.

3.5.3 On-line process analyzers

3.5.3.1 Pour point analyzers. A typical example is the Precision Scientific general pour point analyzer Model 44670, shown schematically in Figure 3.13, which is designed to measure the pour point of

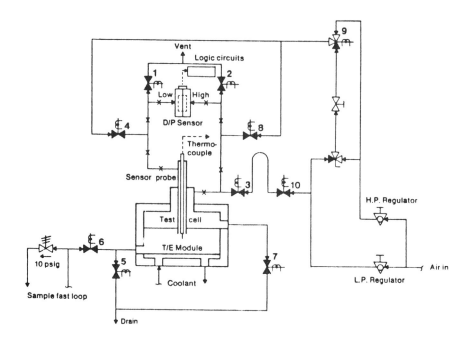

Figure 3.13 Precision Scientific Model 44670 pour point analyzer. (Courtesy of Precision Scientific Inc.)

hydrocarbons streams having viscosities up to 45 cSt at 100°F and pour points down to −58°F. The analyzer has a cyclic operation consisting of five sequences and is programmed to reset to sequence #1 on start-up or upon any interruption. Referring to Figure 3.13, the sequences are as follows:

(1) *Forced drain sequence*: Valves 4, 5 and 8 are energized, applying air to the test cell. T/E module current is reversed warming the test cell and allowing it to drain. Valve 5 is de-energized to terminate the sequence. Valves 4 and 8 remain energized.

(2) *Fill sequence*: Valves 6, 7 and 9 are energized, allowing sample to flow into the test cell. HP or LP air is applied to prevent sample from entering the sensing lines. Valve 6 is de-energized to terminate the sequence.

(3) *Level sequence*: Valves 4, 7 and 8 remain energized pressurizing the surface of the sample and forcing excess sample to drain. T/E module current is applied at normal polarity to cool the test cell. At the termination of this sequence, all valves are de-energized and the sensor probe is submerged below the surface of the sample.

(4) *Equalizing sequence*: Valves 1, 2 and 10 are energized, the test cell, DP sensor and all connecting lines are vented to atmosphere establishing a condition of zero DP in the sensing system. All valves close at the end of this sequence. Valve 10 in closing traps LP air in the tubing between valves 3 and 10.

(5) *Pulse sequence*: Valve 3 is energized. LP air trapped between valves 3 and 10 expands into the test cell pulsing the liquid surface outside the sensor probe and against the HP diaphgram of the DP sensor and the output voltage of the sensor rises. The pressure pulse also tends to push sample liquid up into the sensor probe, compressing the air inside the probe. As the sensor probe is connected to the LP side of the DP sensor, this results in a reduction of the DP sensor output voltage.

The control logic of the analyzer has a built-in time delay to give the sample time to compress the air inside the sensor probe after each pressure pulse. The output of the DP sensor is then compared with an adjustable set point voltage. If the output is below the set point, another equalizing and pulse sequence occur. As the sample cools, its resistance to flow increases and its ability to compress the air inside the sensor probe decreases, resulting in an increase in the DP sensor voltage. Eventually the output exceeds the set point, tripping a peak holding relay and resetting the analyzer to the forced drain sequence to begin another cycle. A thermocouple inside the sensor probe measures sample temperature continuously, the lowest temperature reached during the measuring cycle being the pour point.

3.5.3.2 Cloud point analyzers. A typical example is the Rotork Analysis Ltd. (Hone) cloudar, the operating principle of which is shown in Figure 3.14, and which is the latest microprocessor-based analyzer of a product range originally manufactured by Hone Instruments (UK) since the late 1960s. The analyzer operates on an optical principle for cloud point detection, with the detector mounted at 90° to the light beam so that there is no output initially. When wax crystals form in the sample, light is reflected from the cloud and received by the detector. The operating sequence has three steps:

(1) *Fill cycle*: The test cell is flushed with process sample at normal temperature for 20–45 s.

(2) *Cool cycle*: The cell is cooled by a T/E module and the heat extracted by a cooling water system.

(3) *Read cycle*: The cloud point is detected by a photocell system. This intiates a 1 s read cycle during which the cloud point temperature is read and stored. If no cloud is detected within 25 min, the analyzer is automatically recycled and the sequence of operations repeated.

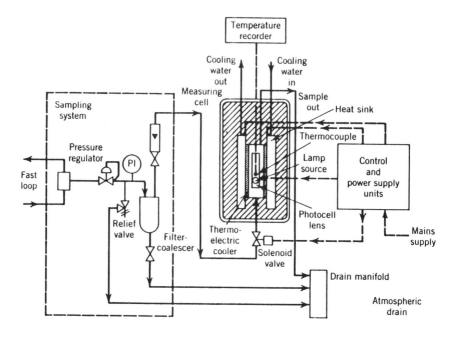

Figure 3.14 Rotork Analysis (hone) cloudar cloud point analyzer. (Courtesy of Rotork Analysis Ltd.)

The minimum cloud point temperature for the analyzer is 140°F below the temperature of the cooling water used. Analyzer cycle time is from 1 to 4 min and repeatability is quoted as better than ±1°F.

3.5.3.3 Freeze point analyzers. Process freeze point analyzers are usually adaptations of the cloud point analyzer described earlier. A typical example is the Rotork Analysis Ltd. (Hone) Freezar, a microprocessor-based instrument with essentially the same design as the cloudar except that it has a slightly different operating sequence. In this case the sample is cooled until the presence of wax crystals is detected by the photocell. The test cell is then warmed by reversing the current in the T/E module, and the freeze point recorded as the temperature at which the wax crystals disappear, as detected by the photocell.

3.5.3.4 Cold filter plug point analyzers. A typical example is the Precision Scientific Model 44675 CFPP analyzer, which is an adaptation of the company's pour point analyzer described earlier and is designed to measure CFPP over the range −58 to +50°F with a repeatability of ±2°F. Referring to Figure 3.15, in this case the test probe has a 45 mm filter screen mounted at its base. As for the pour point analyzer, the

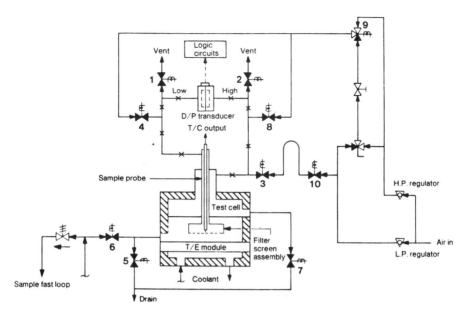

Figure 3.15 Precision Scientific Model 44675 CFPP analyzer. (Courtesy of Precision Scientific Inc.)

CFPP analyzer has five operational sequences for each measurement cycle. In the test sequence, when the air pressure pulse is applied to the sample surface, a small quantity of sample is pressurized through the filter screen, compressing the air inside the probe. The DP transducer, sensing pressure drop across the filter screen, detects the point at which wax crystals plug the filter screen, preventing sample flow. At this point, the DP transducer trips and initiates another measurement cycle and the lowest temperature achieved during the measurement cycle is reported as the CFPP of the sample.

3.6 Vapor pressure measurement

3.6.1 Background and definitions

It is a well-known fact that a gas fills any space to which it has access, whereas a liquid takes the shape of the vessel into which it is placed and a liquid surface is formed. If a volatile liquid is placed in a closed vessel, the liquid will evaporate and the pressure above the liquid will increase until an equilibrium is attained. A point is reached where the system appears to be static (there is no change in liquid level or vapor space pressure). However, the gas molecules are in constant motion, colliding with the surfaces of the vessel (causing pressure) with the surface of the liquid, losing kinetic energy and sometimes returning to the liquid. When the rate at which molecules leave the liquid is the same as the rate at which they return to the liquid, the system is said to be in equilibrium. It is a dynamic equilibrium; although no overall change is observed, two exactly equal and opposite changes are taking place.

The pressure of the gas molecules above the liquid at equilibrium is known as the *vapor pressure* of the liquid at the temperature specified.

Vapor pressure is an important measurement parameter, particularly in the petroleum and petrochemical industries. The relative volatilities of the components of a hydrocarbon mixture forms the basis of all fractionation processes (see section 3.3). Since the vapor pressure of a product is an indication of its purity, vapor pressure is often an important specification and quality parameter. The vapor pressure of a process stream will increase as the concentration of the low-boiling components increases. The concept of vapor pressure is also important in the design of sample handling systems, where a liquid must be vaporized prior to analysis.

Another parameter related to vapor pressure is the *vapor–liquid ratio* (*V/L*). This is the volume of vapor produced at a given temperature and an absolute pressure of 760 mm HG from a given volume of liquid. The *V/L* ratio is sometimes used by oil companies as a specification parameter for gasolines (some companies require a 20:1 ratio). Alternatively, the vapor temperature at a given *V/L* ratio may be measured.

3.6.2 Laboratory methods of measurement

Correlation and validation of process vapor pressure analyzers must be made against the ASTM standard laboratory test methods described below.

3.6.2.1 Reid method ASTM D323. This method was first introduced in 1930 and since then has been the universally accepted standard for the petroleum industry. It covers determination of the vapor pressure of volatile crude oils and volatile non-viscous petroleum products, except LPG which is covered by ASTM method D1267.

The test apparatus consists of two vessels, the air vessel and the sample vessel. The ratio of the volumes of these vessels must be held within the limits 3.8 to 4.2. The sample vessel is filled with chilled sample and connected to the air vessel and the complete apparatus immersed in a constant temperature bath at $100 \pm 0.2°F$ and shaken periodically until equilibrium is reached. The reading of a precision pressure gauge attached to the apparatus is recorded as the *Reid vapour pressure* (*RVP*) of the sample.

From this description, it would appear that this is a relatively simple and straightforward method. In fact it is a method requiring meticulous attention to detail and preparation. The most important part of the test is not the test itself but the collection, handling and transfer of the sample to the test apparatus. Measurement of vapor pressure is extremely sensitive to losses by evaporation or to slight changes in sample composition. Therefore the utmost care must be taken in sample handling. The precise procedure is laid down in ASTM D323 which makes reference to ASTM D270 'Sampling of Crude Petroleum and Petroleum Products'. Today, commercially available automatic RVP measurement systems are often used in the laboratory.

3.6.2.2 Other methods. ASTM standard test methods D-5190 and D-5191 are methods for measuring the true (absolute) vapor pressure of volatile liquid petroleum products, using commercially available automatic test apparatus. The chilled sample cup of the instrument is filled with chilled test sample and coupled to the instrument inlet fitting. The sample is then forced from the sample chamber to the expansion chamber where it is held until thermal equilibrium at 100°F is reached. In this process, the sample is expanded to five times its volume (4:1 vapor/liquid ratio). The vapour pressure is then measured by a pressure transducer.

3.6.3 On-line process analyzers

3.6.3.1 Reid vapor pressure analyzers. A typical example is the Precision Scientific Reid monitor Model 44770, shown schematically in Figure 3.16. This analyzer is designed to duplicate the requirements of ASTM D-323, namely, that the measurement is conducted at 100°F using an air-saturated sample with a vapor/liquid ratio of 4:1. The analyzer readout is made directly in RVP units. Sample enters the analyzer via a feed pump (one head of a duplex pump assembly) at a constant flowrate of 100 ml/min. The exhaust pump (second head) is designed to maintain an extremely accurate flowrate of 500 ml/min of two-phase product, thus establishing the require V/L ratio. As sample flows into the saturation chamber it lifts the float and needle valve assembly away from the orifice. The exhaust pump creates a vacuum in the lower section of the chamber, causing the sample to flash across the orifice into the 4:1 vapor/liquid state. The pressure in the lower section of the chamber is sensed by a pressure transmitter. Instrument measurement range is 2–19 RVP and repeatability is quoted as ±0.1 RVP. The analyzer response time is 60 s for a 63% response to a step change in sample RVP at the analyzer inlet.

3.6.3.2 Kinetic vapor pressure analyzers. A typical example is the Rotork Analysis Model 1354 vapor pressure analyzer, shown schematically in

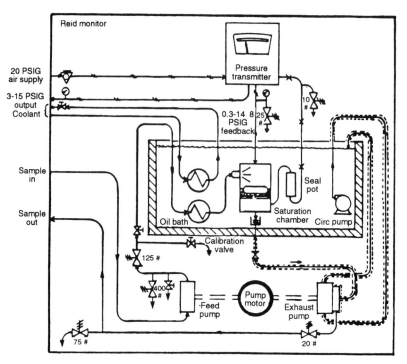

Figure 3.16 Precision Scientific Model 44770 Reid monitor. (Courtesy of Precision Scientific Inc.)

Figure 3.17. The operating principle is similar to that of the common laboratory ejector pump. The sample is forced through a jet orifice in which the increase in fluid velocity produces a pressure drop sufficient to induce incipient vaporization of the sample.

Process sample enters the constant temperature bath (controlled at 100°F) through a pressure regulator, then passes through a heat exchanger and nozzle assembly and out of the analyzer through a second (back-pressure) regulator. As the process sample passes through the nozzle assembly, its velocity increases and the pressure downstream of the nozzle is the effective vapor pressure of the liquid. The nozzle position is readily adjustable. The vapor chamber of the ejector is connected to an absolute pressure transmitter which can be supplied with either pneumatic or electronic transmission. Analyzer response is quoted as 43 s and repeatability as ±0.1 psi

3.6.3.3 Vapor/liquid ratio analyzers. The only known instrument of this type is the Core Laboratories **VOLRAC** vapor/liquid ratio analyzer, shown schematically in Figure 3.18, originally manufactured by Ethyl

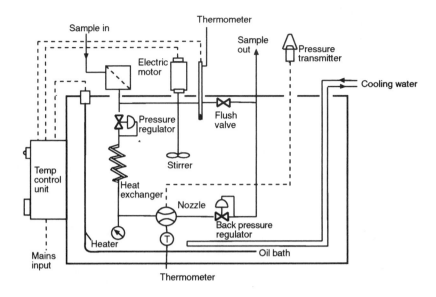

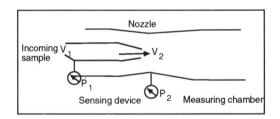

Figure 3.17 Rotork Analysis Model 1354 vapor pressure analyzer. (Courtesy of Rotork Analysis Ltd.)

Corporation. This analyzer is designed to measure the V/L ratio of in-line blended gasolines and readings may be correlated with ASTM D-2533 results.

Gasoline sample first enters the fuel conditioning system and then passes through mechanical chiller at the top of the unit, where its pressure is reduced to atmospheric and excess sample passed to the waste outlet. The fuel metering pump delivers a constant flowrate of chilled sample to an orifice located at the inlet to the vapor/liquid separator. This device is the key component of the system. Here the liquid sample is quickly heated and an extremely thin film dispersed over a relatively large surface area, resulting in an equilibrium V/L relationship. Liquid is removed to waste and vapor continuously removed from the side

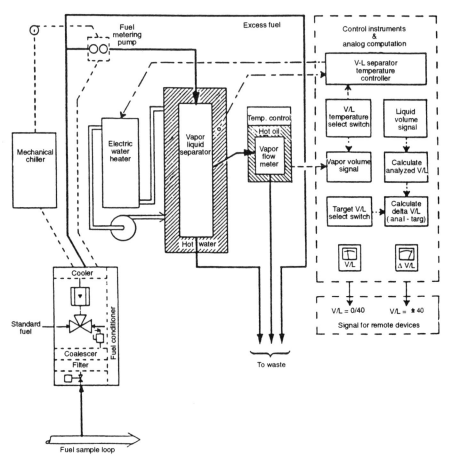

Figure 3.18 Core Laboratories VOLRAC V/L ratio analyzer. (Courtesy of Core Laboratories Inc.)

connection and passed through a highly accurate laminar flowmeter, using a differential pressure, reluctance type transducer. Since sample liquid is introduced at a constant flowrate, calculation of the V/L ratio is a direct function of the vapor flowrate.

Two forms of compensation are provided. One corrects the vapor volume signal for temperature, the other corrects for variations in barometric pressure. The analyzer provides a delta V/L signal where:

$$V/L = \text{analyzed } V/L - \text{target} V/L$$

and target V/L is the desired quality for the fuel.

The standard V/L measurement range is 10–30 and temperature values may be selected from within the range 100–150°F.

3.6.4 Recent developments

As part of its new VISTA on-line analytical products, ABB Process Analytics recently introduced the microprocessor-based Model 4100 Reid vapor pressure analyzer, designed to measure the RVP of gasoline, crude oil and other volatile petroleum products. Readings may be correlated with ASTM D-323 results.

The measurement cell is designed so that the *V/L* ratio will meet the requirements of ASTM D-323. The cell temperature is controlled at $10\pm0.1°F$. The control section contains the power supply, single-board microprocessor, LCD display and operator switches. The measurement section contains the measurement cell with its associated components such as solenoid valves, heater, level transmitter, etc. The cycle time of the analyzer is 7.5 min. During this time, the measurement cell is purged, zeroed, and filled with sample. After sample equilibration, the final

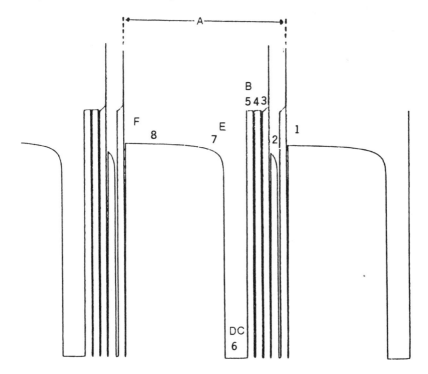

Figure 3.19 ABB process analytics model 4100 RVP analyzer. **Transducer output**: *Once cycle*: 1 Purge cell; 2 False fill; 3,4,5 3 each 20 s purges of cell; 6 Zero pressure reading; 7 20 s sample filling cell; 8 4 min equillibration and final pressure reading; *Diagnostics during one cycle*: A Temperature and temperature runaway monitoring and over pressure protection; B Air purge pressure test; C Zero reading pressure transducer test; D Level sensor test; E Sample fill test; F Cell leak test and RVP temperature confirmation. (Courtesy of ABB Process Analytics Inc.)

pressure reading is obtained. Throughout the cycle, the microprocessor performs self-diagnostics for temperature, purge pressure, zero reading, level sensor, sample filling, cell leaking, as well as over-pressure protection. A typical analysis sequence is shown in Figure 3.19. The status of each diagnostic routine is shown in sequence on the control unit front panel diagnostics display. The 4–20 mA DC output is held at the current value and updated after each cycle. Measurement range is 0–20 psi with a repeatability of ±0.05 psi.

3.7 Density and specific gravity measurement

3.7.1 Background and definitions

The *density* of a substance is defined as the mass (M) per unit volume (V) of the substance at a specified temperature, and may be measured in a variety of different units, such as g/cm^3, lb/ft^3, lb/gallon and kg/m^3.

The *specific gravity* (or *relative density*) of a liquid is defined as the ratio of the density of the liquid at a specified temperature to the density of water at the same temperature. Specific gravity is often referred to a base temperature of 60°F (15.5°C) or 39.2°F (4°C). Since the density of water at 4°C is 1.000 g/cm^3, the specific gravity is numerically equal to its density at this temperature.

The *specific gravity* of a gas is defined as the ratio of the density of the gas to the density of air at a specified temperature and pressure. The conditions are usually specified as *standard temperature and pressure* (*STP*), 0°C and 14.7 psia.

Temperature, pressure and the nature of the fluid all affect the density. Except for fluids such as LPG, liquids can be considered as incompressible fluids and pressure variations have a minimal effect on the density. However, temperature variations have a considerable effect on the density and it is generally necessary to provide temperature compensation if accurate measurement is required.

Although specific gravity is a dimensionless quantity, there are several units related to specific gravity in common industrial use. For example, *degrees API* is used widely in the petroleum industry and *degrees Baume* in the chemical industry. Hydrometers calibrated in these units are available for laboratory measurement of specific gravity.

The density of a gas is dependent upon temperature, pressure and the nature (composition) of the gas, and may be described in terms of the *ideal gas equation* and its deviations.

$$PV = nRT$$

where P is the absolute temperature, V is the volume, n is the number of moles of gas, R is the universal gas constant, and T is the absolute temperature.

For real gases, the ideal gas equation only applies at low pressures and relatively high temperatures. One parameter commonly used as a measure of the deviation of a gas from ideality is the *compressibility factor* (Z) where:

$$Z = PV/nRT$$

The density of a gas (p) is given by:

$$\rho = Mn/V$$

where M is the molecular weight. Substitution in the non-ideal gas equation gives:

$$\rho = MP/ZRT$$

Therefore, the density of a gas may be calculated from this equation, ensuring of course that all units used in the equation are compatible, with special attention given to the value of R used. This basic principle is often used for preparation of standard gases of known density for analyzer calibration.

3.7.2 Laboratory methods of measurement

3.7.2.1 Liquids. Traditional laboratory methods for measurement of density and specific gravity of liquids use two different techniques:

- Methods involving the use of pycnometers (weighing principle)
- Methods involving the use of hydrometers (buoyancy principle)

The pycnometer, often known as the 'density bottle' in high-school jargon, is a small borosilicate glass vessel of accurately defined volume with a mass not greater than 30 g. It is cleaned, dried and weighed empty to the nearest 0.1 mg. It is then filled with the sample under test and immersed in a constant temperature bath at the test temperature. When the sample has equilibrated to the test temperature, the pycnometer is removed, its exterior cleaned and dried, allowed to cool and then weighed again to the nearest 0.1 mg. The density of the sample is calculated from the expression:

$$\text{Density} \, (\text{g.cm}^3) = \frac{W_2 - W_1}{V}$$

where W_1 is the mass of the pycnometer empty (g), W_2 is the mass of the pycnometer and sample (g), and V is the volume of the pycnometer (cm^3).

Based on Archimedes Principle, the hydrometer is usually made of glass and weighted so that it will float upright in the liquid. The upper stem incorporates a graduated scale. When immersed in the liquid under test, the hydrometer sinks to a depth dependent upon its weight in the liquid and therefore upon the density or specific gravity of the liquid. Many different types of hydrometer are available as covered by ASTM standard E100.

3.7.2.2 Gases. Laboratory methods are based on commercially available instrumentation and the various types are recommended in ASTM method D1070. The majority of these instruments are similar in design to process analyzers. Many are direct reading and require no calibration subsequent to that done initially. These instruments may be classified into the following types according to operating principle:

> Pressure balance
> Kinetic energy
> Displacement balance
> Centrifugal force

The following instruments are commercially available:

> Ac-Me gravity balance
> Arrco-Anubis gas gravitometer
> Ranarex gas gravitometer
> Ac-Me recording gravitometer
> Kimray gravitometer
> UGC gas gravitometer

The Arrco-Anubis, Ranarex and UGC instruments are virtually the same as the process analyzers manufactured by these companies.

3.7.3 On-line process analyzers

There is a wide variety of process analyzers available for on-line measurement of density and/or specific gravity and these may be conveniently classified according to their operating principles as follows:

- Buoyancy (Archimedes principle)
- Vibration principle
- Balance beam (weighing) principle
- Radiation absorption principle
- Kinetic energy principle
- Aerostatic weighing principle

Commercially available equipment is listed in Table 3.5 and typical examples of the various types are described below.

Table 3.5 Commercially available process density and specific gravity analyzers

Manufacturer	Model number	Measurement principle	Sample
Anton Paar (Austria)	DPR 2000 Series	Vibrating tube	Liquid
Arrco Inst. Co. (USA)	Anubis Series	Weighing	Liquid
		Aerostatic Weighing	Gas
Automation Products (USA)	Dynatrol CL-10TY	Vibrating Tube	Liquid
EG & G Chandler (USA)	Ranarex Series	Kinetic Energy	Gas
UGC Industries	278	Vibrating Tube	Liquid
	241	Buoyancy	Gas/Liquid
	GR6-01	Balance Beam	Liquid
ITT Barton (USA)	Series 660	Vibrating Vane	Liquid
Kay-Ray (USA)	3600	Radiation	Liquid
Nuclear Research (USA)	Gammatrol Series	Radiation	Liquid
Ohmart (USA)	Densart Series 3400	Radiation	Liquid
Princo Insts. (USA)	Densitrol W-700	Bouyancy	Liquid
Ronan Eng. (UK)	Series X-91	Radiation	Liquid
Rosemount, National Sonics (USA)	Sensall Series 4900	Sonic Attenuation	Liquid
SPRI Ltd. (UK)	Gravitymaster	Balance Beam	Liquid
Sarasota Automation (USA)	FD-810/820	Vibrating Tube	Liquid
	FD-700 Series	Vibrating Spool	Gas
Solartron Transducers (USA)	Series 7810	Vibrating Spool	Gas
	Series 7830/7840	Vibrating Tube	Liquid
TN Technologies (USA)	E-Z Cal Series	Radiation	Liquid
Toray Industries (Japan)	DFL Series	Bouyancy	Liquid
Valmet (Finland)	Dens-EL	Balance Beam	Liquid
Yokagawa (Japan)	VD6D	Vibrating Tube	Liquid

3.7.3.1 Bouyancy principle. A typical example is the Model 241 electronic densitometer manufactured by UGC, EG & G Chandler (USA), shown in Figure 3.20, which is designed for vapor or liquid service.

The measurement system consists of a beam suspended at a pivot point, with a float attached to one end of the beam. The bouyant force exerted on the float by the sample provides a moment around the pivot point that varies in direct proportion with the density of the sample surrounding the float. Attached to the end of the beam opposite the float are two 'slugs' forming part of an electrical force balance system. One senses the position of the deflected beam, the other is coupled to a restoring coil that drives the beam to the null-balance position. The electronic circuitry produces a high-frequency AC signal which is applied to a differential transformer detector that senses the position of the float beam. As the sensing slug moves in or out of the magnetic field of this detector, the amplitude of the signal is changed. The resultant voltage output is rectified and controls the current passing through the restoring coil. The current required to achieve balance, which is proportional to the density of the process sample, is highly dampened to give an instrument response of about 30 s.

The analyzer is suitable for pressures up to 2500 psig and has an operating temperature range of 0–140°F. Minimum range is 0–1 lb/ft^3. Repeatability is quoted as ±0.25% of full scale.

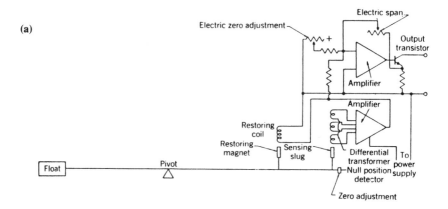

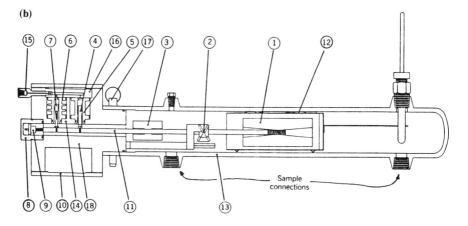

Figure 3.20 UGC Model 241 electronic densitometer. (a) Operating principle; (b) sensing unit. 1, float; 2, pivot; 3, counterweight; 4, restoring magnet; 5, restoring coil; 6, sensing coil; 7, sensing slug; 8, calibration plug; 9, zero calibration weight; 10, knurled sleeve; 11, balance beam; 12, float shield; 13, barrel chamber; 14, terminal strip (not shown); 15, vent; 16, coil cover; 17, closure pin; 18, plug-in circuit board (not shown). (Courtesy of EG & G Chandler Engineering Inc.)

3.7.3.2 Vibration principle. A typical example is the Sarasota Automation series 700 density meter, shown in Figure 3.21, which is designed for vapor or liquid service. The sensing element is a thin-walled cylinder or 'spool' of a magnetic stainless alloy, maintained in circumfernetial oscillation by an electromagnetic circuit. The spool is totally surrounded by the process sample and is mounted in a meter body, secured so that one end is fixed and the other free. The drive coil sets the spool into oscillation and the movement is detected by the pick-up coil. The resultant signal is amplified and fed back to the drive coil to maintain

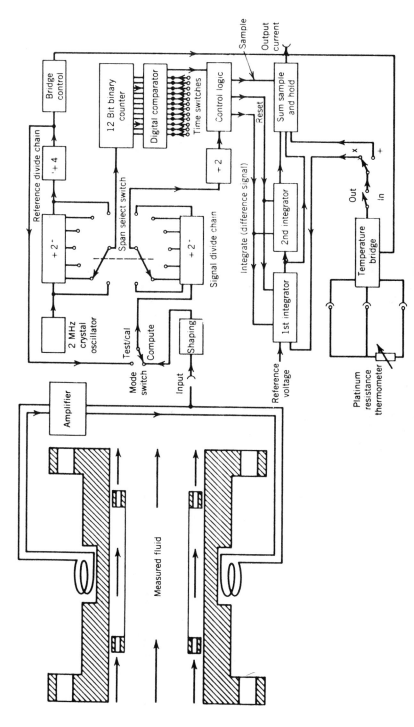

Figure 3.21 Sarasota Automation series 700 density meter. (Courtesy of Sarasota Automation Inc.)

the spool in oscillation. The resonant frequency is is dependent upon the total mass of the vibrating system. Therefore, any change in the density of the process sample will result in a change in oscillation frequency. This change is measured very accurately by the density converter, which provides an output proportional to sample density. The Model FD700 is designed for use in bypass (on-line) installations for gas and clean, non-viscous liquids, whereas the Model ID700 is designed for direct insertion in the process line and may be fitted with a retraction device to allow removal without the need for process shutdown. The latter design has the advantage that the sensing element is at the process temperature and pressure, which is particularly important for high-accuracy gas measurements. The meter body is designed for operating pressures up to 2000 psig.

3.7.3.3 Balance beam (weighing) principle. A typical example is the Arrco Anubis liquid gravitometer, which is essentially a spring balance that measures the weight of a fixed volume of a process liquid in terms of its specific gravity. The closed system consists of a bulb of 1000 ml volume, inlet tube and outlet tube. These tubes also act as the spring elements of the weighing mechanism. The bulb is suspended by suitable links and fulcrums at the left of the main beam. At the opposite end is a counter weight and a vernier weight (used for adjustment). As the specific gravity of the process liquid increases, the bulb tends to move downward, this movement being resisted by the spring action of the inlet and outlet tubes. The motion of the bulb is therefore proportional to the change in specific gravity of the sample liquid. This motion is transferred to the main beam and from there by suitable linkages and fulcrums to a recorder, indicator or transmitter.

3.7.3.4 Radiation absorption principle. A typical example is the Kay-Ray Model 3600X digital density system, which is designed for liquid and slurry service using a non-invasive technique. A caesium-137 radioactive source housed in a lead-shielded holder emits a highly collimated beam of γ-energy which passes through the pipe wall and the process fluid to an ion chamber detector mounted directly opposite the source holder. The detector generates an electrical signal proportional to sample density which is transmitted to a remotely located control unit for amplification, scaling and calibration in density units.

   The source holder and detector housing may be bolted to any existing process pipe with a simple mounting saddle. Since the system is non-contacting, it is unaffected by variations in sample temperature, pressure, corrosivity, abrasiveness, viscosity and solids build-up, and installation does not require process downtime. The microprocessor-based control unit continuously and automatically performs input signal linearization, scaling, compensation, etc. as well as internal diagnostic routines to check system performance. In addition, the system has integral three-mode and proportional controllers for single-loop automated control and can communicate with or be controlled by a host computer for data logging and report generation. A digital display is provided for readout of process parameters.

Measurement accuracy is quoted as ± 0.0005–0.001 SG units and response time is adjustable between 5 and 60 s. The Nuclear Regulatory Commission (NRC) permits sale and installation of Kay-Ray density systems under a general license.

3.7.3.5 Kinetic energy principle. A typical example is the Ranarex gas specific gravity analyzer, shown schematically in Figure 3.22. The instrument consists essentially of two cylindrical, gas-tight measuring chambers, each with separate inlet and outlet sample connections. Each chamber contains an impeller and an impulse wheel, both with axial vanes. These wheels are mounted on separate shafts, facing each other but not touching. An electric motor, and drive belt rotate the impellers at the same speed and in the same direction.

The impellers draw continuous flows of the process gas sample and dry reference air into their respective chambers and rotate them against the vanes of the companion impulse wheels. They create torques, proportional to the density of the sample gas and reference air, respectively, which are transmitted from the chambers by pivot shafts on which wheels are mounted. The wheels are restrained from continuous motion by a flexible tape that is wrapped over the wheel rims in the crossed direction.

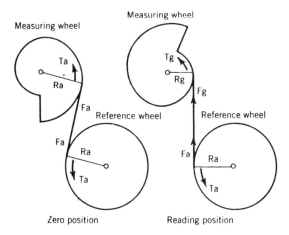

Figure 3.22 Ranarex gas specific gravity analyzer. (Courtesy of EG & G Chandler Engineering Inc.)

The torques, which are proportional to the densities, are thus opposed to each other. Therefore the measuring system divides the torque of the sample gas by the torque of the reference air. Since

$$\frac{\text{Torque of sample gas}}{\text{Torque of reference air}} = \frac{\text{Density of sample gas}}{\text{Density of reference air}}$$

the instrument measures the specific gravity of the sample gas. The rim of the measuring wheel is contoured so that it rotated through equal angels of equal changes in specific gravity. The uniform motion of the measuring wheel is transmitted by a linkage to the readout element, which may be an integral indicator, recorder, pneumatic or electronic transmitter. Ranarex analyzers are available as portable types or for permanent installation.

3.7.3.6 Aerostatic weighing principle. A typical example is the Arrco Anubis gas gravitometer, shown schematically in Figure 3.23, which is a direct weighing instrument designed to produce a temperature, barometric pressure and humidity-compensated recording of gas specific gravity.

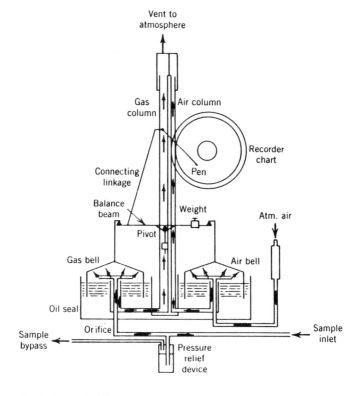

Figure 3.23 Arrco Anubis gas gravitometer. (Courtesy of Arrco Instrument Co.)

The equipment consists of two inverted thin-wall 'bells' suspended on knife-edge links with the lower bell edges submerged in oil to form a gas-tight seal. The bells are of equal size and weight, and are placed at equal distances from the fulcrum of a balance beam that drives the recording mechanism. One bell contains the sample gas, the other dry reference air. The difference in weight produces an unbalance of the beam which is proportional to the sample gas specific gravity. This type of instrument is used for both laboratory and on-line analysis.

3.8 Heating value measurement

3.8.1 Background and definitions

Heating value is an important measurement parameter in the Natural Gas industry, in other industries where gas is burnt for heating purposes and for domestic heating appliances where uniform heat flow at a given setting is critical. With today's extensive use of gas as a heating medium there is a need to measure and control the heating characteristics of a fuel gas to obtain optimum combustion control. In addition, in order to meet the objectives of energy conservation and smoothly functioning heating control systems, it is necessary that fuel gases be blended to provide uniform heating value and that variations be compensated for in the control system.

The *calorific value* (H) of a fuel gas is a measure of the heat produced by unit mass or volume of the gas. Typical industrial units are calories/g, BTU/lb and BTU/SCF.

The *British Thermal Unit* (*BTU*) is defined as the heat required to raise the temperature of 1 lb of water from 60 to 61°F. Another commonly used unit of measurement in the Gas industry is the *Centigrade Heat Unit* (*CHU*) which is defined as the heat required to raise the temperature of 1 lb of water from 15 to 16°C.

One frequently hears the terms *gross calorific value* (H_g) and *net calorific value* (H_n) of a fuel. Calculation of H_g is based on the condensation of water produced in burning the fuel, so that the latent heat of vaporization of water is recovered. In practice, this recovery is often uneconomic and the net calorific value must be calculated. The relative difference between these two units of measurement ranges from about 2% for some coals to up to 10% for natural gas.

Another well-known parameter is the *Wobbe index* or *Wobbe number* (W_o), which is defined as the ratio of the net calorific value and the specific gravity of the fuel (H_n/G_s). The significance of Wobbe index is that it relates fuel gas heating characteristics in a manner that is useful for blending fuel gases or obtaining a constant heat flow from a gas of varying composition. Table 3.6 lists the heating characteristics of some typical fuel gases. The Wobbe index generally groups these gases in a more usable format than classification by calorific value alone. Thus, for wide variations in specific gravity and calorific value due to composition variations, the Wobbe index varies over a much narrower range, providing a more realistic parameter by which to control the heating value.

Table 3.6 Typical fuel gas heating characteristics

Type of gas	Specific gravity	Calorific value	Wobbe index
Blast furnace gas	1.000	100	100
Producer gas	0.878	129	138
Coal gas	0.431	527	803
Reformed gas (refinery)	0.497	530	752
Reformed gas (natural)	0.472	530	771
Natural gas (I)	0.635	1013	1271
Natural gas (II)	0.650	1145	1420
Refinery gas (high olefin)	0.890	1468	1556
Refinery gas (high paraffin)	0.990	1644	1652
Propane (commercial)	1.523	2550	2066
Butane (commercial)	1.950	3200	2292

Whereas Wobbe index has advantages from a combustion control standpoint, calorific value (or heat of combustion) is still used as a quality control and specification parameter for gas sold on a custody transfer basis. The mass heat of combustion of a fuel is particularly important to weight-limited vehicles such as airplanes, whereas the volume heat of combustion is important to volume-limited vehicles such as automobiles and ships

3.8.2 Laboratory methods of measurement

Methods used in the laboratory for measurement of calorific value (or heat of combustion) generally fall into two categories:

- Methods using the continuous recording calorimeters
- Methods using the bomb calorimeter

ASTM standard test method D1826 covers measurement of the calorific value of gases in the natural gas range by continuous recording calorimeter. The recommended instrument is the Cutler–Hammer calorimeter which is essentially a process analyzer and is described in section 3.8.3. An extensive investigation of the accuracy of this instrument when used with gases of high heating value was conducted by the National Bureau of Standards (NBS) in 1957, under a research project sponsored by the Americal Gas Association (AGA).

ASTM standard test method D2382 covers measurement of the heat of combustion of hydrocarbon fuels by the bomb calorimeter, and is designed specifically for use with avaiation turbine fuels where the permissible difference between duplicate determinations is of the order of 0.1%. It can also be used for a wide range of volatile and non-volatile materials where slightly greater difference in precision can be tolerated. This is a high-precision test requiring meticulous attention to detail and strict adherence to all procedures. Details of the required apparatus is covered in the Annex to the test method.

3.8.3 On-line process analyzers

3.8.3.1 Traditional gas calorimeters. The Cutler–Hammer recording gas calorimeter Model 16310 is the latest version of this instrument manufactured by Fluid Data and is used as a standard by the US and UK gas industries. Once a rather bulky instrument, several modifications have been made over the years to make it more compact and amenable to plant installation and it is now microprocessor controlled. A full description of this analyzer is given in ASTM D1826.

The sample gas is burnt under controlled conditions and the heat generated by combustion is imparted to a stream of heat absorbing air. The increase in temperature of the air is directly proportional to the calorific value of the sample gas. Flowrates of sample gas, combustion air and heat absorbing air are maintained in a fixed volumetric ratio by three separate wet-gas meters that are motor driven, geared together and work in a common water seal. Readings are independent of atmospheric pressure, temperature and humidity variations, since metering of the gas and air streams is carried out under identical conditions. The temperature measuring devices are calibrated and the measuring circuit designed so that the recording instrument can be calibrated directly in calorific value units (e.g. BTU/SCF at 30 in Hg and 60°F saturated). Overall instrument range is 75–3600 BTU/SCF and repeatability is quoted as ±5% of full scale. An available option is the SMART-CAL digital display system which computes average BTU and has printer and host computer data links, together with a solenoid-operated calibration gas switching panel. Since this is an open-flame device, precautions must be taken regarding its installation.

3.8.3.2 Other open-flame type analyzers. A typical example is the Fluid Data series 7000 high-speed calorimeter, shown schematically in Figure 3.24, which is microprocessor based and designed to measure either calorific value (gross or net) or Wobbe index. The quantities of sample gas and combustion air are accurately metered and burned under closely controlled conditions. The exhaust gas temperature is measured by a precise sensing system and the amount of combustion and cooling air to

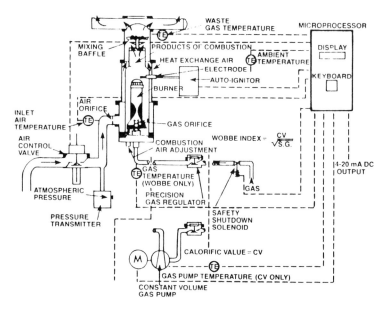

Figure 3.24 Fluid Data series 700 high-speed calorimeter. (Courtesy of Fluid Data Inc.)

the burner controlled to maintain the exhaust gas at a constant temperature as the heating value of the sample gas varies. The flowrate of air, which is proportional to the heating value of the sample gas, is an input to the microprocessor, which provides a digital display, a 4–20 mA DC output and facilities for autocalibration.

For measurement of Wobbe index, a precision gas regulator is provided on the sample gas supply. For calorific value measurement, a constant-volume pump compensates for variations in sample gas specific gravity. Standard analyzer range is up to 3000 BTU/SCF and repeatability is quoted as ±3% of upper range value. Response time is quoted as 6–8 s, but experience has shown that it is somewhat higher (30 s).

3.8.3.3 Other methods of measurement.

3.8.3.3.1 Gas chromatography. Since the introduction of microprocessor-based GC systems, on-line measurement of heating value by GC has been made easier by virtue of the microprocessor's computational ability. The technique involves separation and measurement of the concentration of all components present, followed by normalization and computation of the average heating value from pure component mole fractions and heating values. The technique is satisfactory for relatively simple gas mixtures, such as natural gas, but has difficulties with more complex mixtures such as refinery fuel gases, where composition variations are considerable:

- The GC column must separate all components for measurement.
- The column separation must take account of all expected variations in stream composition.
- The more complex the analysis, the longer the GC cycle time.

In the majority of refineries and petrochemical plants, measurement of fuel gas heating value usually forms part of a BTU control system, where long cycle times are unacceptable. Consequently, the process GC has found only limited application in these services. However, the technique has been used extensively in the natural gas industry and some GC systems are designed specifically for this application. The major microprocessor-based process GC manufacturers all offer their equipment for heating value measurement.

3.8.3.3.2 Correlative techniques. Traditional infrared absorption (NDIR) has been used extensively in industry for many years in applications involving measurement of a single component in a gaseous or liquid mixture. Its success depends upon the correct selection of measurement and reference wavelengths and therefore selectivity is an important consideration because of interference problems and background composition variations. Consequently, NDIR has not been very successful in heating value measurement applications and several analyzers designed specifically for this service are no longer in production. With the introduction of new spectroscopic techniques such as Fourier-transform infrared spectroscopy (FTIR), which involve full spectrum scanning with microprocessor control and chemometrics software, the problems associated with NDIR have been eliminated and FTIR is now finding application in fuel gas heating value measurement.

3.9 Measurement of octane number

3.9.1 Background and definitions

Since the early 1930s the petroleum industry has used standard ASTM laboratory procedures to evaluate the expected performance of gasoline in internal combustion engines. The so-called octane rating system for motor gasolines measures the tendency of the fuel to 'knock'. Knocking is the production of a series of high-pressure waves that bounce off the cylinder walls of the engine. This is evident as a metallic 'pinging' noise and occurs when the engine is subjected to heavy load, such as rapid pulling uphill in high gear. The more resistant the fuel is to knocking, the higher its octane number and the better its performance in high-compression engines. The laboratory octane rating system uses a special internal combustion engine, affectionately known as the knock engine, exclusively manufactured by Waukesha Motor Company (Waukesha, WI, USA). The octane scale is defined by two pure hydrocarbons. Iso-octane (2,2,4-trimethylpentane) is assigned an octane number of 100 and n-heptane is assigned an octane number of 0. Therefore a 50:50 blend of these two pure compounds will have an octane number of 50 by definition. Mixtures of these pure compounds are known as primary reference fuels (PRF). The octane number of any fuel may therefore be defined as the percentage of iso-octane in a known mixture that knocks at the same compression ratio and at the same knock intensity as the fuel under test.

Based on global supply and demand, motor gasoline production is generally the most important process in any petroleum refinery and is probably the most complicated blending system in the plant. The feedstocks for gasoline blending come from several refinery processes, the most important of which are:

- Catalytic cracking processes
- Catalytic reforming processes
- Alkylation processes

Other petrochemical processes such as steam cracking also produce significant quantities of good-quality gasoline blendstock.

As the demand for high-quality gasoline increased, together with the need for more efficient production, the traditional batch blending process was replaced by the in-line blending process. With the introduction of on-line octane analyzers, savings of considerable magnitude could be obtained from a reduction in octane giveaway, thus improving operating efficiency.

3.9.2 Laboratory methods of measurement

The ASTM 'Manual of Engine Test Methods for Rating Fuels' was first published in 1948. Since that time, there have been numberous revisions, including supplementary publications. The current edition, volume 5.04 of the *ASTM Manual of Test Methods*, gives valuable information on installation, operation and maintenance of the test equipment, detailed information on the equipment itself and data on reference materials and accessories for use in blending standard fuels. The following ASTM standards are included:

- ASTM D2699   Knock characteristics of motor fuels by the research method

- ASTM D2700 Knock characteristics of motor and aviation fuels by the motor method
- ASTM D2885 Research and motor method octane ratings using on-line analyzers

The test equipment, known as the ASTM-CFR (cooperative fuel research) engine, consists of a single-cylinder engine of continuously variable compression ratio (4.0 to 18.0), together with all control instrumentation and accessories. The engine is connected to a reluctance type sychronous motor, capable of starting the engine, absorbing the power developed and maintaining the specified engine speed in accordance with the test method. Knock intensity is measured by a detonation pick-up located in the cylinder head, a detonation meter and a knock intensity meter.

The procedure for rating the sample fuel involves 'bracketing', or matching the knocking tendency of two reference fuels (PRFs) of accurately known octane number with that of the sample under test. The octane number of the sample is then calculated by interpolation to the nearest 0.1 octane number. For detailed information on this procedure, the reader is referred to the ASTM manual.

3.9.3 On-line process analyzers

The development of on-line octane analyzers presented quite a challenge. This development, which began in the early 1960s, proceeded along two separate routes; analyzers based on the ASTM-CFR engine and analyzers based on correlative techniques. Three companies chose the first route:

- Ethyl Corporation (USA)
- E.I. DuPont de Nemours (USA)
- Associated Octel Company (UK)

Two companies chose the second route:

- Universal Oil Products (USA)
- Gulf Oil Research & Development (USA)

3.9.3.1 Analyzers based on the ASTM-CFR engine. The Refinery Systems Division of Core Laboratories (Princeton, NJ, USA) recently took over the marketing and production of both the Ethyl Corporation and DuPont Octane Comarator Systems. The Model 8154 octane analyzer is available in two versions:

(1) *Compression ratio method*: This is an updated version of the original Ethyl Corporation analyzer. This microprocessor-based

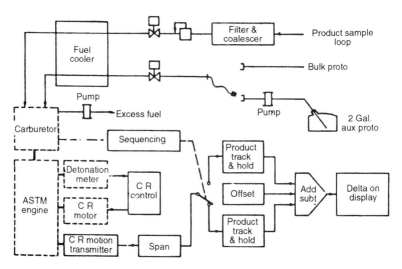

Figure 3.25   Core Laboratories Model 8154 octane comparator (CR Method). (Courtesy of Core Laboratories Inc.)

system determines the difference in octane number between the process sample and a prototype fuel of accurately known octane number. Changes in compression ratio at constant knock intensity are related to octane number difference which is indicated and recorded as delta octane number. A block diagram and schematic of the equipment is shown in Figure 3.25 The engine is operated on proto and sample fuels in an automatic alternating sequence, 5 min. for proto and 25 min. for sample.

(2) *Knock intensity method*: This is an updated version of the original DuPont analyzer. This microprocessor-based system determines the octane number difference between the process sample and a prototype fuel of accurately known octane number. With a constant compression ratio setting, the knock intensity signal produced by a standard ASTM detonation meter is monitored and a delta octane number is calculated. A block diagram and schematic of the equipment is shown in Figure 3.26. The engine is operated on proto and sample fuels in an automatic alternating sequence, 5 min for proto and 5 min for sample.

3.9.3.2 Analyzers based on correlative techniques.   The manual and automatic knock test engines are mechanically complex, require frequent and extensive maintenance, an artifically dampened device is used to measure knock intensity and the equipment is very expensive to purchase and install. Considerations such as these led UOP and Gulf Research in the

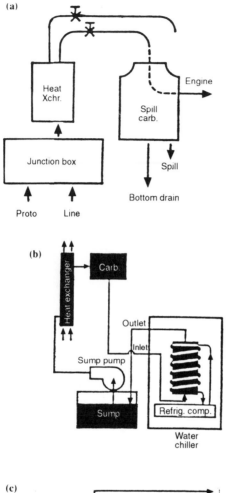

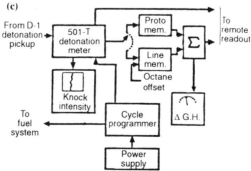

Figure 3.26 Core Laboratories Model 8154 octane comparator (Kl Method). (a) Comparator fuel system; (b) fuel cooling system; (c) console functions. (Courtesy of Core Laboratories Inc.)

1960s to conduct research into finding an alternative technique for measuring octane number. It was established that knock is associated with preignition of 'end-products', defined as that part of the fuel–air mixture in the engine cylinder that has not yet been ignited by the flame front. It was considered possible that, somewhere in the chain of reactions preceding knock, there are one or more reactions, occurring under milder conditions, that can be correlated with octane number as measured by the ASTM-CFR engine. Investigations confirmed that the reactions related to octane number are partial oxidations that proceed at elevated temperatures without the presence of an ignition source or catalyst (the so-called 'cool flame' technology).

A typical example of the correlative type of octane analyzer is the UOP Monirex, first introduced in 1966, which is designed to monitor the reactions that are precursors to knock. A schematic diagram of the analyzer is shown in Figure 3.27. The heart of the system is a reactor tube maintained at an elevated temperature. The partial oxidation reactions are exothermic and result in a fairly well-defined temperature at a certain distance from the reactor inlet. After the initial temperature rise, the reactions either stop or continue at a much slower rate, resulting in a fall in reactant stream temperature. As the octane number of the sample fuel increases, the temperature peak is displaced *away* from the reactor inlet and vice versa. Furthermore, the temperature peak is displaced *toward* the reactor inlet by increasing the reactor pressure. The basis of the measurement is therefore to maintain the temperature peak in a fixed position by varying the reactor pressure.

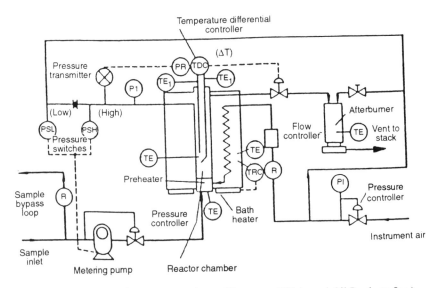

Figure 3.27 UOP Monirex octane analyzer. (Courtesy of Universal Oil Products Inc.)

The temperature peak is located in the reactor by two thermocouples located 1 in apart, as shown in Figure 3.27. The thermocouples are connected to a differential temperature controller the output of which operates a control valve in the reactor outlet line to vary reactor pressure. The reactor pressure is measured by an independent transmitter the output of which is calibrated in units of octane number. The reaction products are a mixture of partially oxidized compounds (e.g. alcohols, aldehydes, ketones, etc.), which are foul smelling and tend to deposit as 'varnish'. An afterburner is provided downstream of the reactor to complete the oxidation of these products and discharge them to atmosphere via a small stack.

Initial calibration of the analyzer involves establishing the response curve slope which represents the relationship between octane number and reactor pressure. Three primary reference fuels and a prototype fuel are required. Two of the PRFs should have octane numbers that 'bracket' the expected octane range of the process stream and the third should have the same octane rating as the proto-fuel. Reactor temperature is varied until the same pressure readings are obtained for the proto-fuel and the PRF with the same octane rating. The three PRFs are then run to establish the slope of the response curve. Thereafter, it is unnecessary to run the PRFs unless there is a major shutdown, but the proto-fuel should be run periodically as a validation check.

3.9.3.3 Recent developments. The two types of analyzers described above are used today in most refienry applications for octane measurement. The ASTM-CFR engine-based comparator systems are generally preferred for use in gasoline blending applications, whereas the correlative type analyzers are limited to monitoring of the feed streams for gasoline blending.

Two other analytical techniques, gas chromatography and refract've index, have also been used, but not with any great success, and are therefore of historical interest only. A third analytical technique, NIR spectroscopy, is a recent development and worthy of a more detailed discussion.

The first experimental work on the correlation of octane number with NIR spectral characteristics was conducted by Dr James Callis *et al.* at the University of Washington CPAC and presented in a paper published in *Analytical Chemistry* in early 1989. This work covered the wavelength range 660–1215 nm. Multivariate analysis of the spectra of 43 unleaded gasolines yielded a three-wavelength prediction equation for pump octane that gave excellent correlations with ASTM engine determined octane numbers. Eight other quality parameters of gasoline were also examined; Reid vapor pressure, API gravity, bromine number, lead, sulfur, aromatics, olefinics, and saturates. A second paper was published in 1990, covering estimation of the classes of hydrocarbon constituents in finished gasolines using the same technique.

This work has generated much interest in the petroleum refining business. As a result, toward the end of 1989, several manufacturers of on-line NIR spectrometers, together with the major oil companies, became involved in extensive field evaluation work. Several other papers on this subject were published in the next two years, culminating in the introduction by Guided Wave of its Spectron Octane Monitor based on NIR fiber-optic spectrometry. Since that time, several other NIR spectrometers have been offered for measurement of octane number and other gasoline parameters and much discussion is taking place at present with the ASTM Committee on Octane Measurement.

There are obvious and significant benefits in using NIR spectrometry for octane monitoring. Current methodology for gasoline blending involves the use of two ASTM engines, one for research octane the other for motor octane. As mentioned earlier, these engine-based systems are very expensive and maintenance-intensive. NIR analysis is faster, can provide information on both RON and MON and is much less expensive. Therefore, the next few years should be highly interesting ones for on-line octane monitoring.

3.10 Refractive index measurement

3.10.1 Background and definitions

When light strikes a medium other than the one through which it is transmitted, a portion of the incident light is reflected and a portion refracted into the new medium, as shown in Figure 3.28. Although the laws of reflection were known to Euclid, and refraction was studied by Ptolemy in the second century AD, it was not until 1621 that Snell discovered the exact relationship between the angles of incidence and refraction.

He considered the two media air and water and established that the sine of the angle of incidence divided by the sine of the angle of refraction is a constant:

$$\frac{\sin i}{\sin r} = n$$

The factor n for the two given media and is called the *refractive index* (RI) of water relative to air. Since refractive indices are usually measured relative to air, the value n is normally referred to as the refractive index of water, which is 1.33.

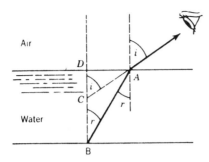

Figure 3.28 The laws of refraction (I). (Reproduced from Ref. 1, by permission of John Wiley & Sons Inc., New York, USA.)

The laws of refraction may be stated as follows:

(1) The incident and refracted rays are on opposite sides of the normal at the point of incidence and all three are in the same plane.
(2) The ratio of the sine of the angle of incidence to the sine of the angle of refraction is a constant (Snell's Law).

These laws can be extended to the refraction of light between any two media. Thus, if a ray of light is incident on a plane surface separating two media of refractive indices n_1 and n_2, respectively, the ray in the second medium will bend toward the normal if n_2 is greater than n_1, and:

$$\frac{\sin i}{\sin r} = \frac{n_2}{n_1} \quad \text{or} \quad n_1 \sin i = n_2 \sin r$$

When light passes from one medium to a more optically dense medium, there will be both reflection and refraction for all angles of incidence, whereas this is not always so when light passes from one medium to a less optically dense medium. Referring to Figure 3.29, consider a ray of

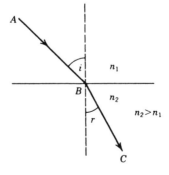

Figure 3.29 The laws of refraction (II) (Reproduced from Ref. 1, by permission of John Wiley & Sons Inc., New York, USA.)

light passing from glass to air. Starting with small angles of incidence, we get a weak internally reflected ray and a strong refracted ray. As the angle of incidence increases, the angle of refraction also increases. At the same time, the intensity of the reflected ray gets stronger and that of the refracted ray gets weaker. Finally at a certain *critical angle of incidence* (*c*), the angle of refraction becomes 90°. Since it is impossible to have an angle of refraction greater than 90°, it follows that for all angles of incidence greater than the critical angle *c*, the light undergoes what is known as *total internal reflection.* Therefore

$$\frac{\sin c}{\sin 90°} = \frac{1}{n} \quad \text{or} \quad n = \frac{1}{\sin c}$$

For crown glass of RI = 1.5, the critical angle is about 42°. For water of RI = 1.33, it is about 49° as calculated from the above equation.

3.10.2 *Laboratory methods of measurement*

Measurement of the RI of liquids in the laboratory involves the use of commercially available precision refractometers. The ASTM has established a number of standard test methods for industrial use. Two that apply specifically to the petroleum industry are:

- *ASTM D1218*, which covers measurement of the RI of transparent or light-colored liquids with an RI in the range 1.33–1.50 at temperatures from 68 to 86°F. The Bausch & Lomb precision critical angle refractometer is the recommended instrument.
- *ASTM 1747*, which covers the RI of transparent or light-colored viscous liquids and melted solids with an RI in the range 1.3–1.60 at temperatures from 176 to 212°F. The recommended instrument is the Abbe precision refractometer, with a measurement accuracy of 0.0002 RI units.

3.10.3 *On-line process analyzers*

3.10.3.1 Industrial applications. Refractive index measurement of a process stream is rarely made to obtain knowledge of that parameter alone. Usually we use changes in RI to infer changes in composition. RI analyzers are therefore non-selective instruments, in the same way that thermal conductivity analyzers are non-selective (see section 3.11). The RI analyzer measures the average RI of the sample, with each component contributing to the average RI. However, refractometers are very sensitive instruments. The Average RI of a mixture may be calculated from the equation:

$$N_m = C_a N_a + C_b N_b + C_c N_c + \dots C_n N_n$$

where C_a, C_b, etc are the mole fractions of components a, b, etc. and N_a, N_b etc. are the pure component RIs. Thus, when used as a process analyzer, the refractometer should be ideally limited to binary mixtures, although more complicated mixtures can be measured provided that the component of interest has an RI significantly different from the other sample components.

In spite of some limitations, there are numerous applications where on-line refractormeters are used in industry. One major area is in the food and beverage industry where RI is used for measurement of sucrose concentration in jams, preserves, sauces, fruit juices, milks and sugars, coffee and soft drinks. The weight percent concentration is known as °Brix and process refractometers are often calibrated directly in this unit. Another important application is in the measurement of iodine values of edible oil during the hydrogenation cycle. The brewing industry uses on-line refractometers for monitoring alcohol mixtures, and in the pulp and paper industry they are used for measurement of percent solids in black liquor.

In the petroleum industry, on-line refractometers have been used in lube oil extraction, and dewaxing processes, gasoline octane measurement, phenol production, aromatics production, and in styrene, formaldehyde and butadiene production. In many cases, however, a better analysis technique has been found.

In the chemical industry, on-line refractometers are used for monitoring phosphoric acid, sulfuric acid, nitric acid, urea/ammonium nitrate, brine concentration, glycol and glycerin, and caustic soda production.

The majority of on-line process refractometers operate on the critical angle principle.

3.10.3.2 Commercially available equipment. A typical example is the Anacon Model 47 process refractometer, shown schematically in Figure 3.30. This is a critical angle refractometer, which is available with two types of sensing unit designs, a weir-type valve design and an in-line probe design. A spinel prism is mounted in the process line in contact with the sample. A conical beam of light is directed through the prism at the interface with the process liquid. A portion of the light is transmitted into the liquid and a portion is reflected, as determined by the critical angle. The reflected light is directed to a dual-element solid-state photodetector. One element is in the constant light area (reference), the other at the light-to-dark edge (sample). As the RI of the process liquid changes, the critical angle also changes, moving the shadow edge on the sample detector. The detector system thereby provides an output proportional to the RI of the process liquid.

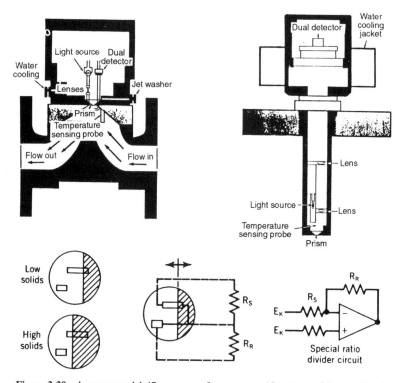

Figure 3.30 Anacon model 47 process refractometer. (Courtesy of Anacon Inc.)

3.11 Thermal conductivity measurement

3.11.1 Background and definitions

All gases have the ability to conduct heat, but to varying extents. The rate of conduction is dependent upon a property of the gas known as the *thermal conductivity* (λ). This difference in the ability to conduct heat may be used to determine quantitatively the composition of gaseous mixtures. Since all components in the mixture contribute to the average thermal conductivity of the mixture, this property is more applicable to the determination of the composition of binary mixtures (it is basically non-selective). However, more complex mixtures may be analyzed adequately provided that all components of the mixture have approximately similar thermal conductivities except the component of interest, which must have a significantly different thermal conductivity.

Hydrogen has a much higher thermal conductivity than hydrocarbons and other gases, except Helium, as shown in Table 3.7 which gives the thermal conductivities of various gases relative to hydrogen = 100 at

100°C. It is therefore not surprising that one of the most common industrial applications of thermal conductivity analyzers is the measurement of hydrogen in hydrocarbons and other gases, or measurement of the purity of hydrogen-rich process streams.

Table 3.7 Thermal conductivities of gases relative to hydrogen

Acetone	7.8
Acetylene	12.9
Air	14.3
Argon	10.4
Benzene	8.2
n-Butane	10.6
Carbon dioxide	9.8
Carbon monoxide	13.7
Cyclohexane	8.2
Ethane	13.9
Ethyl Alcohol	10.0
Ethylene	13.2
Helium	83.5
Hydrogen	100.0
n-Hexane	9.5
Hydrogen sulfide	7.2
Methane	20.8
Methyl alcohol	10.2
Methyl chloride	7.5
Neon	26.4
Nitrogen	14.2
Nitrous oxide	10.4
Oxygen	15.1
n-Pentane	10.1
Sulfur dioxide	5.4
Water vapor	11.1

3.11.2 On-line process analyzers

3.11.2.1 Principal components. A typical thermal conductivity analyzer consists of three principal components, the measuring cell, a Wheatstone bridge circuit with power supply, and a temperature controlled enclosure, or 'oven', in which the measuring cell is mounted.

The measuring cell and its component parts must of course be compatible with the process stream, and at the same time have a high thermal conductivity coefficient. It consists of a relatively large mass of metal to provide a stable heat sink. Stainless steel is used as standard. The metal block is machined to provide flow passages for the sample and a reference gas, and recessed cavities for the sensing elements. These may be hot-wire filaments or glass-bead thermistors. Hot-wire filamants are usually tungsten or platinum alloy materials. Thermistor sensors were popular prior to 1965, but suffered from frequent failure in hydrogen-rich streams. Considerable improvements in design of hot-wire filaments have taken place over the years and they are now the preferred choice.

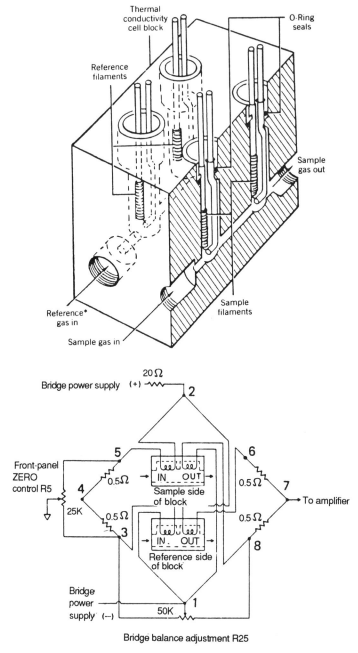

Figure 3.31 Rosemount analytical model 7C thermal conductivity analyzer. Cell block sectioned through sample side. Section through reference side is similar. *Reference ports capped if cell uses sealed-in, non-flowing reference gas. (Courtesy of Rosemount Analytical Inc.)

The Wheatstone bridge, the basic sensing system, includes a high-quality regulated power supply to provide maximum stability of operation. Many different versions are in use today.

Since just about *any* change in operating conditions will result in a change in sensing element temperature, it is important to maintain a constant sample flowrate and install the measuring cell in a constant temperature enclosure. Early developments of the TC analyzer in the UK led to the term 'katharometer' to describe instruments of this type, a term that is still widely used in Europe today.

3.11.2.2 Commercially available equipment. Commercially available process thermal conductivity analyzers are listed in Table 3.8. A typical example is the Rosemount analytical Model 7C, which is designed to measure the concentration of a single component of interest in a simple mixture. The thermal conductivity detector block, bridge circuit and typical analyzer modules are shown in Figure 3.31.

Table 3.8 Process thermal conductivity analyzers

Manufacturer	Model number	Country of origin
Anacon Inc	88	USA
Anarad Inc.	AR-110 Series	USA
Rosemount Analytical	7C	USA
Comsip-Delphi Inc.	B	USA
Hartmann & Braun	Caldos Series	Germany
Horiba Instruments	TCA-31/31F Series	Japan
ABB Kent Taylor	6600 Series	UK
Leeds & Northrup	7866	USA
Maihak, H, AG	Thermor 6N	Germany
Mine Safety Appliances	Thermatron T-3	USA
Schlumberger-Sereg	Type HCD-5	France
Siemens AG	Type TCA	Germany
Teledyne Analytical Insts.	225	USA

The measuring cell is a 316 stainless-steel block with separate passages for sample and reference gases. The reference passage can be flowing or sealed depending upon the application. Both reference and sample passages each contain a matched pair of resistive tungsten filaments. The filaments are connected to form the arms of a Wheatstone bridge circuit. The interior of the analyzer case is maintained at 130°F by a heater and solid-state controller. Repeatability is quoted as ±0.5% of full scale and analyzer response time is 30 s for 90% of final reading with a sample flowrate of 250 ml/min. A single range is standard, but a switch-selectable dual or triple range feature is available as an option.

Reference

1 K. J. Clevett, *Process Analyzer Technology.* John Wiley & Sons, New York, USA, 1986.

4 Process chromatography

G. L. COMBS and R. L. COOK

4.1 Historical perspective of process chromatography

Chromatography has become one of the most widely used analytical techniques in the world, both in the laboratory and for on-line process analysis. The first reported use of chromatography occurred in 1903 by the Russian scientist Mikhail Tswett. He used a simple chromatographic technique to separate plant pigments into different colored bands when solutions of the pigments were washed through a glass column filled with calcium carbonate [1]. The pigments could be separated into distinct colored bands; hence the term *chromato* (color) and *graphy* (graphics) or simply, *color writing*. Many years followed before the full potential of the technique began to be realized. In 1952, James and Martin [2] presented work performed utilizing a gaseous mobile phase instead of a liquid phase. Since 1952, the technique of chromatography has witnessed a tremendous growth, providing a simple solution to otherwise very complex analyses both in the laboratory and in the process. Chromatography is now recognized as one of the most important analysis tools available to the chemist, offering the advantages of separation and quantitation of components in a sample, all performed in a single operation. The rapid growth of chromatography techniques in the laboratory and process plants has been fostered by the development of new separation methods, sophisticated detectors, and the application of microprocessors for control and data manipulation.

The various types of chromatography available today all rely on the basic principles first discovered by Tswett. Under carefully selected conditions, a portion of a mixture is introduced to a stationary phase from whence it is then transported through the stationary phase by way of a moving carrier, or mobile phase. The various components in the mixture are subsequently separated based upon their unique interaction with the mobile and stationary phases. The various types of chromatography are actually classified by the nature of the stationary and mobile phases utilized and the particular types of analytes being measured. Today, there are a number of techniques available, all based on basic principles of chromatography. Some examples of these are listed in Table 4.1.

Table 4.1 Examples of chromatographic techniques

	Mobile phase	Stationary phase	Analytes
Gas chromatography	Gas	Liquid, solid	Volatile compounds
Liquid chromatography	Liquid	Solid	Non-volatile compounds
Super critical fluid	Supercritical fluid	Liquid, solid	Volatile/non-volatile compounds
Ion chromatography	Liquid	Solid	Inorganic compounds

Currently, all of these approaches are in various stages of development and maturity. Nevertheless, all four are currently being utilized, to one degree or another, for process monitoring. This chapter is concerned with the application of these four types of chromatography in process plant analyses.

4.2 A simple chromatography model

All present-day process chromatographs, regardless of the manufacturer, share several basic components as depicted in Figure 4.1. The *sample injection valve* is used to introduce a precise volume of the sample into the *mobile phase* (either gas, liquid, or supercritical fluid). Sample valves are specialized valves which must be carefully designed to allow repeated, reproducible injections with minimal maintenance.

The mobile phase continues to flow through the *column* where separation of the sample into its various components takes place. Within the chromatography column reside various materials (stationary phase) which promote the separation of the components contained in the mixture. These materials vary depending on the type of chromatography being performed and the particular components to be separated. Columns are typically constructed of various types of tubing of widely varying diameter and length.

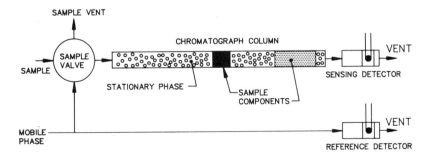

Figure 4.1 Simplified schematic of the basic components of chromatographs.

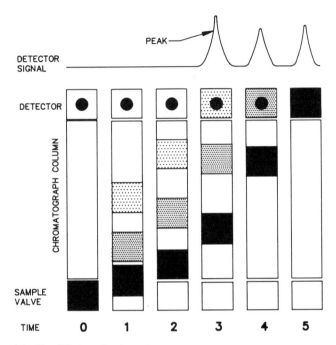

Figure 4.2 Simplified mechanism of separation in a gas chromatograph column.

A simplified mechanism of separation in a gas chromatograph column is illustrated in Figure 4.2. In this example, a gas mixture containing three components is injected using a sample valve at the input of a column and is swept through the column by the mobile phase or carrier gas. The migration rate of each component along the column depends on the carrier velocity and the degree to which each component interacts with the stationary phase within the column. The mixture is separated as it elutes through the column since each component is traveling at a different rate and, therefore, emerges at a different time. At the output of the column are a series of pure components (diluted in carrier) separated by regions of pure carrier gas.

From here, the mobile phase continues the elution process, eventually transporting the separated components to the *detector* wherein a signal proportional to the concentration of the individual component is generated. A graph of this signal versus time produces the characteristic *chromatogram*.

As mentioned earlier, the various chromatography techniques utilized today vary in the particular mobile and stationary phases utilized. For example, whereas gas chromatography utilizes a vapor mobile phase or carrier, including examples such as helium, hydrogen and nitrogen, liquid chromatography utilizes liquid mobile phases, examples being methanol, acetonitrile, water, and tetrahydrofuran. Supercritical fluid chromatography, on the other hand, utilizes vapors operated under supercritical conditions for the mobile phase. Ion chromatography, which in many respects is a subcategory of liquid chromatography, uses aqueous mobile phases, often containing buffers for the purpose of maintaining a particular ionic strength or pH.

The remainder of this chapter will provide more details of the particular hardware and applications which are unique to process chromatography techniques.

4.3 Process gas chromatography

4.3.1 Overview of system hardware

The satisfactory performance of any process gas chromatography (PGC) will depend on a number of factors. Unlike laboratory gas chromatographs, PGCs must be designed to withstand sometimes harsh environmental conditions, hazardous process areas, and continuous operation with minimal attention. Since the PGC is often located close to the actual process, sample transport to the analyzer and sample conditioning is another important aspect when using PGC. Looked at in another way, whereas the basic principles and component operation of a PGC are very similar to the laboratory gas chromatography, there are significant differences which tend to separate the two into distinct analyzer classes. For example, a PGC end-user expects an annual instrument up time in excess of 97%. This means that the instrument will be on-line providing successive analyses around the clock while requiring less than eleven days a year to calibrate and maintain the unit! Most laboratory chromatographs are not designed for this type of service nor are they designed to deal with the dust, vibration, power variations, ambient temperature changes, electromagnetic interferences or potentially corrosive environments that are common in PGC applications.

Conversely, part of the inherent success of any PGC is the ability to 'transfer' a desired laboratory application directly to the process application. Although this goal has been achieved to some extent, further efforts in this regard are continually being pursued by the PGC manufacturers and end-users. This challenge is often exacerbated by the requirement that process analyzer components do not generate a source of ignition under normal conditions or during a fault condition when operated in hazardous plant environments. Placing analyzers near a process often expose them to surrounding atmospheres containing explosive gas mixtures. In addition, the sample to be analyzed very often contains explosive gas concentrations. A number of electrical codes have been developed which address the actual safety features as well as maximum allowable temperatures which must be adhered to in the design and installation of analyzers and other equipment which may be utilized in a plant environment. Each design must be tested thoroughly to insure safety requirements are met.

In the sections that follow, several of the components which make up a PGC will be examined in order to better understand the actual operation of the PGC as well as some of the particular design constraints necessary.

4.3.2 Oven design

Gas chromatography is predicated upon precise temperature control of the components which will be in contact with the sample. This is particularly true of the sample injection valve, columns, and detectors. Most PGC manufacturers have dealt with this issue in a manner analogous to that used by laboratory GCs; that is, the temperature sensitive components are placed in a constant temperature oven or enclosure where the temperature is controlled to within ±0.1°C. This must occur despite the fact that the PGC, unlike the laboratory GC, may be subjected to wide ambient temperature ranges, perhaps as much as −20°C to 55°C. The heating mechanism often chosen is air-bath heating, since a plant air source is almost always available. Air-bath heating offers several advantages, including rapid and uniform heating of the components, and is easier to control than other types of heating schemes.

Present-day ovens often employ electric cartridge type heaters which operate by passing compressed air through the heater and into the oven (Figure 4.3). Temperature is controlled using RTDs and other temperature sensing devices which provide heater temperature control in the event of failure of the control mechanism. The disadvantage to this type of heating mechanism is the lack of ability to achieve elevated temperatures (>250°C), often a desirable parameter for many difficult applications. Whatever heating mechanism is employed, it must be capable of operation in hazardous environments.

4.3.3 Sample injection and column valves

Today, there exist a wide variety of time-proven sample injection valves. The valve chosen must be capable of reliable and reproducible injection of a precise sample volume, representative of the stream being analyzed. The choice of sample valve must be made with due consideration of the type of stream and the required operating conditions of the stream sample as well as the actual column selected for the particular application. Additionally, the sample valve must be durable. For a typical

Figure 4.3 A typical process gas chromatograph oven with electronics enclosure (top). Oven heating is established with the electric cartridge heater shown behind the FID detector block. The RID probe sensor is located directly below the detector block. (Courtesy of Applied Automation/Hartmann & Braun.)

process chromatograph application with a 10 min analysis time, the sample valve will operate over 50 000 times in 1 year!

Sample injection into the column must occur in a time period which is only a small fraction of the time of the narrowest peak width expected in the application. Slow injection times lead to peak band broadening and poor analysis repeatability. Under typical conditions, sample from the sample conditioning system continually flows through the sample valve and returns to the sample system. The sample valve acts to trap a consistent volume of sample and introduce this to the carrier at the next cycle start time. The sample can be either a vapor or liquid, but, must always be a single phase. In some cases, specially designed sample valves are incorporated in the PGC to provide higher temperature operation in order to vaporize a liquid sample prior to injection.

Among the available sample valves are the rotary valve, slider valve, and the diaphragm valve. The rotary valves were among the first sampling valves used in process GCs and have also become popular for many laboratory applications. Typically, a rotary valve is used in a process GC when the need exist to inject very small amounts of liquid sample. Typical injection volumes for a process chromatograph are on

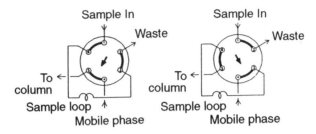

Figure 4.4 Diagram of a typical rotary valve utilizing an external sample loop. Alternatively, an internal sample loop, consisting of a machined groove or slot, may be used when sample volumes less than $0.1\,\mu l$ are required. (Reproduced by permission of Valco Instruments Co. Inc.)

the order of $200\,\mu l$ for a vapor sample and less than $2\,\mu l$ for liquid samples, although the exact volume can vary widely for a particular application. Rotary valves are available with injection volumes of less than $0.1\,\mu l$. A diagram of a rotary valve with an external sample loop is shown in Figure 4.4. In an inactive state, sample flows through a groove or slot machined in the rotor, while carrier flows to the column. Energizing the valve causes it to rotate 60°, placing the sample slot in-line with the carrier, allowing transfer to the column. De-energizing the valve allows the sample to refill the slot for the next injection. Disadvantages to the rotary valve are a decreased life expectancy, presumably due to the sealing surfaces being exposed to sample. Particulate in the sample can cause wear on the rotor surfaces, resulting in premature failure and/or alterations in the sample volume.

Slider valves operate in a manner analogous to the rotary valve. Operation occurs via a moving plate or slide, transferring an aliquot of the sample from the sample flow to the carrier gas, through a small slot machined in the plate. This is a very popular valve used in process chromatography. Its disadvantage is that, as larger sample volumes are needed, the slot volume begins to resemble a cavity, thereby increasing the time to effectively 'sweep' the sample from the valve into the carrier. For most applications, this time frame is acceptable, but with ever-increasing column efficiencies through the use of capillary columns, the time required to transfer sample to the column can become critical.

A diaphragm valve does not contain any sliding surfaces in contact with the sample, but relies on plungers, arranged in a circular configuration shown in Figure 4.5. The plungers operate by opening or sealing the passage between adjacent ports. A Teflon diaphragm is used as a seal and prevents process sample from contacting other components of the valve. Alternating sets of plungers are operated simultaneously through a spring-loaded, air-actuated piston so that one set closes three passages

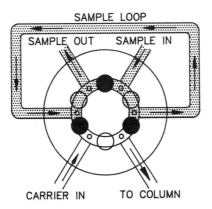

Figure 4.5 Typical diaphragm valve configuration. (Reproduced by permission of Applied Automation/Hartmann & Braun.)

between ports while the other set opens three passages. Since the plungers move only thousandths of an inch, the valve is considered to be not only relatively fast, but highly reliable. Failure of the valve is normally the result of excessive diaphragm wear and eventual leakage.

The majority of process chromatography valves are remotely actuated by pneumatic actuators and require between 30 and 60 psig to operate, depending on the valve type. Deactivation is normally accomplished via spring return. Stainless steel valves are the most common, but other materials are often used when corrosive samples or environments are present. In those cases, valves constructed of Teflon, hastelloy or monel are used. Valves designed for vapor samples generally are equipped with an external sampling loop which may be adjusted in dimension to accommodate different sampling volumes. Liquid sampling valves typically have very small internal sample loops, allowing injection of minute quantities of sample. In some cases, a liquid sample valve may be used for vapor samples when ultrasensitive detectors or microbore capillary columns are used. Vapor sample valves normally operate at low sample pressures whereas liquid sample valves must be capable of handling sample pressures up to 300 psig.

4.3.4 Detectors

The chromatograph detector provides the signal which is used to indicate the presence and amount of sample eluting from the column. Although there are a variety of gas chromatograph detectors available for laboratory GCs, in practice, only three are used extensively in process gas chromatographs. Many of the specialized detectors which work well in the laboratory environment are difficult to adapt to the requirements of a process instrument. Here, long-term stability, minimal maintenance requirements, and broad applicability to a wide range of components are of utmost importance. In addition, it is desirable that the detector construction be kept simple while meeting all electrical safety codes in force at the process plant locale.

The *thermal conductivity detector* (TCD) is currently the most popular detector used in process GCs, owing to its universal response to compounds and its simple construction and operation (Figure 4.6). The TCD is a *bulk property* detector since it responds to a change in some overall property of the sample in the detector; in this case, the physical property of thermal conductivity. Compounds vary in their ability to conduct heat, with lower molecular weight components generally having much better heat transfer capabilities. For this reason, carriers such as helium and hydrogen are generally used with a TCD. When a component is present in the carrier as it passes through the detector, the thermal conductivity will decrease and a signal is generated. The TCD operates by applying an electric current to a filament. More recent versions of these detectors operate in a constant temperature mode by controlling the filament current to maintain constant wire resistance. The energy supplied is then the measured variable. Among the advantages to this approach is the avoidance of filament burnout in the event of loss of carrier pressure and improved response time due to elimination of thermal lags in heating and cooling the filament. The TCD is useful for component concentrations from 0.1 to 100%.

Thermal conductivity detectors come in two types depending on the thermally responsive element used. A *filament* TCD utilizes a coil of

Figure 4.6 Process GC thermal conductivity detector (TCD) with three sensors—sense, reference and ITC (inter-column detector). With the cap in place, the detector constitutes an explosive-proof design. (Courtesy of Applied Automation/Hartmann & Braun.)

resistive wire while a *thermistor* TCD replaces the filament with a heat-sensitive semiconductor or thermistor. Thermistor detectors are made of metal oxides constructed in the form of a bead which is then coated with glass to enhance inertness. As a result of their inherently large negative temperature coefficient, thermistors are very sensitive detectors. Due to their limited temperature range, however, they are used primarily for lower temperature applications. The filament (or hot-wire) type detector has found use where slightly increased sensitivity is desired and with elevated temperature applications. This type of TCD detector uses a thin coil of fine wire, commonly tungsten or tungsten–3% rhenium.

Among the factors which influence the maximum achievable sensitivity of a TCD are carrier flow perturbations and detector temperature stability. To minimize the detector sensitivity to these factors, most TCD detectors today are equipped with two filaments (or thermistors) mounted in close proximity to one another. One filament is exposed to pure carrier gas from a reference line (hence, the term *reference detector*), while the other is connected to the exit of the analytical column (*sense or measurement detector*). The difference in resistance between the two filaments in the detection circuit produces the signal. Reference elements should be identical to the measurement elements and exposed to the same conditions.

The *flame ionization detector* (FID) is often the detector of choice when ppm levels of hydrocarbon are to be measured. The FID utilizes an oxygen-rich hydrogen flame to combust organic solutes as they elute from a column, producing both positive and negative carbon ion species. In most designs, the negative ions migrate through an electric field, produced by applying a high potential (300–1000 volts) across the inlet jet and a collecting grid, shown in Figure 4.7. The negative ions discharge electrons to the positive grid, creating an electrical current. The number

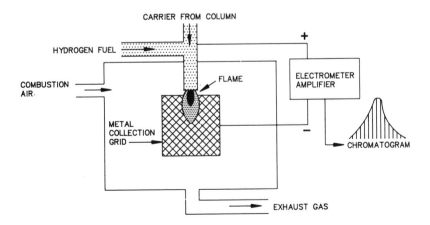

Figure 4.7 Schematic diagram of a flame ionization detector.

of ions formed, and thus the current conducted, is roughly proportional to the number of carbon atoms present in the flame. The FID is quite sensitive, allowing measurement of low ppb level of solutes. Only carbon-containing components solicit a response, with carbon oxides producing little or no signal.

FID response relies on the proper flow settings of the air and hydrogen as well as carrier. In most designs, hydrogen fuel is mixed with the carrier gas and solutes from the column. This mixture is then fed to the FID jet, from whence it is enveloped in air to support the flame. Instrument air of high quality is necessary for proper operation, partly the result of inefficient ionization (approximately 0.0001%) [3]. Most PGC suppliers provide *air treaters* for use with the FID. Plant source air is passed through a catalyst at elevated temperatures in order to convert any residual hydrocarbons to carbon dioxide, which does not respond appreciably in the FID. Use of air treaters can save the user the cost of high purity, bottled air. An FID is shown in Figure 4.8.

The *flame photometric detector* (FPD) is used to measure trace quantities of sulfur and, in very rare instances, phosphorous-containing compounds. The FPD is similar to the FID in that a flame is used to generate the particular species to be monitored. The carrier gas, exiting from the column, is mixed with hydrogen and air and burned. Sulfur-containing compounds are converted to an excited state of elemental

Figure 4.8 Process GC flame ionization detector (FID) shown mounted in the analyzer oven. (Courtesy of Applied Automation/Hartmann & Braun.)

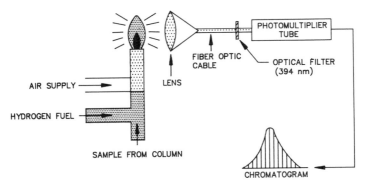

Figure 4.9 Schematic diagram of the flame photometric detector.

sulfur, S2*, which quickly loses a photon (chemiluminesces) in the region above the flame and returns to the ground state. The released energy (blue light) is collected using a lens and, as illustrated in the example shown in Figure 4.9, transmitted via a fiber optic bundle to a photomultiplier tube equipped with a narrow band-pass (394 nm) filter. The photomultiplier tube produces an output proportional to the square of the sulfur concentration. The hydrogen/air flow ratio is adjusted at approximately 1:1 to provide a reducing atmosphere. The phosphorous detection mode is similar in nature. As with the FID, a flame-out sensor and ignitor coil are located within the detector.

Typical applications for an FPD detector include measurement of trace sulfur compounds in fuel gases for EPA monitoring as well as the measurement of sulfur containing impurities to avoid catalyst poisoning [4].

4.3.5 Carrier gas considerations

In gas chromatography, the carrier gas serves as the *mobile phase*, transporting the components of interest from the sample valve, through the column and eventually to the detector. For most types of chromatography, the mobile phase serves as an inert transport and detection medium. For process gas chromatography, the particular choice of carrier is the result of consideration of cost, the detector to be used and, in some applications, the effect of the carrier on column efficiency and pressure drop. The carriers typically chosen are helium, hydrogen and nitrogen. Helium carrier is commonly chosen since it is readily available, non-flammable and relatively inexpensive. Hydrogen is used in areas where helium is not readily available. Nitrogen is often used as the carrier gas when hydrogen is to be measured due to the large difference in thermal conductivity between the two gases. Otherwise, for thermal conductivity detectors, helium and hydrogen are the most popular due to their high thermal conductivities.

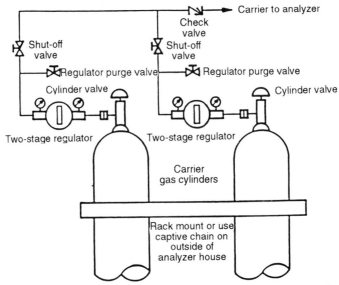

Figure 4.10 Typical carrier gas manifold system using two carrier gas cylinders.

Most gas chromatographs rely on accurately controlled carrier gas flowrates through the column and detector in order to maintain reproducible retention times and detector zero and span settings. Stable, inert pressure regulators must be used for carrier pressure control. Otherwise, unacceptable detector noise and varying component retention times may result. Manifolding two carrier gas cylinders together as illustrated in Figure 4.10 ensures continuity of supply and allows one cylinder to be replaced without affecting the operation of the GC. Each cylinder is fitted with a pressure regulator, with one set at a slightly higher pressure than the other. In this way, the cylinder with the higher pressure is consumed first, with the second automatically switching in when the pressure begins to decrease. All carrier transport tubing must be clean, inert and leak free. Even slight leaks can provide the opportunity for introduction of impurities such as water vapor or air into the system. Very often, drying tubes containing charcoal or molecular sieves are utilized to assist in removing trace impurities in carrier gases.

4.3.6 Columns

The inherent success or failure of a particular chromatography application depends upon how well the various components in the sample have been separated. This separation will, in turn, depend in large part upon the correct choice of analytical column for the application. Today, there are literally hundreds of choices of chromatography columns and considerable excellent reference literature exist on this subject [5]. Columns may vary in the type and amount of packing material, length, internal diameter, and material of construction.

Modern process gas chromatographs utilize either packed columns or open tubular (capillary) columns, the differences generally being the column internal diameter and the method of suspending the stationary phase. Packed columns tend to have larger internal diameters (2–4 mm), shorter lengths (1–6 m) and are typically constructed of metal tubing whereas capillary columns typically have very small internal diameters (0.2–0.75 mm), very long lengths (5–150 m) and, until recently, have been constructed primarily of fused silica glass. Technological advancements have permitted the construction of capillary columns using metal capillary tubing which has been rendered inert through various processes (Figure 4.11). These columns are much more durable than their fused silica counterparts and, as such, lend themselves well to the process environment. Although the wide use of capillary columns is a relatively recent event in process gas chromatography (1985), it is expected that the majority of applications will be performed with capillary columns in the near future.

Typically, a particular stationary phase is chosen based upon its ability to effect widely differing partition coefficients between the various

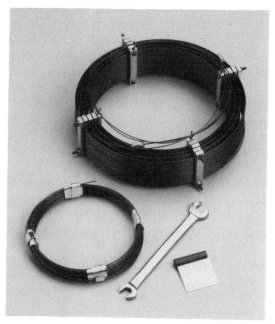

Figure 4.11 Current metal capillary columns shown in two configurations. (Courtesy of Restek Corporation.)

components to be separated in a sample. As a sample is transported through a GC column, the components within the sample will partition themselves between the stationary and mobile phases. Recall that the mobile phase is the carrier gas and the stationary phase is the packing material of the column. The partition of a component between the two phases depends upon interactions that occur between individual component molecules and stationary phase molecules. This interaction can be based on differences in molecular size, polarity or boiling point. The stationary phase will selectively retard components of the mixture based on these differences.

Most stationary phases rely on differences in compound polarity to achieve separation. As a general rule, as the polarity of the stationary phase increases, the retention of non-polar compounds decreases and the retention of polar compounds increases. For example, olefins, which contain a polarizable double bond, can be easily separated from saturated hydrocarbons, which tend to exhibit non-polar characteristics. The majority of stationary phases exist in the liquid phase under chromatograph conditions, and sample components become partially dissolved (or partitioned). Other phases, such as molecular sieves, remain a solid and rely on surface adsorption differences for separation. Molecular sieves are often used for the separation of oxygen, argon, nitrogen, hydrogen and methane.

4.3.7 Separation theory

Earlier in this chapter, a simple chromatography model was presented which described how a sample containing several components could be separated into its individual constituents using process chromatography. In practice, achieving component or peak separation requires not only the correct choice of column and stationary phase, but also the correct adjustment of a number of parameters which can impact column efficiency.

4.3.7.1 Relationship of liquid phase solubility (partition coeff K) to retention time. A fundamental principle of gas chromatography is that different components will dissolve to different extents in the same liquid. Chromatography can be thought of as a series of discontinuous steps, each step representing a zone wherein the sample component and stationary phase liquid are in equilibrium. That is, the ratio of the concentration of the sample component in the liquid phase and the gas phase is a constant:

$$K = X(\text{liq})/X(\text{gas})$$

where $X(\text{liq})$ is the concentration of the sample component in the liquid stationary phase, and $X(\text{gas})$ is the concentration of the same sample

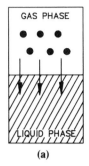
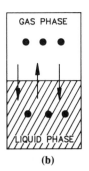

Figure 4.12 Simple model of the chromatography principle. (a) Pre-equilibrium conditions exist at initial introduction of the sample; (b) equilibrium conditions established, $K = 1$.

component in the mobile gas phase. This constant K is known as the partition coefficient. This is illustrated in Figure 4.12. At equilibrium, if equal portions of the sample component are in the mobile gas phase and liquid stationary phase, then K will equal 1. As long as the pressure and temperature are kept constant, there will always be equal portions of the sample component in each phase. Although in a gas chromatograph, we know that the carrier is continuously moving over the stationary phase, this model is useful for understanding what occurs in the system.

Although this discussion has dealt only with gas–liquid systems, gas–solid chromatography follows essentially the same mechanism, with sample components in contact and attracted to a solid of high surface area. Equilibrium conditions are achieved through surface attractions between the components and the solid stationary phase. These interactions are quite strong and this type of stationary phase is used mainly for fixed gas and light hydrocarbon separations.

4.3.7.2 Theoretical plates/column efficiency. Figure 4.13 illustrates the effect of having two components present in the sample, each component having its own unique partition coefficient. If we again consider the system as a series of discontinuous steps, each step representing an equilibrium condition, we begin to see how separation occurs and peaks are formed. Each step or equilibrium is known as a theoretical plate, similar to the plates of a distillation column. The number of theoretical plates for a particular column can be calculated from the actual chromatogram peaks obtained, as shown in Figure 4.14 and the following relationship:

$$\text{Number of plates } (N) = 5.54 \, (t_R/W_{0.5})^2$$

where t_R is the retention time from injection and $W_{0.5}$ is the peak width at half height. Increasing the number of theoretical plates for a chromatographic

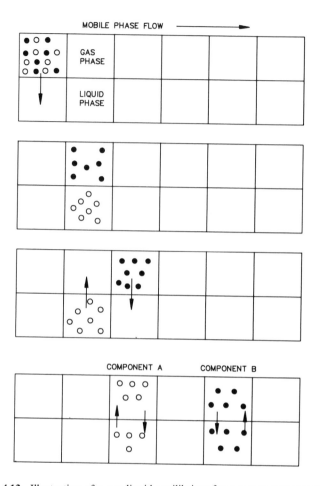

Figure 4.13 Illustration of a gas–liquid equilibrium for a two-component mixture.

application will result in narrower peaks and more efficient separation. Hence, N is used to measure the efficiency of a particular column.

4.3.7.3 Resolution. Referring again to Figure 4.13 we see that component A, being more soluble in the liquid stationary phase, is retained longer than component B. The resulting chromatogram may resemble Figure 4.15. Here, the two components A and B elute from the column at 11 and 9 min, respectively, after sample injection. Correcting for the column dead time of 0.5 min, we can calculate a relative solubility (α) for the two components as follows:

α = solubility of A/solubility B = 11-column dead time/9-column dead time

For Figure 4.15, $\alpha = 10.5/8.5 = 1.24$.

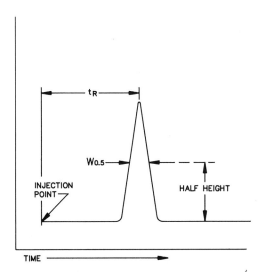

Figure 4.14 Determination of column efficiency (N) requires the data illustrated in this chromatogram.

Another parameter which is useful to calculate is the time spent by the component in the liquid phase, calculated from the expression K is equal to the peak time after column dead time/peak time from injection. For component A, $K = 95\%$ and for component B, $K = 94\%$.

Resolution is defined as peak separation/average peak width. Improvements in resolution will result from increasing peak separation and/or

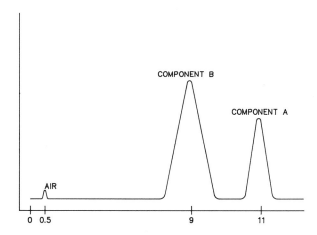

Figure 4.15 Two-component chromatogram illustrating retention times used for calculating relative solubility (α).

reducing peak width. We have seen how the separation of two peaks on the chromatographic column is completely dependent upon the liquid phase efficiency, with peak width completely dependent upon column efficiency or number of theoretical plates. Liquid phase efficiency can be optimized by choosing a liquid which provides the largest difference in solubility between the two components to be separated. α should be as large as possible. In addition, one should strive to choose a liquid phase which will result in the greatest residence time in the liquid phase. A general rule of thumb is 50–95% of the time. Maximum use of the liquid phase effect will result.

4.3.7.4 Factors which affect partitioning.   Other factors which can affect the separation of two peaks are column temperature, column pressure, carrier flowrate and amount of liquid phase (or liquid phase film thickness). For column temperature, theory predicts that lower column temperatures favor better separation. In general, this is what is found in practice and one chooses the lowest possible temperature which is appropriate for the application. Most columns, however, have a lower temperature limit which must not be exceeded. Additionally, for samples which contain multiple components to be separated, often the column temperature is chosen which optimizes the separation of the two most difficult components. This can result in unacceptable analysis times due to the elution times of the other, more strongly retained components. A technique which is often used to circumvent this problem is programmed temperature GC (PTGC). Here, the temperature of the column is increased linearly with time during the analysis, after some initial time period at isothermal conditions. The result is shorter elution times for the strongly retained components with resultant shorter analysis times.

Carrier flowrate affects analysis time, is easily adjusted and has a direct effect on column efficiency. When choosing a flowrate, consideration should be given to the type and size of column being used. Generally, optimum flowrates are found to be near 20–40 cm/s. Since this is a linear velocity specification, the optimum volumetric flow rate will vary depending upon actual column ID and whether it is packed or capillary. Table 4.2 lists four sizes of capillary columns available along with typical manufacturer's recommended flow rates.

Table 4.2 Typical column characteristics (Courtesy of Restek Corporation)

Column ID	0.18 mm	0.25 mm	0.32 mm	0.53 mm
Sample capacity	10–20 ng	50–100 ng	400–500 ng	1000–2000 ng
He flow at 20 cm/s	0.3 cm^3/min	0.7 cm^3/min	1.0 cm^3/min	2.6 cm^3/min
H$_2$ flow at 40 cm/s	0.6 cm^3/min	1.4 cm^3/min	2.0 cm^3/min	5.2 cm^3/min
TZs	40	30	25	15
Theoretical plates/m	5300	3300	2700	1600
Effective plates/m	3900	2500	2100	1200

Increasing the film thickness results in increased column sample capacity. Generally, greater film thicknesses serve to improve the resolution of early eluting components, increase analysis times and, in some cases, can help reduce the need for subambient cooling in programmed temperature GC applications. Also from Table 4.2 note the effect of decreasing the column ID. Smaller ID columns can result in improved resolution of early eluting compounds, but at the expense of longer analysis times and limited component concentration ranges on the column. Larger ID columns often result in less resolution for early eluting compounds but, generally shorter analysis times, and greater dynamic ranges. In most cases, the larger ID capillary columns offer sufficient resolution for complex mixtures found in process applications.

4.3.8 Multidimensional chromatography

In practice, process gas chromatography applications must be capable of meeting a number of often conflicting goals. For example, attempts to achieve baseline separation with a single column and simple hardware configuration may not allow one to achieve the analysis or cycle time desired for the application. Further, consideration must be given to the possibility of additional components being present in the sample from time to time, possibly overlapping the peaks of interest. In addition, sample component concentrations may vary widely in a particular process, creating the necessity of large dynamic range capabilities for the application. These problems, as well as others, have resulted in the development of techniques utilizing a number of columns in various configurations and switching between these columns during analysis to address particular requirements. This technique of column switching is known as multicolumn or multidimensional chromatography.

Column switching techniques provide a means to achieve several chromatography functions which are particularly important in process analysis:

- *Eliminate undesirable components from the analysis*—Rapid analysis cycle times are often very critical for process control. In a process chromatography application, rapid analysis of only the components of interest is of vital importance. Various column switching techniques are available which allow venting and eliminating from the measurement undesirable components.
- *Shorten analysis cycle times*—Various switching techniques are often used to shorten the analysis time. Rather than wait for all components to elute from a single column, very often different separating columns are applied to different groups of components in the same analysis.

- *Simplify an analysis*—Although column switching techniques generally add to the complexity of the analysis hardware, and as such are not necessarily desirable when they can be avoided, the vast majority of applications in process chromatography use some form of column switching technique. Many separations are not achievable by a single injection onto a single column. Use of separate columns, optimized for different portions of the analysis, can result in an easier solution to the analysis problem, while providing the desired speed of analysis.
- *Column cleaning*—Most process streams will contain more components than are desired to measure. Quite often, these components are more strongly adsorbed and, if not removed between each analysis, can accumulate and modify the characteristics of the column. In other cases, these components may elute during a later analysis cycle and interfere with the measurement.

Two of the most widely used column switching techniques are backflush and heartcut. Back flushing is accomplished by reversing the direction of carrier gas flow through the column during the analysis. Any components which were still in the column will be eluted in the opposite direction from the injection (Figure 4.16). Back flushed components can then be: (i) taken to vent for discarding, (ii) taken to the detector for measurement, (iii) taken to a second column for separation, or (iv) taken to some combination of these three configurations.

Among the many uses of back flushing are ensuring that the columns do not accumulate impurities, reducing analysis time by flushing un-

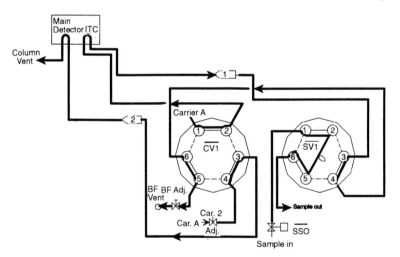

Figure 4.16 Backflush-to-vent column/valve arrangement with sample valve ($\overline{SV1}$), column or backflush ($\overline{CV1}$), backflush column (1) and analytical column (2).

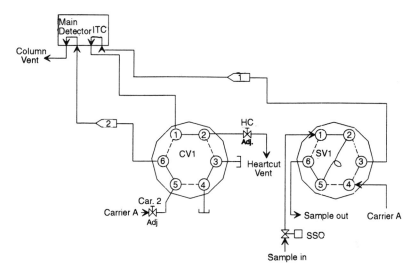

Figure 4.17 Heartcut column arrangement.

desired components to vent, or separate components in a second column which are not separated on the first column. Very often, back flushing is not used alone, but rather with some other column switching technique.

The hardware configuration used for the heartcut column switching technique is depicted in Figure 4.17. Generally, heartcutting is used to separate a minor component for measurement from a major component, such as a solvent peak. Two columns are used for heartcutting, the heartcut column and the analysis column. The heartcut valve can be used to direct the eluent from the heartcut to either vent or to the second column for separation and quantitation.

Combining various column switching configurations results in an almost limitless number of possible analysis approaches, but often at the expense of an increasingly complicated system. System complexity and reliability must be weighed to justify the value of each component to be measured, since increasing the number of components to be monitored will contribute to the complexity of the measurement. Relatively simple column switching systems can be utilized for one or two components, whereas separation of four or more components will very often result in significantly more complex column designs. The user must weigh the value of each component to be measured against its contribution to overall system complexity and reliability. Considering the analyzer up-time expectations, previously mentioned, it is highly desirable to accomplish the measurement in the simplest possible way that meets the application requirements.

4.3.9 Practical process gas chronatography

4.3.9.1 Historical perspective of process applications. Historically, PGC has provided an input to control schemes for simple refinery processes [6]. For example, applications for overheads, bottoms, and feed stock samples for distillation towers in the processing of light hydrocarbons, hydrocarbon solvents and aromatics have been and remain common. Distillation towers are often used to process light hydrocarbon, hydrocarbon solvent and aromatic streams. A distillation (or fractionation) tower generally separates the lower boiling point components in a feed stream overhead while the higher boiling point components exit the tower through the bottoms stream. For example, in a demethanizer tower, the desired operation of the distillation tower might maximize the amount of methane in the overhead stream (while maintaining the least amount of ethane and heavier component impurities) and conversely minimize the amount of methane present in the bottoms stream. A PGC could sample both of these streams to provide the analytical information for the tower control to an automated computer control system or to a human operator in the control room. The information can then be used to adjust parameters of the tower operation to control the energy usage, throughput capacity and product quality.

While laboratory applications still provide many of the analyses not available with PGC, the migration of chromatographic techniques and technology from the laboratory to on-line use continues to be the direction of PGC. As equipment continues to become more capable, additional applications are possible and the demand by end-users to increase their productivity and reduce their product costs molds the driving force for successive generations of PGC equipment. While the quality of the analytical result is comparable to laboratory equipment, the time to obtain the result has improved between one and two orders of magnitude. In addition, transferring the sample to the analyzer by mechanical means eliminates human errors in collecting, storing and sampling the processes.

More recently, applications have been developed for the petrochemical and chemical industries. These include olefins such as ethylene, propylene, and butadiene, as well as applications for chemicals such as freon, vinyl chloride and styrene processes. Improved correlations to certain product impurities and catalyst poisons result as more information and tighter process control becomes available to the end-user. Applications to monitor the levels of key impurities have allowed controllers of the process to reduce levels of catalyst poisons and prolong the usefulness of costly catalysts. More recently, applications for processes such as isoprene, linear olefins, oxo alcohols, and simulated distillation applications for motor gasoline or diesels have shown value in the market place [6].

Societal changes have also impacted the direction of some applications now in wide use. Changes in motor gasoline formulations to reduce the environmental impact of petroleum fuels have generated application needs for methyl tertiary butyl ether and other oxygenated fuel additive processes. Many applications today deal with regulatory requirements. For instance, water and air quality monitoring applications are becoming more commonplace. Water analyses for a variety of pollutants as well as stack gas monitoring for sulfur emissions are more recent examples [7].

Today, PGC applications can generally be divided into three general types: process control, mass balance and monitoring. Process control applications typically focus on the measurement of a few 'key' components and require a short cycle time. Here, the term *cycle time* refers to the span of time between successive analytical analyses. A control application would normally be expected to provide a cycle time of 1–10 min and seldom would a single analyzer sample more than one stream.

A mass balance application normally requires from 5–30 min per cycle, measures most of the individual components and groups the remaining components. In this case, a single analyzer might sample one to five streams. The information provided by this type of analysis can be coupled with process flow information, thus providing the end-user with information that is valuable both to the process engineer, regarding the efficiency of a given process, and to the accounting department, concerning the production, quality, and inventory of a product. A common application in this category is the measurement of pipeline gas to calculate BTU value, which is primarily used to address custody transfer issues regarding product quality and quantity [8].

Finally, monitoring applications generally meet a regulatory need. These require widely varying cycle times from 15 s to 30 min and may sample anywhere from one to 60 streams with a single chromatograph.

In actual practice, the preceding application types may not be so clearly discernible, as the economics involved in decisions made by both the end-user and the equipment supplier often create hybrids of these three types.

4.3.9.2 General application considerations. Generally, the development of a PGC application requires knowledge of the desired analysis cycle time, the hazardous area classification required, the components to be analyzed as well as their corresponding concentrations at normal operating conditions, as well as in upset conditions, for each stream to be measured. Additionally, information regarding the sample phase, pressure and temperature (with minimum and maximum values) will be required for each stream.

PGC application considerations begin with the requirements mentioned above and a preliminary review of the components in order to identify any special material requirements for the sample wetted paths as well as any unusual safety design considerations due to the presence of highly toxic stream components. Either of these issues can add significant cost to the chromatograph system. The presence of sample components which might strongly adsorb or absorb must be given consideration in the choice of sample wetted materials. Components of this type are typically quite polar, or, are appreciably soluble in a sample wetted material. In practice, most application needs can be addressed with hardware that utilizes materials such as stainless steel combined with appropriate polymeric materials such as Teflon, Rulon, Viton and Kalrez. On occasion, some applications will require extraordinary measures to allow for the presence of highly toxic or corrosive components.

As has been previously stated, PGC analyses are commonly applied to the refinery, chemical and petrochemical production and distribution industries. Samples are typically liquids or vapors having little or no dissolved solids. A primary concern is the pressure and temperature of the sample. First, the sample delivered to the chromatograph must remain single phase in order to obtain precise, repeatable injections. The sample system must be capable of continuously supplying a single-phase vapor or liquid sample at a known range of temperature and pressure variations.

There exists a variety of commercially available software programs capable of estimating the bubble points and dew points of chemical streams. The minimum and maximum values for sample pressure, temperature and composition must be reviewed for each stream to determine the boundary estimates of bubble point and dew point for the particular application. Additionally, it is important to consider the melting point in some applications. Sample melting points are rarely a problem, but the consequences of a liquid sample unexpectedly solidifying in a chromatograph system are less than convenient. In practice, with available references and a working knowledge of phase diagrams, an individual can make adequate estimates without the aid of commercial software packages.

Chromatograph oven temperatures that may be required for a given application are largely determined by the sample temperature and pressure. Vapor samples must be maintained at a temperature that will prevent condensation of any components during process upsets in composition or pressure. Liquid samples require additional consideration. A liquid sample containing components with boiling points as great as 100°C above the operating oven temperature can often be adequately vaporized for PGC analysis. PGC oven temperatures are generally operated at temperatures of at least 60°C in order to insure a stable oven temperature while located in an environment with widely varying ambient temperatures. Generally, oven temperatures in excess of 150°C begin to adversely affect the maintenance and reliability aspects of the application and hardware. The issue of temperature is far less critical in a laboratory wherein the laboratory chromatographer is rarely concerned about determining the most moderate temperature at which a given application can be accomplished, the exception being analyses involving thermally labile species.

Locating a sample valve external to the analyzer oven is a technique often used when dealing with samples having elevated boiling points, or in cases wherein sample polymerization is a concern. Some externally mounted liquid sample valves have a temperature controlled vaporization chamber which overcomes the problem of a sample with high boiling components or a sample pressure insufficient to maintain a liquid phase at the chosen oven temperature. This type of configuration allows vaporization of high boiling point components into the carrier stream while allowing the oven temperature to remain below the point capable of efficient vaporization of the sample.

The concentrations of components of interest often determine the type of detector chosen. Thermal conductivity detectors are primarily selected for components at tenths of percent concentrations and larger. FID detectors normally require additional utilities of fuel and hydrocarbon-free air, but provide good performance from less than one ppm concentrations to percentage levels. The total analysis of a mixture of high and low percentage level components may restrict useful detection of the lowest ppm range peaks, but in practice this is an acceptable situation and is little different from the normal laboratory approach. An FPD generally is required for ppm levels of sulfur species. Detector choice and good calibration practices can be complicated by the presence of streams of widely varying component concentrations. This situation should be avoided, if possible.

The measurement of inert or fixed gases at low ppm concentrations represents a situation of some interest. Inert or fixed gases are not detectable by an FID. One approach is to choose conditions that extend the detectable range of a TCD. Sample sizes can be increased to the limits of what the column is capable of separating as a means of extending the lower detectable concentration range for a given detector. In some cases, the selection of an appropriate TCD detector temperature can more than double the detector sensitivity. Another approach is to chemically convert the gas to an FID sensitive specie by the use of a catalyst and appropriate conditions. As an example, in the presence of hydrogen carrier, carbon monoxide and carbon dioxide can be converted by a catalyst at moderate temperatures to form methane, which is then detectable at sub-ppm levels.

Column selection involves consideration of the oven temperatures available coupled with a review of the component species, concentrations, and separations required. Usually, identification of a single column that will address all of these issues, and still meet the required cycle time, is seldom possible; hence, the common use of multiple valving schemes in PGC analyses. In order to make a first approximation selection of the column, information reflecting the relative retention times of sample components for specific columns at a given temperature must be obtained through laboratory data, references, calculation or experience. In practice, fewer than a dozen liquid phases or uncoated adsorbents (such as graphitized carbon black, porous polymers and molecular sieves) are capable of addressing the application needs of most packed column PGC applications. The correct column length or, as appropriate, percent of liquid load can then be more easily determined. Capillary columns typically require fewer permutations to adequately address most applications.

Finally, while there are over 50 possible ways for one to six valves of the six port type to be connected to one detector, ten configurations are commonly used to accommodate most applications. Optimizing an application for all the previously mentioned considerations is the desired goal and often requires iterative attempts.

4.3.9.3 Programmed temperature GC.   Process applications utilizing programmed temperature gas chromatographs (PTGCs) more closely resemble a laboratory application and have only become widely available since 1987 [9]. Here, the sample is injected onto a column which resides in a smaller, temperature programmable oven, located within the isothermal oven (Figure 4.18). This is then followed by programming the column temperature to effect the timely separation of the components of interest, normally on a single column. The useful temperature range for this type of application approach is greater than isothermal applications, covering the range of 0–250°C. Cycle times are generally longer, carrier purity is of greater concern, and the technique generally requires more maintenance than a more traditional isothermal oven approach. However, process PTGCs do provide a powerful solution for complete analyses of complex samples containing components with a wide range of boiling points.

4.3.9.4 Calibration considerations.   Practically all PGC applications utilize the absolute calibration method. This method compares the area of a known sample component to its known concentration to derive a response factor. Using this response factor, subsequent component areas are then related mathematically (and in some cases, normalized) in order to reduce the sensitivity of the results to minor instrumental variations, such as sample size.

Other calibration techniques, such as standard addition or internal standard, add a known amount of a separate component to the sample,

Figure 4.18 Temperature programmable oven (containing analytical column) mounted within an isothermal PGC oven. The isothermal oven is also equipped with a detector and sample valve. (Courtesy of Applied Automation/Hartmann & Braun.)

which has previously been correlated to a given response of other components to be measured. By measuring the added peak of known concentration, the previously established correlation can be used to determine the component concentration in the stream. However, these techniques are seldom utilized in PGC calibration.

Calibration standards, containing the components of interest at concentrations that approximate the expected or normal measured component value, are typically blended with an accuracy of plus or minus 2% of value. Important considerations for calibration standards include compatibility with the application parameters. For example, single-phase samples must be maintained over the range of the specified ambient temperature environment, as well as during sample transport to the analyzer or sample system and in the analyzer sample valve. The standard must also remain chemically stable and homogeneous during its specified life.

4.3.10 Process gas chromatography applications

The next section will describe three PGC applications, commonly encountered in on-line analysis. The intent is to make the reader aware of the general concepts and the important considerations for PGC applications needed in order to make a useful and reliable process measurement.

4.3.10.1 FID, PTGC oven application. Motor gasoline analysis is a recent PTGC application of importance to the refinery industry. The analysis of some low boiling point components is utilized for blending control while the complete sample analysis provides additional information related to product quality. The sample is composed of a complex mixture of chemical classes and concentrations, with widely varying boiling points. The components are carbon based and range in concentrations from moderate percent levels and lower. Individual separation of components, other than control components, is not required; but, automatic integration of all eluting peaks is electronically stored for post-analysis processing. Elution of components in order of increasing boiling points is an important chromatographic requirement. The elution times of peaks are mathematically related to the boiling point temperatures and are used to calculate initial and final boiling points, percent off at a temperature, temperature at a percent off, Reid vapor pressure and the vapor to liquid ratio for the process sample.

The hardware used in this application is shown schematically in Figure 4.19. This application represents a relatively simple hardware configuration insofar as a sample is provided to a single sample valve which injects onto a non-polar column followed by FID detection of all eluting components. The PTGC oven is indicated in Figure 4.19 by the dashed lines and, is surrounded by a constant temperature, isothermal oven. The sample valve and detector are located in the isothermal oven which is at a temperature sufficient to provide a rapid vaporization and transport of the sample inject to the column as well as transport of the eluted components from the PTGC oven to the FID detector. Figure 4.20 represents a typical chromatogram of motor gasoline with elution times increasing from the right hand side of the figure to the left. The cycle time of this analysis is about 15 min.

Figure 4.21 indicates the elution, identity, and concentration of components in the calibration blend. These are components of known boiling point which allow calibration of the retention times of eluting hydrocarbons to a specified temperature.

4.3.10.2 FID, isothermal oven application. This application is one of several employed in the operation of an ethylene plant. The main focus of this application description is the use of valving to accomplish a

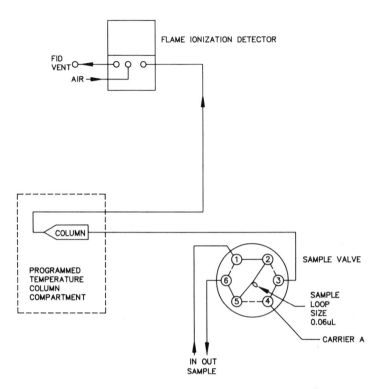

Figure 4.19 Chromatographic hardware configuration utilized for a typical motor gasoline analysis.

'heartcut-backflush-to-vent' arrangement using six port valves. A heart-cut is primarily utilized to aid in quickly separating minor components from one or more high concentration components. The sample is an ethylene product stream containing two analytes, methane (C1) and ethane (C2), in ppm concentration ranges. Figure 4.22 indicates the arrangement of valves, columns, detectors and adjustable restrictors. An internal-column thermistor detector (ITC) is indicated between columns one and two. The ITC is a convenient maintenance tool and provides the user with useful information such as selecting critical valve actuation times. Initially the valves are set in an orientation to establish the carrier pressure required for a specified flow, allowing adjustment of the restrictors for maintaining a given flow to the vents for each valve orientation utilized. Piping diagrams such as Figure 4.22 depict the valves in the deactivated state and, it is useful to understand that, with six port valves, each operation (i.e. sample backflush and heartcut) is primarily associated with a single valve operation. Figure 4.23 is useful for visualizing how the values operate to accomplish the transport of components to the

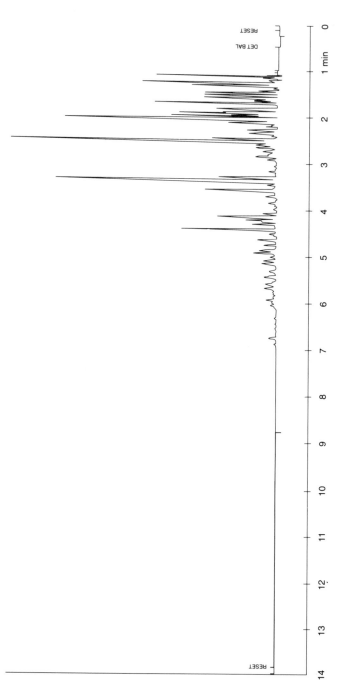

Figure 4.20 Typical chromatogram of a motor gasoline analysis. Elution times increase from right to left. (Courtesy of Applied Automation/ Hartmann & Braun.)

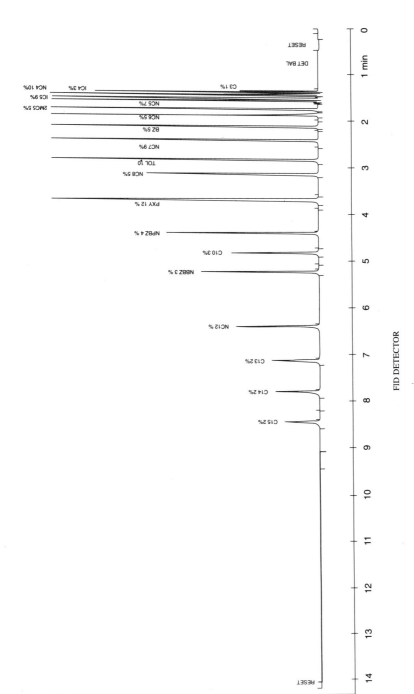

Figure 4.21 Motor gasoline calibration blend chromatogram. (Courtesy of Applied Automation/Hartmann & Braun.)

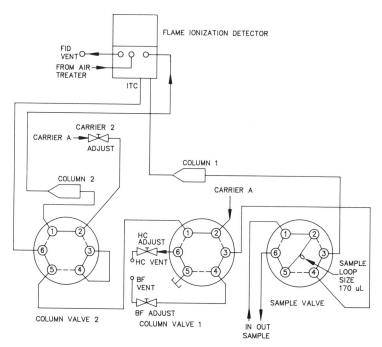

Figure 4.22 'Piping' diagram for a heartcut-backflush-to-vent column/valve arrangement. (HC, heartcut; BF, backflush; ITC, internal column thermistor detector).

detector or vents. The columns separate the components indicated in the order methane, ethylene, and ethane.

The analysis begins with the valves in position three, indicated in Figure 4.23. The new cycle first actuates the valves to position 1 to allow the flows to stabilize prior to the sample inject. The sample valve will be actuated and sample injected into the carrier stream flowing toward the HC vent in this position. At the appropriate time, the valves will change state to accomplish position 2 and allow the methane peak moving through column one to pass to column two. As soon as the methane peak is on to column two, the valves are activated to return to position one. Most of the ethylene exits through the HC vent. Again, at the appropriate time, the valves return to position 2 and ethane (mixed with small traces of unseparated ethylene) flows onto column two. As soon as the ethane has passed to column two, the valves are switched to return to position 1. After the last HC, the valves are switched to position 3 to flush any remaining higher boiling components and residual traces of ethylene to the BF vent.

Finally, Figure 4.24 depicts the chromatograms from the detectors in this application. The valve manipulations in this application result in a

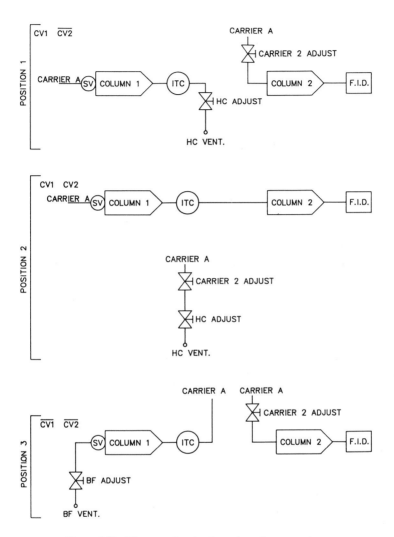

Figure 4.23 Diagram showing how the valves operate.

separation at the FID detector of methane, residual ethylene and ethane in about 6 min. The ITC chromatogram indicates the relationship of the heartcuts to the main process stream component, ethylene.

After the first heartcut, methane continues through column two and is sensed by the detector. The second heartcut is a mixture of traces of ethylene and ethane, which column two separates into what is referred to as the HC ramp and the ethane peak. The window of a heartcut is the time between moving to position 2 and returning to position 1 or 3. If the HC window for the ethane peak is made equally wider on the front

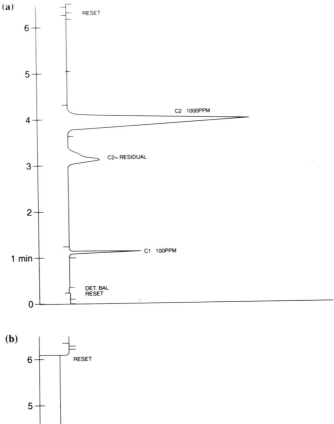

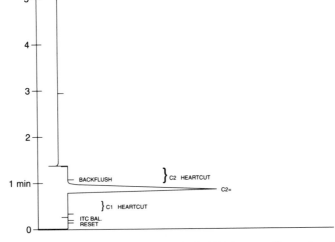

Figure 4.24 Chromatograms from (a) FID and (b) ITC detectors illustrating the use of a valving technique; heartcut, backflush to vent, as applied in an ethylene purity analyzer (Courtesy of Applied Automation/Hartmann & Braun.)

and backside, the ethylene heartcut ramp will continually increase in size until the ethane is no longer separated. Conversely, if the heartcut window is too narrow, some of the component(s) of interest will not be sent toward the detector, which is a situation to be avoided.

4.3.10.3 TCD, isothermal oven application. The main focus of this application is the use of valving to accomplish a backflush-to-vent, trap bypass (BV, TB) arrangement using six port valves. The backflush/trap arrangement is primarily used to aid in the quick separation of components of varying boiling points or polarities. The sample is a cracked gas stream of widely varying percent level concentrations of light hydrocarbon gases, all of which are suitable for thermal conductivity detection. Figure 4.25 indicates the arrangement of valves, columns, detectors and adjustable restrictors. Initially, the valves are set in a particular orientation to establish the carrier pressure required for a specified flow. Restrictors are adjusted to maintain a given flow rate to the vents for each valve orientation utilized. Each operation (i.e. sample backflush or trap bypass) is primarily associated with a single valve operation. Figure 4.26 is useful in visualizing how the valves operate to accomplish the

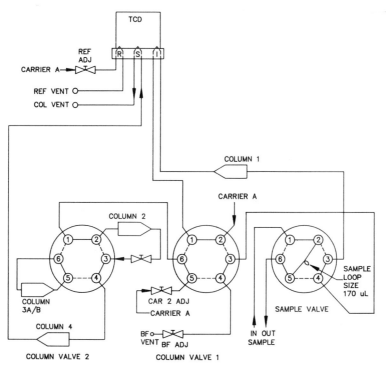

Figure 4.25 'Piping' diagram for a backflush-to-vent, trap bypass column/valve arrangement. R = reference detector; S = sample detector; I = intercolumn detector.

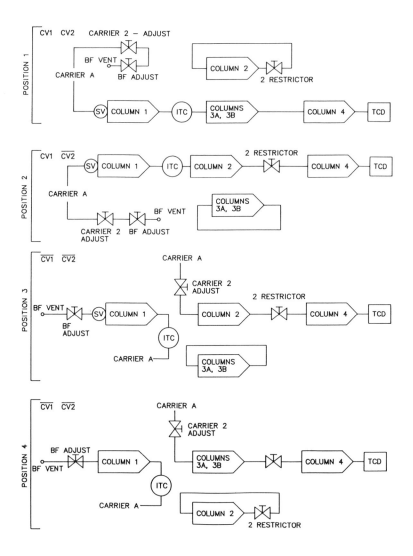

Figure 4.26 Diagram showing how the valves operate.

transport of components to the detector through one of the columns, or, to the BF vent.

The analysis begins with the valves in position four, as indicated in Figure 4.26. The new cycle first actuates the valves to position 1 to allow flows to stabilize prior to the sample inject. The sample valve will be actuated and sample injected into the carrier stream flowing through columns one and three toward the TCD. At the appropriate time, the valves will change state to accomplish position 2, which traps the lowest boiling point components in column three, allowing the higher boiling point components to pass through column two toward the TCD. As soon as the last component of interest passes onto column two, the valves are operated to accomplish position three, which backflushes the unwanted sample to the BF vent. Just after the last component allowed onto column two has passed through the TCD, the valves are placed in the correct states to effect position 4. This action 'untraps' the lowest boiling point components and forces these remaining components through the TCD, after which the entire cycle can be repeated. In order of elution (indicated in Figure 4.27), the following components are separated and detected in this application in 7 min: propane, propene, air, methane, ethene, ethane, and ethyne. The backflushed components include, propyne, allene, C4 paraffins and olefins as well as some C5s. The order of components detected from lowest to highest boiling point is as follows: air, methane, ethyne, ethene, ethane, propene, propane, propyne, allene. The difference between the elution order and the boiling point order is due to the valving scheme and the polarity of the columns eluting those components. The ITC chromatogram indicates the groupings of peaks that take different paths during the analysis.

4.3.11 Future developments in process gas chromatography

Considerable research efforts continue to be undertaken in order to expand the application and use of process gas chromatography. More recent efforts have centered on the speed of analysis, improved technology and overall process chromatograph size [10]. The advent of capillary columns in the 1980s has resulted in remarkable improvements in column separation power and speed, with some applications now completed in seconds rather than minutes. Improvements in capillary column technology will allow more durable metal capillary columns to continue to replace the fused silica columns introduced only a few years ago, while the use of conventional packed columns will continue to diminish.

Improved chromatograph detector technology will serve to enhance the capabilities of the process chromatograph. More recent efforts have been directed at the design of more universal detectors with increased sensitivity. While specific detectors will continue to be widely used in laboratory environment, process gas chromatographs will benefit from more universal detector technology. Examples of these types of detectors are pulsed discharge helium ionization and mass spectrometry.

Silicon micromachining will play an increasingly significant role in efforts to reduce the overall size of chromatography systems, as further improvements are made in the manufacture and cost of micromachined valves and detectors [11]. Improvements to man–machine interfaces will provide users increased hardware and application flexibility while reducing

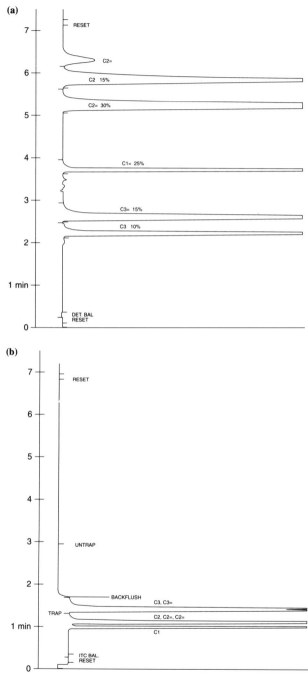

Figure 4.27 Chromatograms from (a) FID and (b) ITC detectors illustrating the use of a valving technique; backflush to vent, trap bypass, as applied in a cracked gas analyzer (Courtesy of Applied Automation/Hartmann & Braun.)

the amount of training and knowledge required for the basic operation of the process chromatograph. Additionally, increased use of system diagnostics will continue to play a role in advancing ease-of-use and reducing analyzer downtime and maintenance requirements.

4.4 Liquid chromatography

4.4.1 Overview of system hardware

4.4.1.1 Comparison to PGC. Process liquid chromatography (PLC) was first introduced in 1974 [12]. Although similar to gas chromatography in terms of its fundamental aspects, it differs in a number of significant ways. As previously mentioned, liquid chromatography utilizes *liquid* mobile phases rather than vapors. In addition, liquid chromatography is normally performed at much lower temperatures as compared to GC. Generally, the application of PGC is focused on compounds with relatively high vapor pressure, a characteristic of the majority of compounds processed in the petroleum and chemical industries. In addition, the technique of PGC is compatible with many industrial processes in that the sample is often a vapor, and actual sample preparation is often minimal. Nevertheless, there are a large number of high boiling compounds which either cannot be analyzed by GC or, because of the conditions required for GC analysis, namely elevated temperatures, one would not choose to perform with a process GC. Operating a PGC under these types of conditions results in greatly reduced life span of the valves and columns. For these reasons, PLC has been found to be a useful complementary tool to PGC. In cases where a process stream could be analyzed by either method, PGC would normally be the preferred method since it can usually provide an analysis with less cost and effort. Other considerations include long-term maintenance. Although prior experience with PGC will be useful, technical personnel will require training in the specific characteristics of PLC in order to be effective.

Besides high boiling compounds, other applications utilizing PLC are focused on compounds which are thermally labile or react when heated [13]. When heated, these compounds will often decompose or polymerize, thereby preventing their analysis by PGC.

Finally, PLC employs detectors which, in some cases, provide PLC increased selectivity for a given analysis relative to that obtainable using PGC [14].

4.4.1.2 Comparison to laboratory LC. There exist a number of differences between PLC and laboratory LC instrumentation, primarily as a result of different objectives. Whereas laboratory LC analyses are generally accomplished by manual sample injection followed by analysis on a versatile LC system to provide fully resolved analyte peaks, PLC requires robust automated sample preparation and injection, automated analysis and rugged instrumentation for placement in hazardous process locations. Applications must be developed in order to deliver highly reproducible results and in a timely manner, unlike laboratory LC applications which may take up to 1 hour or more. Further, just as in process gas chromatography, the PLC must frequently be capable of operation in environments with widely varying ambient temperatures.

Figure 4.28 illustrates a typical PLC schematic [15]. Among the key differences between PLC and laboratory LC is the liquid mobile phase pump. While piston-type reciprocating pumps are popular for laboratory systems, PLCs typically utilize pneumatically operated pumps for pressurizing the liquid mobile phase. These pumps generally have lower maintenance requirements and are suitable for use in hazardous areas. PLC applications often utilize multiple columns and valve switching in order to obtain acceptable analysis times. Timely results are of vital importance for PLC applications. Also, the life expectancy of a process column must exceed that of a typical laboratory column. Typical life expectancies for a laboratory column are on the order of a thousand injections. In PLC, this would provide only 1 week of life. Measures must be taken to prevent the accumulation of trace components, which act to change the characteristics of the column and shorten its life.

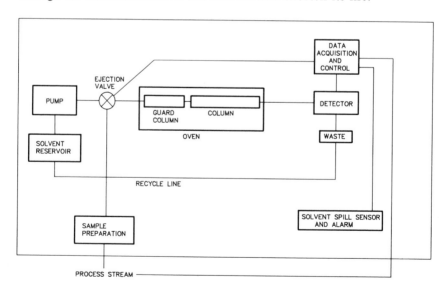

Figure 4.28   Schematic diagram of a process liquid chromatograph. (Reproduced from Ref. 15, by permission of International Scientific Communications Inc.)

4.4.1.3 PLC detectors.   Three detectors are currently utilized for process LC. The most popular is the ultraviolet (UV) or optical absorbance detector [16]. Reference and sample flows are through individual cells so that the absorption characteristics of the sample stream are compared with those of the reference stream. A typical optical path length for the detector is 1 cm, with a cell volume of 18 μl. The detector is rated for operation to 500 psig with discrete UV wavelengths from 214 to 350 nm. UV detection in liquid chromatography is generally the preferred method if the sample components to be measured have a chromophore with suitable absorbance.

The refractive index detector detects the presence of a component by the change of a refractive index of the column effluent [17]. The deflection type of detector used for PLC uses a single light beam which passes through the detection and reference cells twice. The disadvantage to a refractive index detector is the sensitivity to temperature changes. These effects are somewhat negated through design considerations which minimize temperature variations. For a PLC detector, this is accomplished by separate temperature control of the detector or by placing the detector in an explosion-proof housing with significant mass to act as a heat sink and stabilizer.

Finally, the dielectric constant detector measures small changes in the dielectric constant of a flowing liquid stream [18]. The detector contains both a reference and sensing cell, each cell containing adjacent, parallel-plate capacitors, with the sample component and mobile phase serving as dielectric material between the plates. Each cell forms a portion of a parallel inductance/capacitance resonant circuit, where the frequency of each cell oscillator is determined by the capacitance of the cell. The output frequency of the two oscillators is fed to a mixer that senses the frequency difference and generates a signal proportional to this difference.

4.4.2 Types of liquid chromatography

4.4.2.1 Liquid–liquid chromatography.   Liquid–liquid chromatography is based on the partitioning of the sample between the liquid carrier and a liquid stationary phase that coats a solid packing material making up the column. Normally, the liquid stationary phases are chemically bonded to the packing material to prevent problems of *bleeding*, or washing off.

Liquid–liquid chromatography is further divided into normal and reversed-phase systems. Normal phase refers to systems wherein the mobile phase is less polar than the stationary phase, whereas reversed-phase systems utilize carriers which are more polar than the stationary phase. Reversed-phase LC separates many organic compounds with water as the major mobile phase solvent. This is attractive for the process application since it is generally less flammable and less costly than a normal phase carrier. Unfortunately, reversed-phase separations also tend to be lengthy. The technique is most useful for separation of non-polar hydrocarbons, particularly those which differ only in their carbon number.

4.4.2.2 Liquid–solid chromatography. Liquid–solid or adsorption chromatography was the first reported type of chromatography [18]. Liquid–solid chromatography arises from a competition between sample and mobile phase molecules for a site on the adsorbent stationary phase surface. For normal phase separations, silica has been found to be an excellent adsorption material for PLC, with alumina used on occasion for particularly unique separations. Liquid–solid chromatography is normally used for intermediate molecular weight organic materials, particularly isomers or mixtures of similar compounds. Typical mobile phases include methylene chloride and hexane.

Reversed-phase separations using liquid–solid chromatography often use non-polar functional groups chemically bonded to silica and utilize carriers such as water/acetonitrile or water/methanol mixtures.

4.4.2.3 Size-exclusion chromatography. The application of size-exclusion chromatography has become the single most important application for PLC, primarily the result of polymer stream analysis, wherein the technique is ideally suited. Size-exclusion chromatography has often been referred to by a variety of other names, such as gel permeation chromatography, gel filtration chromatography, liquid-exclusion chromatography, and steric exclusion chromatography, All of these terms refer to the same general technique of component separation according to molecular size.

Size-exclusion chromatography utilizes column packing materials containing small pores. Components (molecules) which are smaller than the average pore size of the packing material will reside within the pores for longer time periods than molecules which are larger. The result is a molecular size (weight distribution) chromatogram with the larger molecules eluting first followed by the smaller molecules, which will have a longer elution time.

4.4.3 Process liquid chromatography applications

PLC applications can address a wide variety of analytical needs. One of the most prolific commerical applications for PLC is the analysis of polymeric materials [14]. A common example of polymer analysis is styrene-butadiene copolymer product (Figure 4.29). The hardware is straightforward in configuration and includes a sample valve, an analytical column and a UV detector enclosed by an isothermal oven. The sample

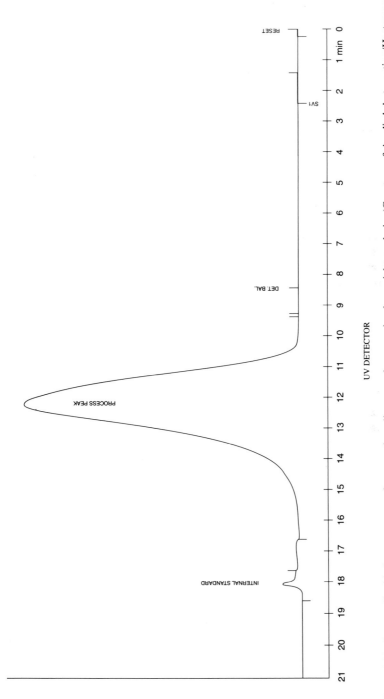

Figure 4.29 Process liquid chromatogram of styrene–butadiene copolymer molecular weight analysis. (Courtesy of Applied Automation/Hartmann & Braun.)

is typically diluted prior to injection. The information derived from the analysis is generally a molecular weight analysis. The results describe the sample in various types of molecular weight results and component concentration. The calculated parameters may all be used to characterize the properties of the polymer. The calculations for this type of an analysis are well documented in existing literature. The results are utilized to tune the process parameters and provide a record of the quality of each batch of product. PLC provides a useful tool for relatively quick analysis of difficult process samples such as polymers, materials with boiling point ranges outside the range of PGC, thermally labile species and inorganic samples.

4.5 Process supercritical fluid chromatography

4.5.1 Overview of system hardware

Supercritical fluid chromatography is a chromatographic technique that uses supercritical fluids for the mobile phase. Heating and compressing a gas above its critical temperature and pressure causes it to exhibit properties which are intermediate between those of gases and liquids. One key property is that the density varies as a function of the temperature and pressure. Increasing the pressure causes the density to increase, and subsequently causes the solvent strength to increase. Changes in solvent strength results in changes in the partitioning of the component between the stationary phase and the supercritical fluid mobile phase. Related to density, the viscosity of the supercritical fluid will be similar to gases and much lower than that of liquids.

PSFC combines to a large extent the favorable attributes of PGC and PLC, but extends the range of possible samples which can be analyzed by chromatography. It is estimated that only 20% of known organic compounds can be satisfactorily separated by GC without prior chemical modification. Generally, compounds which are thermally labile, have high molecular weights, are low UV absorbers, and/or are chemically reactive with certain PLC mobile phases are prime candidates for PSFC. Further, as a result of differences in the mobile phase characteristics as compared to PGC and PLC, increased column efficiencies, resolution and theoretical plates generally results in PSFC [19].

Figure 4.30 illustrates a phase diagram of carbon dioxide (CO_2) showing its triple point, critical temperature (31.1°C) and critical pressure (73 ATM, 1070 psig). As a result of its low critical parameters and intermediate polarity, carbon dioxide is the most widely used solvent in SFC. Also, its lack of toxicity and the absence of an appreciable response in an FID make it an ideal mobile phase for PSFC.

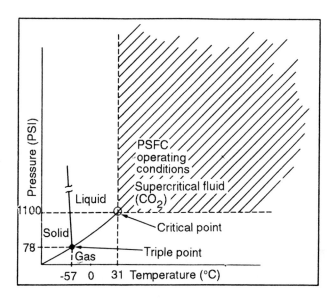

Figure 4.30 Phase diagram of carbon dioxide. (Reproduced by permission of ABB Process Analytics.)

Figure 4.31 shows a schematic diagram of a PSFC. Several components are critical to the operation of the PSFC and differentiate it from PGC and PLC. The pump serves to maintain supercritical conditions in the system and to provide a programmed pressure ramp, similar to a

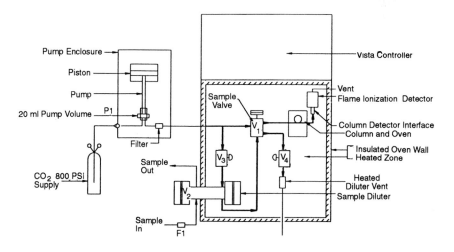

Figure 4.31 Schematic diagram of a process supercricical fluid chromatograph (PSFC). (Reproduced by permission of ABB Process Analytics.)

programmed temperature ramp in PGC. The pump is pneumatically driven from an 80 psig air supply, providing supercritical CO_2 from 1000 to 7000 psig pressure.

Small-bore (0.1 mm ID) fused-silica capillaries, as well as packed columns, are used for PSFC applications. Capillary columns generally exhibit higher resolution, whereas packed columns can handle higher sample-loading capacities, which in turn can yield lower detection limits and higher flow rates, thus shortening analysis time.

Although a variety of detectors have been utilized in laboratory SFC applications, only the flame ionization detector has been used for process SFC. This is the result of its ready availability from PGC in a design suitable for process instruments, as well as its near universal response to hydrocarbons, while having no response to the CO_2 mobile phase.

The final critical component of the PSFC is the capillary restrictor column–detector interface. Much of the difficulty encountered in the development of SFC for the process has been focused on the capillary restrictor. Effluent from the PSFC column must be depressurized prior to introduction to the FID detector. This is achieved by incorporation of a capillary restrictor, typically of the porous frit type. The restrictor has the task of providing for a complete transfer of the non-volatile component to the detector without allowing formation of component particles and subsequent plugging, as well as providing for a pulse-free flow throughout the column.

4.5.2 Process supercritical fluid chromatography applications

To date, actual incorporation of PSFC to address on-line process applications has been limited. Nevertheless, the relatively mild conditions required for PSFC have made it attractive for a number of analyses. For example, simulated distillation analysis by PGC has typically been limited to those petroleum fractions which have a final boiling point (FBP) lower than 400°C. Attempts to operate PGC at elevated temperatures is plagued by premature failure of valves and other PGC hardware components. SFC has been found to be successful for extending this range under relatively mild conditions. Figure 4.32 shows a chromatogram and distillation report of a vacuum gas oil with a FBP of 600°C.

4.6 Ion chromatography

4.6.1 Overview of system hardware

Ion chromatography is a chromatographic technique which traditionally has utilized columns packed with cation and anion exchange resins,

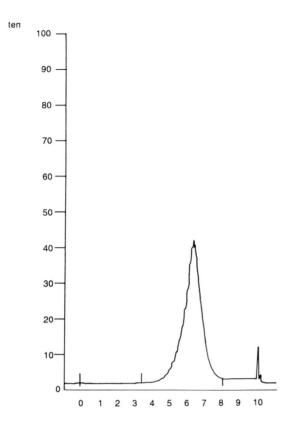

SIM DIS REPORT
CT 1

% Off	Time	Temp (°C)
0.5	243	302
5.0	316	367
10.0	335	387
15.0	346	399
20.0	354	408
25.0	361	417
30.0	367	426
35.0	371	431
40.0	375	437
45.0	379	443
50.0	383	448
55.0	387	453
60.0	390	457
65.0	394	462
70.0	398	468
75.0	403	477
80.0	408	484
85.0	415	492
90.0	423	508
95.0	437	526
99.5	474	589
100.0	484	607

Figure 4.32 Chromatogram and simulated distillation report for a vacuum gas oil analysis. (Reproduced by permission of ABB Process Analytics.)

aqueous mobile phases and conductivity detectors to separate and identify soluble ionic and polar compounds. The technique came into widespread use by 1975 [20] and process ion chromatography (PIC) was first employed as an on-line monitoring technique in 1977 [21] for boiler feed water analysis. The primary benefit of ion chromatography is that analyses which normally take hours to complete via wet chemistry methods can be accomplished in a few minutes with ion chromatography, and with generally better accuracy and reproducibility.

Ion chromatography is very closely related to liquid chromatography and, in many cases, laboratory vendors provide a single set of hardware for both analysis techniques. Nevertheless, several advances in instrumentation, including materials of construction, column and detector technology and process configuration have served to separate these two techniques. In ion chromatography, all sample wetted components must be constructed of inert non-metallic materials in order to avoid contamination of the liquid mobile phase during trace component analysis. More recently, newer polymeric materials have been introduced which are not only highly inert, but are able to withstand the high mobile phase pressures often used in ion chromatography (up to 5000 psig). Ion chromatography has benefited from developments in non-functionalized, polymer based resin columns for use in reversed-phase and ion pairing separations. Latex bonded ion exchange materials have allowed simultaneous separations of ionic and non-ionic compounds on the same column. Typical detectors for ion chromatography include photometric, conductivity and pulsed amperometric detection (PAD). Within the PAD detector, compounds are oxidized or reduced at an electrode surface in the detector cell. After the current is sampled at the appropriate potential, the electrode potential is alternatively pulsed more positive and then negative in order to clean the electrode surface. PAD detection is useful for electroactive species which are neither ionic or chromophoric. Figure 4.33 is a schematic diagram of a PIC.

Figure 4.34 shows a typical process ion chromatograph. At present, PIC instruments will typically consist of laboratory hardware placed in NEMA enclosures, either to avoid dust and corrosive vapors or to comply with local safety codes for electrical devices. Due to the nature of a typical PIC analysis, many applications do not require placement of instruments in hazardous locations. Nevertheless, as PIC becomes a more accepted on-line analysis approach, increased use of instruments in hazardous locations will be predicated upon systems which are designed to be process worthy.

4.6.2 Process applications

A variety of applications have been developed utilizing process ion chromatography. Control of corrosion caused by contaminants such as

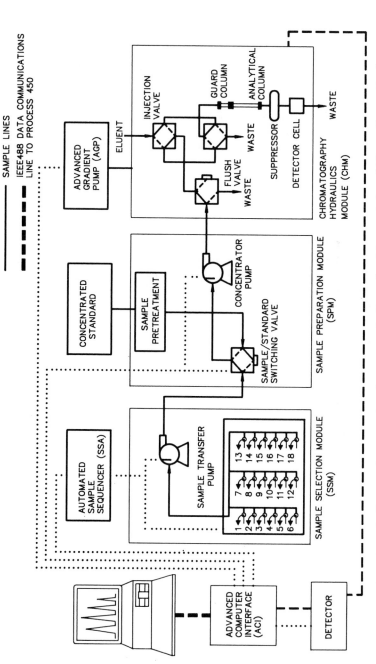

Figure 4.33 Schematic diagram of a process ion chromatograph. (Reproduced by permission of Dionex Corporation.)

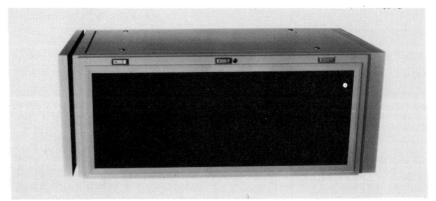

Figure 4.34 Photograph of a typical process ion chromatograph showing molecular components in a NEMA enclosure. (Reproduced by permission of Dionex Corporation.)

chloride, sulfate, sodium, and silica is critical in the power generation industry. Minimizing the concentration of corrosive ions in boiler feed water and effluents can reduce the damage to turbines and nuclear power plant steam generators. Determination of ion concentrations at critical points throughout the power plant operation can provide engineers and water chemists with valuable information regarding possible sources of contamination as well as serve to provide an early detection and warning system for possible equipment or process failure.

The semiconductor industry increasingly relies on ultrahigh purity water sources, since virtually any type of contamination can significantly reduce device yields. On-line ion chromatography has been utilized for monitoring inorganic anions, cations, and silica in high purity water production facilities. Continuous monitoring of these component analytes can reveal when an ion exchange demineralizer requires regeneration.

Plating solutions require the careful control of major constituents as well as the addition of trace amounts of chemicals to brighten a deposit or stabilize a bath. Ion chromatography has been used in conjunction with flow injection techniques to provide a complete profile of plating bath constituents, providing information for control of critical bath parameters.

A number of industries are placing greater emphasis on the continuous monitoring of waste water streams as a result of an increasing number of governmental regulations. Of particular concern are the concentrations of transition and heavy metals in plant effluents. Removal of metals from a waste water effluent usually depends on a complexation or precipitation reaction. Continuous monitoring of metals in the influent to a waste treatment facility can be used to optimize the quantities of treatment reagents added. Optimization of these types of facilities allows plant engineers to reduce cost of expensive treatment chemicals while ensuring that the system operation meets discharge limits.

4.7 Programmer/controller/communication

The programmer/controller section of a process chromatograph generally contains the bulk of the electronics necessary to power the system, including the controller and the data reduction hardware and software. The modern process chromatograph is designed as a stand-alone, micro-processor-based system. Typical input/output (I/O) characteristics of a process chromatograph include analog outputs, digital communications and alarms. Analog outputs include trend outputs via 4–20 mA signals which are assignable to components, chromatogram, and/or bargraphs and are updated after each analysis cycle is completed. Alternatively, digital communications are typically available and include serial ports supporting RS-232C, 20 mA current loop, RS-485 or RS-422. Communications are normally devoted to printer terminals and host computer or DCS systems.

Alarms may include data alarms, system alarms and status alarms, all via display or printer. System alarms monitor the pneumatic, mechanical, and electrical operation of the system. These would include low carrier, low calibrant, low sample flow, detector failure, oven temperature out of range, PROM error, etc. Data alarms monitor the actual analytical results of the system, sometimes flagging either high or low component concentrations. Increased attention is being given to developing within the design of the process chromatograph self-diagnostic features capable of continuously monitoring the functionality of the instrument and providing instrument personnel information regarding potential problems and locations.

References

1. M. Tswett, *Ber. Deut. Bot. Gesellsch.* **24** (1906) 318.
2. A. T. James and A. J. P. Martin, *Analyst* **77** (1952) 915.
3. W. C. Askew, *Anal. Chem.* **44** (1972) 633.
4. S. O. Farwell and C. J. Barinaga, *J. Chromatogr. Sci.* **24** (1986) 483.
5. K. Grob, *Making and Manipulating Capillary Columns for Gas Chromatographs.* Huethig Publ. Co., Heidelburg, Germany, 1986.
6. V. N. Lipavsky and V. G. Berezkin, *J. Chromatogr* **91** (1974) 583.
7. R. Annino and R. Villalobos, *Am. Lab.* **10** (1991) 15.
8. R. Kenter, M. Struis and A. L. C. Smit, *Proc. Control Qual.* **1** (1991) 127.
9. J. Crandall, R. L. Cook and R. K. Bade, *I&CS* (1989) 31.
10. R. Sacks, *et al., Analyst* **116** (1991) 1313.
11. G. Lee, C. Ray, R. Siemers and R. Moore, *Am. Lab.* (1989) 110.
12. R. A. Mowery, Jr and L. B. Roof, *Anal. Instrum.* **14** (1976) 19.

13. R. A. Mowery, Jr *Chem. Engng* **88** (10), (1981) 145.
14. L. V. Benningfield, Jr and R. A. Mowery, Jr, *J. Chromatogr. Sci.* **19** (1981) 115.
15. R. E. Synovec, L. K. Moore, C. N. Renn and D. O. Hancock, *Am. Lab.* (1989) 82.
16. S. E. Walker, R. A. Mowery, Jr and R. K. Bade, *J. Chrom. Sci.* **18** (1980) 639.
17. R. A. Sanford, R. K. Bade and E. N. Fuller, *Am. Lab.* (1983) 99.
18. D. T. Day, *Proc. Am. Philos Soc.* **36** (1897) 112.
19. G. B. Levy, *Am. Lab.* **18** (12), (1986) 62.
20. H. Small, *et al., Anal. Chem.* **47** (1975) 1801.
21. G. J. Lynch, (1991), *Process Control and Quality* (Vol. 1). Elsevier Science Publishers BV, Amsterdam, The Netherlands, p. 249.

Bibliography

R. Annino, *Am. Lab.* **21** (1989) 60.
R. Annino, *Process Gas Chromatography: Fundamentals and Applications.* Instrument Society of America, Research Triangle, NC, USA, 1992.
K. J. Clevett, *Process Analyzer Technology.* John Wiley & Sons, Inc., New York, USA, 1986.
K. J. Clevett, *Bioprocess. Technol.* (1990) 47.
K. Haak, S. Carson and G. Lee, *Am. Lab.* **12** (1986) 18.

5 Flow injection analysis

P. MacLAURIN, K. N. ANDREW and
P. J. WORSFOLD

5.1 Introduction

Flow injection (FI) is now accepted as a robust and reliable means of
sample presentation and on-line sample treatment in analytical laboratories. The term flow injection analysis (FIA) was first used by Ruzicka
and Hansen in 1975 [1] and its subsequent success can largely be
attributed to its simplicity and versatility. FI is an unsegmented continuous flow technique and by manipulating basic operational parameters,
e.g. flow rates and sample volume, a wide variety of analytical conditions
can be established in a highly reproducible manner, allowing the routine
determination of a diverse range of analytes, sample matrices and
concentration ranges. The key to this versatility is 'controlled dispersion'.

Many of the salient features that have established FI as an indispens-
able tool for laboratory analysis are often quoted as prerequisites for
process analysers; e.g. speed of response, high sample frequency and
dependable instrumentation. The number of genuine on-line applications
of FI discussed in the literature remains small in spite of the attractive
features stated above. More encouraging for the future of FI as a process
analytical technique is the fact that reported applications cover a diverse
range of industries including bulk chemical and metal production,
bioprocessing and wastewater processing. The aim of this chapter is to
present the current status of process FI, identify recent developments
most relevant to its continued evolution and define the future role for FI
in process analysis.

5.2 Fundamental principles

The essence of FI is the controlled physical dispersion of an injected
liquid sample into a continuously flowing unsegmented liquid carrier
stream. If the carrier stream contains a suitable reagent, chemical
reaction can also occur, resulting in a transient reproducible detector
response proportional to the analyte concentration. This is illustrated in
Figure 5.1, which schematically represents a simple single-channel FI
manifold. The carrier stream is propelled by a pump through a narrow

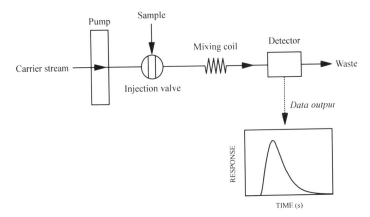

Figure 5.1 Schematic diagram of a single-channel flow injection manifold, showing the transient nature of the signal output. (——) Liquid flow; (.....) Data flow.

bore conduit to an injector, and then via the reaction manifold to the detector and to waste. Each injection yields a single FI peak at the detector, the height, area or width of which is proportional to the concentration of the analyte.

Central to the flexibility of FI is the control that can be exerted on the dispersion process. The coefficient of dispersion (D) has been defined as the ratio of the concentrations prior to (C_o) and after (C_{max}) the dispersion process has taken place in the element of fluid that yields the analytical response [2].

$$D = C_o/C_{max}$$

In FI, dispersion is therefore a measure of the degree of dilution of the injected sample and has traditionally been classified as limited $(D = 1–3)$, medium $(D = 3–10)$ and large $(D > 10)$. The degree of dispersion is governed by the operational parameters of the FI manifold and can be manipulated to achieve optimal analytical performance. In general terms, limited dispersion is used for analysis requiring high sensitivity, and can be achieved through large injection volumes, low flow rates and short lengths of narrow bore manifold tubing. Conversely, combinations of small injection volumes, higher flow rates and longer manifolds will yield greater dispersion. A full treatment of the theory of the dispersion process is given in FI monographs [2,3].

It follows, therefore, that dispersion can be intuitively controlled to suit the application, reaction chemistry and system of detection being utilised. For example, a dispersion coefficient of 1 may be used to present an undiluted sample to the detector for measurement of an inherent property such as pH or conductivity. Dispersion coefficients of > 10 are generally used for on-line dilution prior to detection.

As a technique, FI is often compared with continuous segmented flow analysis (CFA). CFA developed as a result of the pioneering work of Skeggs in the 1950s [4] and became an established means of automating liquid phase analytical procedures (e.g. Technicon AutoAnalyzer®), particularly in clinical and water laboratories. Dispersion was limited by means of air-segmentation, whereby the liquid stream was separated into small compartments by air bubbles. These compartments represent isolated reaction chambers wherein homogenous mixing produces steady state signals. This is in contrast to FI where heterogeneous mixing and the absence of segmentation offer a number of advantages, most notably, improved reproducibility and a faster initial response. In addition, the absence of air-segmentation leads to a manifestly less complex instrument, vitally important for successful unattended process analysis. A direct comparison of the key features of FI and CFA is given in Table 5.1.

Table 5.1 Comparison between the principal features of flow injection and segmented flow techniques

Parameter	FI	CFA
Transportation	Unsegmented	Segmented
Sample introduction	Injection	Aspiration
Sample frequency	~60 h^{-1}	~60 h^{-1}
Sample mixing	Heterogeneous	Homogeneous
Analytical signal	Transient	Steady-state
Analytical response	~10–120 s	~3–15 min
Sample volume	~10–100 ml	~0.5–1.0 ml
Reagent consumption	~0.5–2.0 ml min^{-1}	~5–10 ml min^{-1}
Internal diameter of flow tubing	0.5–1.0 mm	2.0 mm

FI has also been compared with high-performance liquid chromatography (HPLC) (chapter 4). Whilst the hardware used in both techniques is ostensibly the same, describing FI as 'HPLC without the column' fails to convey the merits of reproducible controlled dispersion or the power of on-line sample treatment. In general terms, selectivity in FI is achieved by chemical means and/or selective detection, whereas chromatography relies on physical separation prior to detection. There is however significant overlap; e.g. post-column derivatisation in chromatography whereby components of a column effluent are chemically manipulated in an FI manifold prior to detection. Furthermore, minicolumns are frequently used in FI for sample manipulation and a wide variety of non-chromatographic continuous separation techniques can be utilised [5].

5.3 Instrumentation

The three major elements of any FI system are propulsion, injection and detection. Each of these is discussed in detail below, and a summary of the various options for each of these elements is given in Table 5.2. For process applications automation is also an essential feature, and this is discussed at the end of this section.

Table 5.2 Various options for each of the major FI manifold components

Component	Options	Description
Propulsion system	Peristaltic pump	Set of rollers on a revolving drum, which squeeze flexible tubing to produce a constant, pulsing flow.
	Gas pressurised vessel	Pressurised inert gas vessel connected via a flow regulator to each reagent/carrier reservoir, producing pulseless flow.
	Reciprocating pump	Reciprocating piston pumping fluid through a small chamber, with valves alternately opening and closing to control flow through the chamber. Produces pulsing flow.
	Piston pump	Computer-controlled, cam-driven piston, which produces bidirectional, variable speed, precise and pulseless flow.
Injection system	Rotary valve	Six-port unit incorporating a sample loop, which can be switched between filling and emptying positions. Electric or pneumatic operation.
	Hydrodynamic injection	Involves the selective stopping and starting of a sample pump and a reagent pump, with sample entering the reagent stream while the latter is stopped, then transported into the manifold when it is restarted.
	Multiposition selector valve	Multiport unit allowing sequential selection of a number of flow streams (e.g. sample, standard and reagent streams). Electric operation.
Detection system	Optical	e.g. UV-visible spectrophotometry; solid-state photometry; diode array spectrophotometry; IR spectrophotometry; fluorimetry; chemiluminescence, atomic spectrometry.
	Electrochemical	e.g. Potentiometry (ion-selective and pH electrodes); conductimetry; amperometry; coulometry; voltammetry.

5.3.1 Propulsion systems

The maintenance of consistent flow patterns is critical to FI reproducibility. FI flow rates are typically in the range of 0.5–2.0 ml min^{-1} with between 1 and 8 channels depending on the manifold complexity. A variety of flow patterns can be exploited, the most straightforward and widely used being a continuous linear flow. The stopped-flow technique, as the name implies, requires consistent cessation of flow at preset times, and is generally used to increase sensitivity and/or measure reaction rates. Similarly, intermittent flow approaches require consistent stopping and restarting of flow, and flow reversal requires the stop/start features in addition to gearing for reverse drive. Furthermore, continuous and non-continuous flow and flow reversal can be used with non-linear flow patterns. The particular flow pattern adopted therefore imposes certain requirements on the propulsion system and various approaches to the induction of flow are examined below.

The multiroller peristaltic pump is by far the most widely used FI propulsion system and is capable of delivering a constant volumetric flow that results in highly reproducible residence times. Peristaltic pumps do not however, deliver an entirely pulseless flow and in order to reduce the fluctuations in flow a pump head with 8–10 closely positioned rollers is recommended. Half of the rollers should be in permanent contact with the tubing, through sufficient (but not too much) compression from the pump bridge. The ideal bridge allows independent compression control of each channel. Coupled with an appropriate drive speed (~20–50 rpm) and an element of damping due to the tubing elasticity, peristaltic devices deliver a wide range of flow rates (~0.1–5.0 ml min^{-1}) in up to eight channels; each with a different flow rate if required.

The delivery rate is governed by the internal diameter of the pump tubing and the drive speed of the pump head. Over time the volumetric delivery rate will gradually decrease as a result of pump tube degradation. This is due to reduced flexibility with use and is dependent on the tubing material and the nature of the solution streams. It should be stressed however that the changes occur very slowly and are compensated for through regular calibration and periodic replacement. The use of pump tubing also has the advantage that no other parts of the pump come into direct contact with the liquids being pumped. Furthermore, the tubing is relatively inexpensive, its replacement is straightforward and is ideally changed as part of a maintenance programme. A variety of tubing material is commercially available and the correct choice is important for safe operation and sustaining consistent flow. The most widely used material is poly(vinyl chloride) (PVC) which comes in various grades for different applications. Standard PVC (e.g. Tygon®) is the general choice for aqueous solutions and dilute acids and bases but is not suitable for organic solvents and concentrated acids. Aliphatic hydrocarbons and alcohols can be delivered using modified PVC but this remains unsuitable for other organic solvents and strong acids/ bases. Thermally set fluorine rubbers (e.g. Viton®) are the material of choice for concentrated mineral acids and most organic solvents although they have a limited pump life and are unsuitable for certain organic species such as ketones. Other tubing materials are commercially available and should be used according to the manufacturers' specifications.

A well constructed peristaltic pump will generally have no inherent inertia, allowing instantaneous stop/restart for stopped/intermittent flow and flow reversal procedures, and due to the small internal volume, rapid washout and system start-up is possible. Finally, and very importantly, peristaltic pumps are insensitive to minor variations in the system back-pressure and stream viscosity which is critical to reproducible residence times and hence analytical precision.

Gas pressurised vessels provide an alternative to peristaltic pumps; they offer mechanical simplicity and pulseless flow but suffer from flow rate fluctuation due to back pressure variation. Incorporation of pressure transducers can help overcome back pressure issues through feedback and pressure control systems but flow reversal is not possible.

Reciprocating pumps such as those used for liquid chromatography (chapter 4) provide a further option for FI propulsion. Whilst they are markedly more complex than peristaltic pumps, reciprocating devices can be fabricated from inert materials and provide a suitable means of organic solvent delivery. The major drawback with reciprocating devices is financial; purchase costs are high and each FI stream requires a dedicated pump. In addition, they deliver a slightly pulsed flow, flow reversal is not possible, they can suffer from air locks and are sensitive to particulates.

By far the most attractive alternative is the piston pump, capable of both pulseless flow and flow reversal. For continuous linear flow methods, piston pumps are subject to the same financial restrictions as reciprocating devices but offer real advantages for stopped-flow and flow-reversal. The most obvious differences are the absence of pump tubing and check valves resulting in low maintenance operation. The major drawback is the reduction in capacity due to the time required for reservoir filling.

Piston pumps are currently experiencing a renaissance due to the successful exploitation of sinusoidal flow patterns [6] and sequential injection techniques [7]. Reproducible sinusoidal flow patterns can be generated through the use of a cam-driven piston, powered by a computer-controlled motor.

5.3.2 Injection systems

The delivery of a well-defined portion of liquid into another liquid stream is fundamental to FI practice. This zone is then carried into the FI manifold where dispersion takes place prior to detection. Conventional FI procedures involve the injection of sample but in some cases the injection of a portion of reagent into a sample stream is adopted. This is often beneficial when hazardous or expensive reagents are being used and is referred to as reagent-injection or reverse FI. One of the major factors governing the analytical reproducibility of the system is therefore the precision of the injection process. The various approaches used are discussed below.

The most versatile and widely used FI injectors are low pressure rotary injection valves such as those from Rheodyne®. Typically fabricated from poly(tetrafluoroethylene) (PTFE) with six ports, the rotary device can be rapidly actuated either electrically or pneumatically. The volume injected is dictated by the length and internal diameter of the external sample loop which is readily changed. Two pumps are used in tandem with the valve; the first conveying the carrier stream through the valve, with the second intermittently flushing and filling the injector loop prior to actuation. Rapid switching of the valve ensures minimal flow perturbation as the carrier stream moves through the loop carrying the injected zone into the manifold. Very high precision is possible with rotary valves and provided that the solutions are free from particulates they are capable of long-term maintenance-free operation. A series of rotary injectors can be included in the same manifold to provide a variety of FI configurations. For example, secondary and tertiary valves can be used to extract portions of the dispersed zones from primary and secondary injections, respectively, to provide a high degree of sample dilution. Zone merging techniques require the simultaneous switching of two injectors to allow the merging of sample and reagent zones at a confluence point, and zone penetration requires the switching of two valves in series to allow the mutual penetration as the zones disperse.

Hydrodynamic injection relies on the forces exerted by stationary liquid columns to provide a well defined injection volume. Once again two pumps are used to control the liquid movements but the injector itself includes no moving parts. One pump propels the carrier stream. With this stream stationary, the second pump flushes and fills the portion of conduit defining the injection volume. The pump action is then alternated to transfer the injected portion into the reaction manifold. Compared to rotary valve injection, hydrodynamic injection is generally less precise but the benefits of mechanical simplicity may outweigh this for certain applications. Control of the injected portion may also be achieved by means of propulsion into the conduit for a precisely timed interval. However, time-based injection procedures tend to suffer from poor precision due to their reliance on precise volumetric flow rates.

A much more attractive alternative is the selector, multiposition or directional valve, with six, eight, 10 or 12 ports and a common port to the pump. These systems allow the aspiration of multiple liquid portions in a reproducible preset sequence, prior to dispersion and detection through flow reversal. Akin to rotary injection valves, they are electrically actuated and are capable of long-term maintenance-free operation. The use of piston pumps ensures long-term reproducibility of the injected volumes and reagent aspiration keeps its usage to a minimum. Selector valves are key to the techniques of sequential injection and sinusoidal flow.

5.3.3 Detection systems

The diverse nature of FI applications is reflected in the range of detection systems that are employed, e.g. spectroscopic (chapter 7), and electro-chemical (chapter 8). Many of the detection systems employed in laboratory FI are inherently unsuitable for unattended operation and/or the process environment but it is worthy of note that, for a number of laboratory techniques, flow injection provides an effective means of precise sample presentation.

An obvious prerequisite of any detector for flow analysis is its compatibility with a flowing liquid stream and general analytical requirements such as sensitivity and response time have to be considered from the outset. Of the detection systems that are appropriate for the process environment, the fundamental differentiation between optical and electrochemical approaches is used below.

Spectrophotometry is the most extensively used FI detection system. In the laboratory, it is commonly applied in the form of a 'Z-cell' liquid chromatography detector or a general purpose UV–visible spectrophotometer adapted with an appropriate flow-cell. Filter or monochromator based spectrophotometers may be used. The monochromator systems may use a prism or dispersion grating for single wavelength measurement or a grating for fast scanning instruments and photodiode array devices.

An elegant approach to FI detection for the process environment is provided by solid-state photometers [8] which incorporate light emitting diode (LED) sources, photodiode detectors and manifold tube flow-cells. The costs are a fraction of those for conventional spectrophotometers and as a result of their uncomplicated construction offer a low maintenance, reliable alternative. LEDs are available covering the majority of the visible region of the spectrum and can be used singly or in combinations for multidetection. For certain applications, fluorescence detection can provide a very sensitive alternative when the native or derivatised fluorescence of the analyte can be exploited. Flow injection systems are ideally suited to automated derivatisation procedures which can provide excellent selectivity in addition to the inherent sensitivity of fluorescence. This concept can be extended to the application of chemiluminescence procedures, which can offer detection limits in the nM range [9] provided a stable photomultiplier or photodiode-based [10] system is available.

Interest is also growing in the use of flow injection techniques for both dispersive and Fourier transform infrared spectroscopy with the development of robust flow-cells for both transmission and attenuated total reflectance (ATR) measurements [11].

Fibre-optic technology has had a major impact on process spectroscopy and fibres may also have a role to play in process FI. The potential exists for developing in-process flow cell configurations (or optrodes) connected to a remote spectrometer by fibre-optic cables, with the option of multiplexing a number of flow cells.

In the laboratory one of the most promising developments involving FI is its use for front-end sample treatment and delivery for atomic spectroscopy (e.g. inductively coupled plasma-mass spectrometry). While this is currently restricted to laboratory applications, recent developments in portable atomic spectrometers suggest that this combination has potential scope for the future.

Electrochemical detection systems differ from optical systems in that they respond to species at the sensor interface rather than the bulk of the flow cell contents. The flow cell design therefore must incorporate a means of focusing the dispersed zone onto the electrochemically active surface through a wall-jet or cascade configuration. Many of the developments in electrochemical liquid chromatography detection are directly transferrable to FI detection, although some of the modes of detection differ markedly. For example, potentiometry is widely used in flow injection for pH and selective ion analysis whereas non-selective conductimetry is generally used for ion detection in chromatography. Coulometric, voltammetric and amperometric sensing devices have also found some application in FI.

Deposition of material on optical windows and the active surfaces of electrochemical flow cells will adversely affect detector response and flow cells must therefore exhibit stay-clean properties. The perturbation of flow within the flow cell is often sufficient to prevent accumulation with the analytical stream providing adequate flushing but secondary clean-up can be necessary, e.g. surfactant and solvent addition to reagents, acid/base washes or backflushing with water and/or air.

5.3.4 Instrument control

Fundamental to the practice of FI is reproducible event timing and this is best achieved through automation, a prerequisite for unattended operation. Computer control of the pumps, injection system and ancillary devices such as switching valves is illustrated in Figure 5.2. This diagram also highlights the role of the computer in data acquisition and processing.

In addition to the central processing unit, FI computer units require sufficient memory for program execution and raw data storage and manipulation. In order to acquire data from the sensing device, an

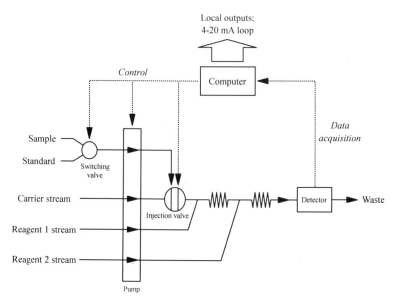

Figure 5.2 Schematic diagram of an automated FI monitor. (———) Liquid flow; (.....) Data flow.

analogue to digital converter is required that offers high resolution and a sampling frequency compatible with transient detector response. Data output facilities such as 4–20 mA loops and/or RS-232 links are also required. To enable rapid and consistent system start-up, especially in the event of a power failure, the incorporation of battery-backed RAM and a programable EPROM is recommended [12].

Long-term accuracy of unattended FI systems can be maintained by the incorporation of automated self-calibration procedures. Frequent recalibration allows the monitor to adapt to temporal changes in reagent sensitivity, detector response and degradation of pump tubing. If the system response is demonstrably linear, a single point calibration is adequate for slope adjustment and this is easily programmed through the activation of a switching valve.

Once accumulated, the data can be manipulated by a variety of algorithms to sustain the quality of the output information. General smoothing functions such as box-car averaging can be used to improve precision in addition to peak-find routines, procedures to determine peak height and width, and area integration methods. Algorithms based on differential rates of change in detector response can be implemented to eliminate detector spikes due to entrained air, and various statistical process control parameters can be monitored to identify rogue results. In the event of an unexpected result, FI systems are readily reset for repeat analysis and/or calibration, and depending on the complexity of the manifold and reaction chemistry, the response time is typically 10–120 s (see Table 5.1).

5.4 FI methodologies

The versatility of flow injection systems is a testament to the flexibility of FI manifolds to accommodate chemical and physical manipulations. The aim of this section is to present the methodologies most suited, and most commonly applied, to operation in the process environment. The applications discussed have been selected on the basis of their actual on-line use or their potential for process implementation.

If a desirable feature of process worthy instrumentation is its simplicity, then *single-channel* manifolds represent the ideal FI configuration. As discussed above, single-channel manifolds can be used for limited dispersion with a selective detection system. However, the incorporation of a reaction coil facilitates the medium dispersion suitable for a variety of reaction chemistries. This is illustrated in Figure 5.3 for the single-channel spectrophotometric determination of thiocyanate in process liquors [13]. Ruzicka and Marshall [7] used this same application and reaction chemistry to demonstrate the practicalities of single-channel sequential injection techniques. The sequential injection manifold is schematically presented in Figure 5.4. Operation is governed by a preprogrammed sequence of piston movements and valve selections, providing reproducible sample–reagent dispersion in the reaction coil followed by transient detection.

Incorporation of a second stream in the FI manifold opens up further options for sample manipulation. Firstly, reagents can be premixed prior to dispersion with the sample zone (necessary for reagents that are unstable when mixed together). Secondly, a reagent line can be merged into the carrier stream post-injection, providing a constant concentration of reagent available for reaction across the entire dispersed sample zone.

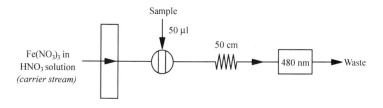

Figure 5.3 Single-channel manifold for the determination of thiocyanate in metallurgical solutions containing free cyanide and metal cyanide complexes [13].

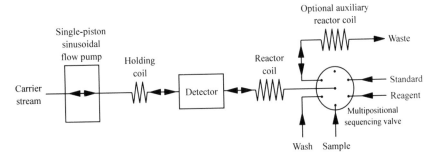

Figure 5.4   Schematic diagram of a sequential injection (SI) manifold [7].

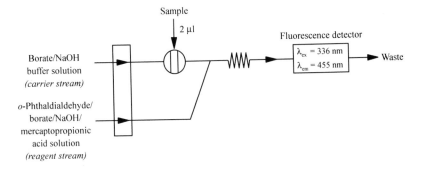

Figure 5.5   Two-channel manifold for the determination of L-phenylalanine in the continous cultivation of *Rhodococcus* sp. MA [14].

A biotechnological application (Figure 5.5) has been used to illustrate the *two-channel* approach with fluorescence detection of an isoindole derivative [14].

A *multichannel* manifold is shown in Figure 5.6 for the determination of aluminium in potable and treated waters [15]. Here, three reagents are added sequentially to a water stream carrying the sample zone prior to detection by visible spectrophotometry. Sequential addition ensures formation of the reaction intermediates during each phase of the analytical scheme prior to each subsequent stage of the reaction. Reagent-injection is an effective means of conserving reagent and is illustrated in another multichannel manifold (Figure 5.7). In this design [16], a portion of reagent is injected into a carrier stream before merging with a preconditioned continuous sample stream. The sample is subjected to dilution and pH adjustment from a buffer stream.

The versatility of FI manifolds facilitates *multideterminations*; the determination of several analytes by one analytical system [17]. Clearly, this can be achieved by combining two manifolds and reaction chemis-

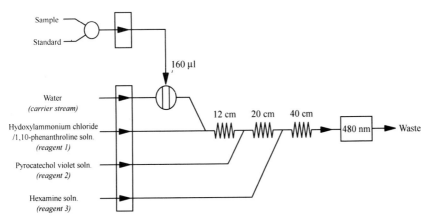

Figure 5.6 Multichannel manifold for the determination of residual aluminium in potable and treated waters [15].

tries into a single instrument. This approach has been adopted by Lynch and co-workers [18] for the speciation of iron in process liquors and the manifold design, given in Figure 5.8, shows the parallel injection and detection systems. A much more elegant approach is that developed by Faizullah and Townshend [19] (Figure 5.9). This type of manifold allows the splitting of a single sample injection for different treatments followed by remerging and detection. This system utilises 1, 10-phenanthroline for the determination of iron(II) in iron speciation. However, the need for a second reaction chemistry and detection system is avoided by the

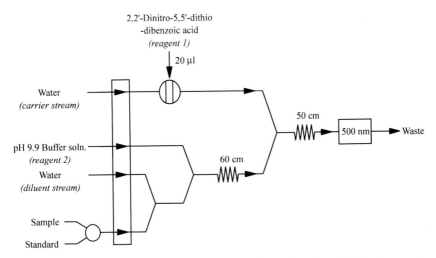

Figure 5.7 Multichannel manifold for the determination of sulfite in high ionic strength potassium chloride brine, using reagent injection [16].

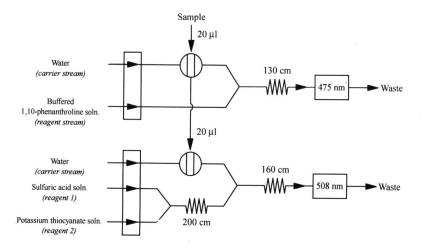

Figure 5.8 In-series manifolds for the simultaneous determination of iron(III) and iron(II) in mineral process liquors [18].

inclusion of a Jones reductor column and holding coil in the split stream. This yields a second dispersed zone and hence a transient response which is completely resolved from the iron(II) peak and is directly proportional to the sum of iron(II) and iron(III) concentrations. The concept of stream splitting is restricted to applications in which a single reaction chemistry yields a distinguishable response for all of the analytes of interest.

Multidetection systems greatly enhance the capabilities of FI to perform multideterminations and offer a number of advantages for process analysis. A multidetection system is a single device capable of recording a number of analytical signals simultaneously, examples of which include electrochemical sensor arrays, multi-LED solid-state photometers and photodiode array (PDA) detectors. PDAs are being increasingly utilised

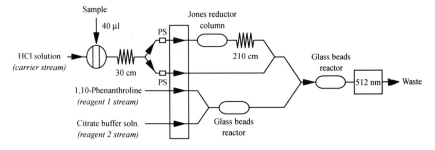

Figure 5.9 Multichannel manifold for the simultaneous determination of iron(II) and total iron, using post-injection sample splitting (*PS* = pulse supressor) [19].

due to their mechanical simplicity, excellent wavelength stability and their ability to collect an entire UV–visible spectrum in as little as one-tenth of a second. Much of the work published using the FI-PDA combination describes selective wavelength procedures for matrix compensation and simultaneous determinations. However, the full potential of FI-PDA is only realised in combination with multivariate calibration. Furthermore, recent developments in chemometrics (chapter 8) render the multianalyte calibration of mutually interfering systems feasible through bilinear modelling techniques such as principal components regression and partial least squares regression. This is demonstrated in Figure 5.10 for the simultaneous determination of phosphate and chlorine in industrial cooling waters with diode array detection and partial least squares calibration [20]. Two well established spectrophotometric procedures have been combined in a basic two-channel manifold and although the visible spectra reveal interference between the reaction chemistries, full selectivity is achieved through the bilinear calibration.

As discussed above, physical manipulation offers a further means of selectivity enhancement that is complementary to chemical and numerical approaches. Two such approaches have been employed in process analysis. Firstly, *packed reactor columns* can be incorporated into the flow stream for interference removal, sample conversion or sample preconcentration and secondly, *gas-diffusion* units can be used to selectively remove the analyte from its matrix. The Jones reductor column shown in Figure 5.9 is an example of an FI minireactor for sample conversion. A similar strategy has been adopted for the reduction of nitrate via a copperised cadmium reductor column [21] (Figure 5.11). In both examples, the life-time of the column is maximised through the injection of small sample volumes and short sample residence times.

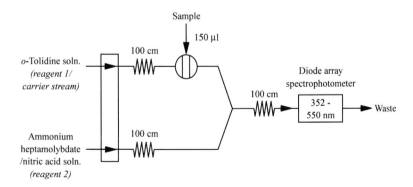

Figure 5.10 Dual-channel manifold for the simultaneous determination of phosphate and chlorine in industrial cooling water, using multiwavelength detection and multivariate calibration [20].

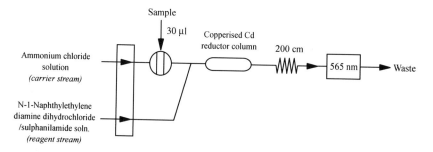

Figure 5.11 Dual-channel manifold for the determination of nitrate in river water, using a packed reactor reaction column [21].

Total sample conversion is not a requirement due to the reproducible timing of the FI system but high rates of conversion will increase sensitivity. Similarly, reproducibility is the key to the non-steady-state transport of gases across the membrane of a gas-diffusion unit, the application of which is illustrated in Figure 5.12. Gaseous ammonia crosses the hydrophobic microporous PTFE membrane into a buffered solution of bromothymol blue, wherein the resulting pH change yields a detectable colour change [22]. Not only is such a system inherently selective, it also maintains a protective barrier between the detection system and the sample and provides arguably the simplest, least expensive and most reliable form of on-channel sample clean-up available.

Of the methodologies summarised above, it is likely that more than one approach will provide a suitable FI solution to a particular application. It is therefore recommended that the application objectives remain foremost during the development programme as mechanical, chemical and operational simplicity will pay dividends after process implementation.

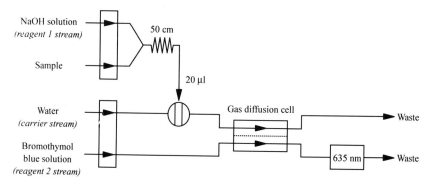

Figure 5.12 Multichannel manifold for the determination of ammonia in industrial liquid effluents, using a gas-diffusion cell.

5.5 Process applications

Conceptually, flow injection techniques are closely allied to liquid phase manufacturing. Many of the procedures carried out in the microconduits of the manifold are miniaturised reproductions of the large-scale manipulations performed in process plants. There are also similarities between the control, acquisition and communication functions of the FI computer and the central process control computer, as illustrated in Figure 5.13. This schematic diagram also highlights the process–analyser interface discussed in chapter 2. Effective sample presentation is vitally important, using a constant head vessel for example, and in many cases some degree of filtration and the control of biofouling is required. The number of reported applications of process FI remains low considering the versatility and potential of the technique, but this is due in part to industrial confidentiality. Nevertheless, the body of literature is increasing, as reported in a recent review article [23]. The reported applications are collated in Table 5.3 and classified according to process areas in Figure 5.14. The available data indicates that most activity is in the areas of chemical production, water quality monitoring and biotechnology. This data also underlines the flexibility of process FI for monitoring diverse analytes and its ability to deal with harsh sample matrices such as dye production liquors and fermentation broths. The long-term operation of an FI system has also been demonstrated by a nitrate monitor that has been running continuously for several years at a remote river water site, a period of which is reported in reference [24].

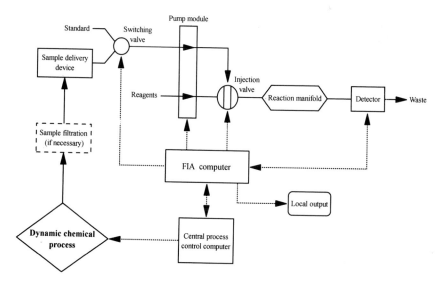

Figure 5.13 Schematic diagram of an automated process FI monitoring system. (——) Liquid flow; (.....) Data flow.

Table 5.3 Process FI applications classified by area and analyte

Area	Analyte	Comments	Reference
Chemical production	Sulfuric acid, ammonia and caustic solutions		25
	Sulfide in di-isopropanolamine solutions		26
	HCl in concentrated hydrochloric acid		27
	Azo dyes		28
	Sulfite in KCl brine	On-line monitoring of industrial process streams	29, 16
	Salicylic and acetylsalicylic acids in pharmaceutical preparations	Continuous monitoring of tablet dissolution tests	30
	Morphine		31
	Hydrogen cyanide in process gas streams	On-line monitoring of industrial process gas streams	32
Metal production	Iron(II) and iron(III) in mineral process liquors		18
	Soluble aluminium in steels		33
	Thiocyanate in metallurgical process solutions		13
	Trace gold in cyanide process solutions		34
Paper production	Calcium in paper machine back water		35
Fish farming	Ammonia	On-line monitoring of tanks containing fish farming plant seawater	36
	Ammonia and nitrite	On-line monitoring of sea and tap water tanks containing suspended fish feed	37
Hydroponic cultivation	Nitrate	On-line monitoring of outflow water from a hydroponic watercress bed	38
Wastewater monitoring	Sulfates and phosphates		28
	Chloride and ammoniacal-N		39
	Phosphate, ammonia and nitrogen	On-line monitoring of a pilot-scale wastewater treatment process	40
	Total phosphorus		41
	Glucose	On-line monitoring of a laboratory-scale waste whey treatment process	42
Treated water monitoring	Fluoride	On-line monitoring of a simulated fluoridation process	43

Area	Analyte	Comments	Reference
	Aluminium	On-line monitoring of potable water	15
	Aluminium and iron	On-line monitoring of potable water	44
Power-plant/ cooling water monitoring	Ammonia, hydrazine, copper, iron, silicon and pH		45
	Phosphate and chlorine		20
Freshwater monitoring	Phosphate		46
	Nitrate	On-line monitoring of river water	47, 48, 24
	Nitrate	On-line monitoring of tap water	12
	Ammonia	On-line monitoring of river water	49
Biotechnology	Protein	On-line monitoring of microorganism cultivation and disruption processes	50
	Formate dehydrogenase and L-leucine dehydrogenase	On-line monitoring of microorganism disintegration and diafiltration processes	51
	L-Phenylalanine	On-line monitoring of microorganism cultivation processes	14
	Glucose, lactic acid and protein	On-line monitoring of lactic acid fermentation	52
	Oxidases	On-line monitoring of enzyme purification LC eluent	53
	Extracellular proteins	On-line monitoring of cellulase fermentation processes	54
	Glucose	On-line monitoring of microorganism cultivation processes	55
	Ethanol	On-line monitoring of bioethanol production	56
	Glucose, DMFase and protein	On-line monitoring of microorganism cultivation processes and enzyme purification LC eluent	57
	Alanine dehydrogenase, formate dehydrogenase and phenylalanine dehydrogenase	On-line monitoring of enzyme purification LC eluent	58
	Cellulase		59
	Ammonium, glucose and proteins	On-line monitoring of fermentation processes	60
	Ammonium and glucose	On-line monitoring of penicillin fermentation processes	61

Table 5.3 (*Contd.*)

Area	Analyte	Comments	Reference
	Glucose	On-line monitoring of microorganism cultivation processes	62
	Acetate and phosphate	On-line monitoring of fermentation processes	63
	Proteins	On-line monitoring of cell-culture and microorganism fermentation processes	64
	β-Galactosidase	On-line monitoring of microorganism cultivation processes	65
	Immunoglobulin		66
	Glucose and ethanol	On-line monitoring of yeast fermentation processes	67
	Total acidity, reducing sugars, ethanol and pH	On-line monitoring of fermentation processes	68
	Penicillin, ethanol, glucose, maltose and sucrose	On-line monitoring of microorganism cultivation processes	69
	Ammonium, glucose, maltose, amino acids, lactose, lactate and glutamine	On-line monitoring of alkaline protease and penicillin fermentation processes	70
	Antithrombin III, immunoglobulin and pullulanase	On-line monitoring of simulated and real (cell-culture and microorganism) cultivation processes	71
	Pullulan and glucose		72
	Serum albumin, immunoglobulin and peroxidase		73
	Amylase, xylanase, polygalacturonase and protease activities		74
	Acetic acid	On-line monitoring of vinegar production	75
	Glucose and lactate	On-line monitoring of cell-culture fermentation processes	76
	Urea and glucose	On-line monitoring of microorganism cultivation processes	77
	Penicillin V	On-line monitoring of penicillin fermentation processes	78
	Glucose	On-line monitoring of microbial gluconic acid production	79
	Pullulanase and immunoglobulin	On-line monitoring of microorganism and hybridoma cultivation processes	80

Area	Analyte	Comments	Reference
	α-Amylase		81
	Ethanol	On-line monitoring of yeast fermentation processes	82
	pH, urea, penicillin V and immunoglobulin		83
	Glucose, disaccharides and β-galactosidase	On-line monitoring of recombinant protein production	84
	Formate dehydrogenase and malate dehydrogenase	On-line monitoring of yeast fermentation processes	85

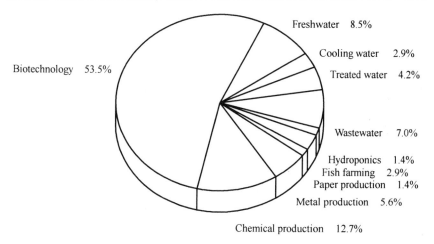

Biotechnology 53.5%

Freshwater 8.5%
Cooling water 2.9%
Treated water 4.2%
Wastewater 7.0%
Hydroponics 1.4%
Fish farming 2.9%
Paper production 1.4%
Metal production 5.6%
Chemical production 12.7%

Figure 5.14 Pie chart showing distribution of published process FI methods by area of application.

5.6 Conclusions

It is clear that the full impact of process FI has yet to materialise and this is primarily due to commercial process instrumentation not being widely available. The applications discussed in the literature have predominantly been implemented using flow injection systems designed and constructed in-house. Such projects therefore tend to have long lead times and require a higher level of on-going support and maintenance than 'off the shelf' commercial instruments. As a result of this process FI is finding a niche as a 'problem-solving' technique, i.e. as a means of providing the required process analytical information in applications where the more traditional approaches such as direct spectroscopy and titrimetry prove inappropriate. This is a further demonstration of the versatility and wide applicability of process FI, however in some cases process FI will out-perform the more established approaches to process measurement. Furthermore, FI has a wider, complementary role to play in sample handling for process electrochemistry and spectroscopy, particularly when complex sampling issues can be resolved through micro-sample manipulation.

Acceptance and utilisation of process FI will continue to grow as the FI technology evolves; the on-line utilisation of sequential injection and sinusoidal flow and the incorporation of quantitative chemometric routines are certain to fuel this growth.

FI has an important role to play as an application-specific tool for resolving process analysis problems. This role will develop in tandem with developments in FI instrumentation and techniques and the wider availability of commercial process worthy systems.

References

1. J. Ruzicka and E. H. Hansen, *Anal. Chim. Acta* **78** (1975) 145.
2. J. Ruzicka and E. H. Hansen, *Flow Injection Analysis* (2nd edn). Wiley, New York, USA, 1988.
3. M. Valcarcel and M. D. Luque de Castro, *Flow Injection Analysis: Principles and Applications.* Ellis Horwood, Chichester, UK, 1987.
4. L. T. Skeggs, *Am. J. Clin. Path.* **28** (1957) 311.
5. M. Valcarcel and M. D. Luque de Castro, *Non-chromatographic Continuous Separation Techniques.* Royal Society of Chemistry, Cambridge, UK, 1991.
6. J. Ruzicka, G. D. Marshall and G. D. Christian, *Anal. Chem.* **62** (1990) 1861.
7. J. Ruzicka and G. D. Marshall, *Anal. Chim. Acta* **237** (1990) 329.
8. P. K. Dasgupta, H. S. Bellamy, H. Liu, J. L. Lopez, E. L. Loree, K. Morris, K. Petersen and K. A. Mir, *Talanta* **40** (1993) 53.
9. D. Price, P. J. Worsfold and R. F. C. Mantoura, *Anal. Chim. Acta* **298** (1994) 121.
10. A. N. Gachanja and P. J. Worsfold, *Anal. Chim. Acta* **290** (1994) 226.
11. M. Guzman, J. Ruzicka, G. D. Christian and P. Shelley, *Vib. Spec.* **2** (1991) 1.
12. N. J. Blundell, A. Hopkins, P. J. Worsfold and H. Casey, *J. Auto. Chem.* **15** (1993) 159.
13. E. A. Jones and M. -J. Hemmings, *S. Afr. J. Chem.* **42** (1989) 6.
14. U. Nalbach, H. Schiemenz, W. W. Stamm, W. Hummel, and M. -R. Kula, *Anal. Chim. Acta* **213** (1988) 55.
15. R. L. Benson, P. J. Worsfold and F. W. Sweeting, *Anal. Chim. Acta* **238** (1990) 177.
16. P. MacLaurin, P. J. Worsfold, A. Townshend, N. W. Barnett and M. Crane, *Analyst* **116** (1991) 701.
17. P. MacLaurin and P. J. Worsfold, *Microchem. J.* **45** (1992) 178.
18. T. P. Lynch, N. J. Kernoghan and J. N. Wilson, *Analyst* **109** (1984) 843.
19. A. D. Faizullah and A. Townshend, *Anal. Chim. Acta* **167** (1985) 225.
20. P. MacLaurin, P. J. Worsfold, P. Norman and M. Crane, *Analyst* **118** (1993) 617.
21. J. R. Clinch, P. J. Worsfold and H. Casey, *Anal. Chim. Acta* **200** (1987) 523.
22. J. R. Clinch, P. J. Worsfold and H. Casey, *Anal. Chim. Acta* **214** (1988) 401.
23. K. N. Andrew, N. J. Blundell, D. Price and P. J. Worsfold, *Anal. Chem.* **66** (1994) 916A.
24. H. Casey, R. T. Clarke, S. M. Smith, J. R. Clinch and P. J. Worsfold *Anal. Chim. Acta* **227** (1989) 379.
25. K. G. Schick, *ISA Trans. Adv. Instr.*, **39** (1984) 279.
26. W. E. van der Linden, *Anal. Chim. Acta* **179** (1986) 91.
27. J. F. van Staden, *Fres. Z. Anal. Chem.* **328** (1987) 68.
28. M. Gisin and C. Thommen, *Trends Anal. Chem.* **8** (1989) 62.

29. P. MacLaurin, K. S. Parker, A. Townshend, P. J. Worsfold, N. W. Barnett and M. Crane, *Anal. Chim. Acta* **238** (1990) 171.
30. J. M. López Fernández, M. D. Luque de Castro and M. Valcárel, *J. Auto. Chem.* **12** (1990) 263.
31. N. W. Barnett, D. G. Rolfe, T. A. Bowser and T. W. Paton, *Anal. Chim. Acta* **282** (1993) 551.
32. D. C. Olson, S. R. Bysouth, P. K. Dasgupta and V. Kuban, *Process Quality and Control* **5** (1994) 259.
33. H. Bergamin, F. J. Krug, E. A. G. Zagatto, E. C. Arruda, and C. A. Coutinho, *Anal. Chim. Acta* **190** (1986) 177.
34. M. J. C. Taylor, D. E. Barnes, and G. D. Marshall, *Anal. Chim. Acta* **265** (1992) 71.
35. J. Nyman, and A. Ivaska, *Talanta* **40** (1993) 95.
36. H. Muraki, K. Higuchi, M. Sasaki, T. Korenaga and K. Tôei, *Anal. Chim. Acta* **261** (1992) 345.
37. A. C. Ariza, P. Linares, M. D. Luque de Castro, and M. Valcárel, *J. Auto. Chem.* **14** (1992) 181.
38. J. R. Clinch, P. J. Worsfold, H. Casey and S. M. Smith, *Anal. Proc.* **25** (1988) 71.
39. J. F. van Staden, *Anal. Chim. Acta* **261** (1992) 453.
40. K. M. Pedersen, M. Kümmel and H. Søeberg, *Anal. Chim. Acta* **238** (1990) 191.
41. R. L. Benson, I. D. McKelvie, B. T. Hart and I. C. Hamilton, *Anal. Chim. Acta* **291** (1994) 233.
42. R. Pilloton, G. Mignogna, and A. Fortunato, *Anal. Letters* **27** (1994) 833.
43. D. Chen, M. D. Luque de Castro and M. Valcárel, *Anal. Chim. Acta* **230** (1990) 137.
44. R. L. Benson and P. J. Worsfold, *Sci. Tot. Env* **135** (1993) 17.
45. M. L. Balconi, F. Sigon, M. Borgarello, R. Ferraroli and F. Realini, *Anal. Chim. Acta* **234** (1990) 167.
46. P. J. Worsfold, J. R. Clinch and H. Casey, *Anal. Chim. Acta* **197** (1987) 43.
47. J. R. Clinch, P. J. Worsfold and H. Casey, *Anal. Chim. Acta* **200** (1987) 523.
48. R. L. Benson, P. J. Worsfold and F. W. Sweeting, *Anal. Proc.* **26** (1989) 385.
49. J. R. Clinch, P. J. Worsfold and F. W. Sweeting, *Anal. Chim. Acta* **214** (1988) 401.
50. A. Recktenwald, K. -H. Kroner, and M. -R. Kula, *Enzyme Microb. Technol.* **7** (1985) 146.
51. A. Recktenwald, K. -H. Kroner and M. -R. Kula, *Enzyme Microb. Technol.* **7** (1985) 607.
52. K. Nikolajsen, J. Nielsen and J. Villadsen, *Anal. Chim. Acta* **214** (1988) 137.
53. W. Künnecke, H. M. Kalisz and R. D. Schmid, *Anal. Letters* **22** (1989) 1471.
54. W. W. Stamm, G. Pommerening, C. Wandrey and M. -R. Kula, *Enzyme Microb. Technol.* **11** (1989) 96.
55. U. Brand, B. Reinhardt, F. Rüther, T. Scheper and K. Schügerl, *Anal. Chim. Acta* **238** (1990) 201.
56. W. Künnecke and R. D. Schmid, *J. Biotechnol* **14** (1990) 127.
57. H. Lüdi, M. B. Garn, P. Bataillard and H. M. Widmer, *J. Biotechnol.* **14** (1990) 71.
58. W. W. Stamm and M. -R. Kula, *J. Biotechnol* **14** (1990) 99.
59. P. J. Worsfold, I. R. C. Whiteside, H. F. Pfeiffer and H. Waldhoff, *J. Biotechnol.* **14** (1990) 127.
60. S. Chung, X. Wen, K. Vilholm, M. de Bang, G. Christian and J. Ruzicka, *Anal. Chim. Acta* **249** (1991) 77.
61. L. H. Christensen, J. Nielsen and J. Villadsen, *Anal. Chim. Acta* **249** (1991) 123.
62. C. Filippini, B. Sonnleitner, A. Fiechter, J. Bradley and R. Schmid *J. Biotechnol.* **18** (1991) 53.
63. L. W. Forman, B. D. Thomas, F. S. Jacobson, *Anal. Chim. Acta* **249** (1991) 101.
64. R. Freitag, C. Fenge, T. Scheper, K. Schügerl, A. Spreinat, G. Antranikian and E. Fraune, *Anal. Chim. Acta* **249** (1991) 113.
65. H. -A. Kracke-Helm, L. Brandes, B. Hitzmann, U. Rinas and K. Schügerl, *J. Biotechnol.* **20** (1991) 95.
66. M. Nilsson, H. Håkanson and B. Mattiasson, *Anal. Chim. Acta* **249** (1991) 163.
67. I. Ogbomo, R. Kittsteiner-Eberle, U. Englbrecht, U. Prinzing, J. Danzer and H. -L. Schmidt, *Anal. Chim. Acta* **249** (1991) 137.

68. M. Peris-Tortajada, A. Maquieira, E. López and R. Ors, *Analusis* **19** (1991) 266.
69. T. Scheper, W. Brandes, C. Grau, H. G. Hundeck, B. Reinhardt, F. Rüther, F. Plötz, C. Schelp, K. Schügerl, K. H. Schneider, F. Giffhorn, B. Rehr and H. Sahm, *Anal. Chim. Acta* **249** (1991) 25.
70. K. Schügerl, L. Brandes, T. Dullau, K. Holzhauer-Rieger, S. Hotop, U. Hübner, X. Wu, and W. Zhou, *Anal. Chim. Acta* **249** (1991) 87.
71. A. Degelau, R. Freitag, F. Linz, C. Middendorf, T. Scheper, T. Bley, S. Müller, P. Stoll and K. F. Reardon, *J. Biotechnol.* **25** (1992) 115.
72. U. Englbrecht and H. -L. Schmidt, *J. Chem. Tech. Biotechnol.* **53** (1992) 397.
73. M. Nilsson, H. Håkanson and B. Mattiasson, *J. Chromatogr.* **597** (1992) 383.
74. H. F. Pfeiffer, H. Waldhoff, P. J. Worsfold and I. R. C. Whiteside, *Chromatographia* **33** (1992) 49.
75. T. Becker, R. Kittsteiner-Eberle, T. Luck and H. -L. Schmidt, *J. Biotechnol.* **31** (1993) 267.
76. T. Becker, W. Schuhmann, R. Betken, H. -L. Schmidt, M. Leible and A. Albrecht, *J. Chem. Tech. Biotechnol.*, **58** (1993) 183.
77. M. Busch, W. Höbel and J. Polster, *J. Biotechnol.* **31** (1993) 327.
78. M. Carlsen, C. Johansen, R. W. Min, J. Nielsen, H. Meier and F. Lantreibecq, *Anal. Chim. Acta*, **279** (1993) 51.
79. B. Gründig, B. Strehlitz, H. Kotte and K. Ethner, *J. Biotechnol.* **31** (1993) 277.
80. C. Middendorf, B. Schulze, R. Freitag, T. Scheper, M. Howaldt and H. Hoffman, *J. Biotechnol.* **31** (1993) 395.
81. M. Nilsson, G. Mattiasson and B. Mattiasson, *J. Biotechnol.* **31** (1993) 381.
82. I. Ogbomo, A. Steffl, W. Schuhmann, U. Prinzing, H. -L. Schmidt, *J. Biotechnol.* **31** (1993) 317.
83. T. Scheper, W. Brandes, H. Maschke, F. Plötz and C. Müller, *J. Biotechnol.* **31** (1993) 345.
84. K. Schügerl, L. Brandes, X. Wu, J. Bode, J. I. Ree, J. Brandt and B. Hitzmann, *Anal. Chim. Acta* **279** (1993) 3.
85. K. Steube and U. Spohn, *Analyst* **287** (1994) 235.

6 Molecular spectroscopy

N. C. CRABB and P.W.B. KING

6.1 Introduction

Molecular spectroscopy is a vast subject encompassing mass spectroscopy, nuclear magnetic resonance (NMR) and the optical spectroscopies. These have all been applied to on-line process measurement but the optical spectroscopies are of particular importance. It is probably fair to say that no other analysis technique has evolved into as many forms. This reflects the versatility that optical spectroscopy offers. Instruments can be devised to detect concentrations from parts per million up to major component levels. The sample itself may be in the gaseous, liquid or solid phases (or even mixtures of these). Its non-invasive nature lets us probe samples at extremes of pressure and temperature and to examine the most toxic and aggressive substances.

The power of this technique has been recognised for a long time. Its applicability to process analysis was obvious and systems began to move from the laboratory to the factory from the outset. Process instruments were introduced as far back as the 1940s leading to the development of a wide range of systems during the 1960s and 1970s. Today, the versatility of optical spectroscopy is enhanced through the use of multivariate calibration techniques.

The aim of this chapter is to provide a practical introduction to optical spectroscopy and the instrumentation used for process measurement. In addition, mass spectroscopy is given brief coverage owing to its increasing importance in environmental monitoring. Many important aspects of molecular spectroscopy in process analysis are not mentioned. Some of these will be covered in chapters 10 and 11.

6.2 Optical spectroscopy

6.2.1 Theory

Molecular optical spectra arise through the interaction of molecular species with electromagnetic radiation. Many important properties of electromagnetic radiation (reflection, refraction, diffraction, interference) show that it has wave-like characteristics. It can be considered as an

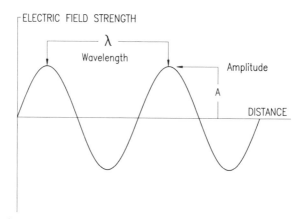

Figure 6.1   Representation of monochromatic radiation.

electrical field force oscillating in a direction that is 90° to the direction of wave propagation. An associated magnetic field is at 90° to both the electric field and direction of propagation. A plot of a time or distance against electrical field strength produces a sine wave as shown in Figure 6.1, electromagnetic radiation is often represented in this way. The distance between maxima or minima is the wavelength (λ) and the number of oscillations per second is the frequency (v, in Hz). The amplitude (A) is the electrical field strength at the maxima, A^2 is related to the beam intensity. A further unit often used in spectroscopy, is the wavenumber. This is the number of waves per cm and is equal to $1/\lambda$ when the wavelength is expressed in cm. The wavenumber has units of cm^{-1}. In Figure 6.1, radiation of a single frequency is represented, such radiation is said to be monochromatic. A beam comprising radiation of several frequencies is said to be polychromatic.

Interaction with matter is better explained if electromagnetic radiation is considered particulate or quantised. This 'wave–particle duality' can be fully rationalised by wave mechanics. The number of possible energy levels in a molecule is vast reflecting a multitude of electronic, vibrational, rational and nuclear spin states. Although vast in number, molecular energy states are not continuous, only certain energy states are permissible. This quantum theory, proposed by Plank in 1900 and subsequently widely accepted, explains the discontinuous nature of energy absorbed or emitted by matter. It assumes there to be a minimum, indivisible quantum of energy that can be exchanged and that all transitions involve whole integral units of this energy quantum. The quantum of energy is directly proportional to the frequency of radiation ($E = nhv$ where E is the energy in joules, v is the frequency of the radiation and h is Planck's constant). High-energy transitions (e.g. inner shell electrons) are characterized by the absorption or emission of high frequency radiation (X-rays, around 10^{17}–10^{19} Hz in this case) while low energy transitions (e.g. nuclear spin) result in the absorption or emission of low frequency radiation (radio waves, around 10^6–10^8 Hz in this case). These two types of transition involve interaction with radiation at extremes of the electromagnetic spectrum.

This chapter is concerned primarily with a relatively small part of the electromagnetic spectrum spanning from around 10^{13}–10^{16} Hz. This region can be divided into the mid-infrared (MIR), near-infrared (NIR), visible and ultraviolet spectral regions. These arbitrary divisions (Figure 6.2) are based largely on instrumental requirements although generalisations on the types of energy transitions involved can also be made. The Beer–Lambert law is of major importance to quantitative spectroscopy, in all of these spectral regions. It is an expression of two observations. Firstly, the intensity (I) of a beam of monochromatic radiation decreases exponentially as the pathlength of an absorbing species (fixed concentration) is increased (I/I_0 }-log pathlength). Secondly, beam intensity also decreases exponentially as the concentration of an absorbing species

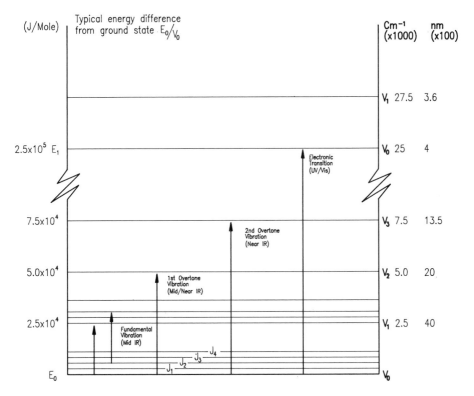

Figure 6.2 Electronic (E), vibrational (V) and rotational (J) energy levels.

(fixed pathlength) is increased (I/I_0 }-log concentration). I/I_0 is known as the transmittance (T), the absorbance (A) is equal to $-\log T (= \log I_0/I)$. A is therefore directly proportional to pathlength and concentration. The Beer–Lambert law is usually expressed as $A = abc$ where b is the pathlength, c is the concentration and a is a constant known as the absorptivity. The MIR, NIR and UV–Vis spectral regions are now considered in turn.

6.2.1.1 Mid-infrared (MIR) spectroscopy. MIR analysis involves the absorption of infrared radiation with consequent vibrational excitation of the sample. The frequencies of fundamental molecular vibrations (and hence those of MIR radiation) fall in the range 10^{13}–10^{14} Hz. It is conventional to calibrate spectrometers in wavenumber units, typically a MIR spectrum covers the range 4000–400 cm^{-1}. To absorb MIR radiation, a molecule must undergo a net change in dipole moment as a consequence of its vibrational motion. In practise, any molecule having an asymmetric charge distribution will absorb IR radiation to some extent. The dipole moment is determined by the magnitude of charge difference and the distance between the two centres of charge. The longitudinal vibration of polar diatomic molecules results in the regular fluctuation of dipole moment. The resultant electrical field can interact with the electrical field associated with the IR radiation. If the frequency of radiation matches a natural molecular vibration, then net transfer of energy occurs (IR absorption) leading to a change in the amplitude of the molecular vibration. Dipole moment changes do not occur on the vibration of homonulcear diatomic molecules, such molecules cannot, therefore, absorb MIR radiation. The vast majority of molecular species do, however, exhibit MIR absorption. Dipole fluctuations also occur on the rotation of asymmetric molecules and again interaction with radiation is possible. The energy differences between rotational levels are very small and consequently rotational transitions occur in the low energy far infrared region. For each vibrational state, however, there are several rotational energy levels. Multiple vibrational–rotational transitions therefore occur resulting in a series of closely spaced lines in the MIR spectra of gases. Rotation is restricted in liquids and solids, here, vibrational–rotational transitions result only in the broadening of the vibrational peak. In Figure 6.2, three vibrational energy states are shown (V_0, V_1 and V_2). Transitions between any of these energy levels are possible but higher energy transitions are successively less likely to occur so V_{1-0} transitions will produce much stronger bands than V_{2-0}, which in turn will produce stronger bands than V_{3-0}. Higher energy transitions are possible but with very low intensities. V_{1-0} transitions are known as fundamental vibrations while V_{2-0} and V_{3-0} transitions are the first and second overtones, respectively. First and second overtone vibrations are observed at around twice and three times the frequency of the fundamental, respectively. This is due to the energy differences between successive vibrational states being approximately equal for the lower vibrational states.

Detailed qualitative interpretation of MIR spectra is difficult for large molecules due to both the number of vibrating centres and interactions between them but correlation charts can be of value. These are based on quantities of data accumulated over several years and show the frequency range within which different functional groups are likely to absorb. Very briefly, absorptions due to stretching vibrations between hydrogen and some other atom occur mainly between 3700 and 2700 cm^{-1} (hydrogen stretching region). Absorption between 2700 and 1850 cm^{-1} is normally due to triple bond stretching although, S–H, P–H or Si–H stretching also occurs in this region. Double bond stretching (aromatic, alkene, carbonyl) results in absorption between 1950 and 1450 cm^{-1}. In the fingerpint region (1500–700 cm^{-1}), minor structural differences between molecules result in significant changes in their IR absorption patterns. This region is therefore excellent for the identification of compounds against standard spectra. With expert interpretation, the MIR spectrum of a pure material can often yield extensive structural information.

MIR has also been widely applied to quantitative analysis. Traditionally this has involved identifying fully resolved peaks associated with the analytes of interest and measuring their absorbances. This approach has been particularly successful in monitoring gaseous emissions. Here the background (air), being composed primarily of homonuclear species, gives minimal absorption. There is a good chance that somewhere in the MIR spectrum a resolved peak due to the analyte of interest, can be identified. High sensitivity is also possible through the use of long pathlengths. The control of chemical processes often requires quantitative, multicomponent analysis, often in complex matrices. To obtain such information from the MIR spectrum would generally require multivariate calibration (spectra of individual components are likely to be overlapped to the extent that resolved peaks for individual analytes cannot be identified). Multivariate calibration involves the use of chemometrics (see chapter 8) to establish correlation between sample spectra and their known analyte values.

6.2.1.2 Near-infrared (NIR) spectroscopy. The NIR region of the electromagnetic spectrum spans from around 780–2500 nm (12 800–4000 cm^{-1}), lying between the visible and MIR regions. Most practical applications are based on the region between 1100 and 2500 nm. In NIR spectroscopy there is a convention for expressing spectral regions in terms of wavelength in nm, this is largely the result of early instruments being extended UV–Vis instruments. NIR absorptions are overtones and combinations of C–H, N–H and O–H fundamental vibrations which occur in the MIR.

Overtones. The relationship between fundamental and overtone vibrations was described under MIR spectroscopy. The first overtone vibration is expected at approximately twice the frequency of the fundamental. The first overtones of fundamental vibrations in the fingerprint or double bond regions will therefore still be in the MIR. The first overtones of vibrations occurring in the hydrogen stretching region and to a lesser extent in the triple bond region, will lie within the NIR spectral region.

Combinations. The simultaneous excitation of two different vibrational modes is possible. The frequency of radiation required to bring about such excitation is the sum (or combination) of the two individual vibrational frequencies. If these occur in the hydrogen stretching region then the combination will be in the NIR. As with the overtones, these higher energy transitions are relatively weak. The number of potential combinations is large and many NIR band assignments are unresolved. This makes detailed qualitative interpretation of NIR spectra very difficulty.

In comparison with MIR spectroscopy, NIR absorptions are weak (low sensitivity), peaks are broad and overlapping (low selectivity) and consequently NIR is a very poor primary characterisation tool. Quantitative analysis by NIR almost inevitably requires the use of multivariate calibration. Interest in NIR spectroscopy arises from the very practical way in which it can be applied. Low sensitivity, for example, allows direct transmittance measurements, even on concentrated process streams. The shorter NIR wavelengths are also reflected more efficiently than MIR radiation. Together with the availability of higher energy sources, this makes NIR superior to MIR spectroscopy for reflectance work. A further advantage is that NIR radiation can be transmitted efficiently through relatively inexpensive, low moisture silica fibres. This allows remote interfacing of process and instrument. These features, together with excellent signal to noise, often makes NIR spectroscopy feasible and attractive even where the analyte of interest has minimal influence on the whole matrix NIR spectrum.

6.2.1.3 UV–Vis spectroscopy. Most UV–Vis spectroscopy concerns the spectral region from around 200–750 nm. Radiation in the range 100–200 nm is still considered to be UV radiation but in this range, atmospheric oxygen is a strong absorber and this region can only be exploited by holding the sample under vacuum. This region is often termed the vacuum UV. In the UV–Vis spectroscopy, the energy transitions associated with absorption are much higher than in IR spectroscopy and correspond to electronic transitions (see Figure 6.2). Absorption peaks tend to be very broad and overlapped, resulting from multiple electronic–vibrational–rotational transitions. Despite this, useful structural information can be derived from the UV–Vis spectrum of a material. A spectrum is best characterised in terms of the position(s) of maximal absorbance (1λ max) and the intensity of that absorbance.

Electronic transitions do occur for sigma (σ) electrons (covalent bonds) but as these are held tightly, excitation requires high energy and such transitions are associated with the vacuum UV region. Excitation of π electrons in double bonds and aromatic systems requires lower energy. Such transitions in simple alkenes still occur in the vacuum UV but conjugation of double bonds lowers the energy required for excitation resulting in absorption in the UV or even visible region. Colour results from multiple double bonds in conjugation. Qualitative and quantitative visible spectroscopy is routinely used in the quality control of dyes and pigments. UV-spectroscopy is widely used as a quantitative tool. Where the analyte of interest is the only absorbing species, direct quantitative analysis can be used. Versatility can be further increased using chemometric techniques. UV spectroscopy is also widely used as a detection technique after matrix isolation (e.g. in chromatography) or following selective derivatisation procedures (e.g. flow injection analysis).

6.2.2 Instrumentation

The most common early techniques used gas filled detectors to provide selectivity and high sensitivity. As technology developed, interference filters were used to isolate spectral bands. This led to the plethora of simple and rugged instruments that still provide a good economic solution to many simple analysis problems. The last 10–15 years has seen the emergence of whole spectrum analysis. Instruments offering high performance are available operating in the near and mid-infrared regions. The use of modern small computers and sophisticated statistical methods allow analysis of the most complex of systems. In this group we can include the large range of NIR instruments currently on the market, MIR Fourier transform IR (FTIR) systems and, of growing importance Raman spectroscopy (see section 6.2.5). Currently, a diverse range of instrument types are available.

6.2.2.1 Types of instrument. As space is limited, no attempt will be made to cover every type of instrument available. It will be more useful to concentrate on the main categories and to merely indicate what other options exist. For the sake of simplicity we can divide instruments into two broad groups. The first we can describe as photometers. These are basically simple, they analyse by interrogating the spectrum at a few selected wavelengths. Such instruments are set up to do one particular job, if the process being analysed changes then the instrument may also have to be changed. Despite this rigidity, photometers are of great use. They are an economic solution to many process problems. The second group we can consider is the spectrometer, systems that use large parts of the available spectrum. These have the great advantage of flexibility. The instrument remains the same irrespective of the measurement. The set-up for a particular task is largely contained in software. This advantage is offset by the fact that such systems tend to be expensive and complex.

Both approaches have their uses, their strengths and weaknesses and both occupy a different niche in the process analysis armoury.

6.2.2.2 Filter instruments.   These form one of the most common groups of instrument. They are available from many sources, each type differing in detail. The basic principle common to all of them is that they use filters as the wavelength selecting element. The technique produces simple rugged analysers that are well suited to process use. Although limited in spectroscopic terms they are surprisingly versatile and successful. A minimum of two filters are used. One is centered on an absorption feature that has been chosen to characterise the component to be analysed; this can be called the measure filter. A second filter, the reference; is placed at some neutral point in the spectrum. The absorbance due to the component of interest can then be calculated from the logarithmic ratios of the energy transmitted at the two wavelengths. A layout of a typical instrument is shown in Figure 6.3. Energy from a source is collimated by a lens or a mirror and projected through the sample cell onto a detector. The energy entering the cell is chopped by a rotating disc carrying the filters. The detector therefore receives alternating pulses of energy at the measure and reference wavelengths separated by dark levels. It is then quite straightforward for the instrument electronics system to extract the signal pulse heights as the difference between the dark level and pulse height, to convert this to a logarithm and then by subtraction to generate the logarithmic ratio. This ratio will be proportional to the absorbance due to the sample and needs only to be offset and scaled to calibrate.

It will be understood that such a simple instrument has great advantages for process use. The configuration gives the system a high degree of stability, any broad band effects due to contamination of the sample cell, changes in the source and small variations in the optical system tend to ratio out. The system will be physically rugged and a high degree of reliability can be expected. The sample is usually isolated from the instrument allowing operation in hazardous areas and the analysis of

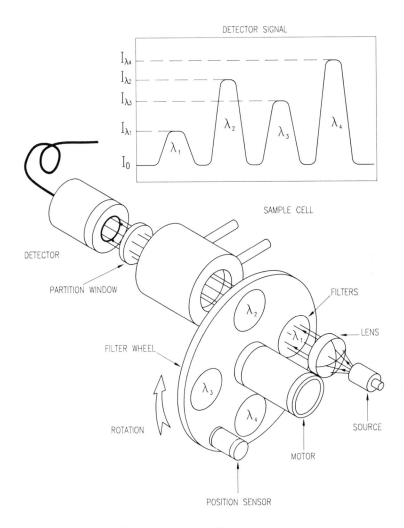

Figure 6.3 Typical filter instrument.

aggressive materials. The key to the use of such instruments lies in the selection of filters. In the simplest cases this can be trivial. As an example, if it was required to measure the concentration of carbon dioxide in nitrogen, then a measure wavelength of 4.2 μm would be chosen with a reference at 3.8. The principle would be to position the measure on the peak of an absorption band, making its bandwidth much narrower than that of the band, say 2% half-height bandwidth. The reference filter should be placed as close is possible to the measure to give the best immunity to any broad band spectral effects such as source temperature changes. More complicated analyses will require much more thought to be given to filter selection. A mixture of several components may mean that it is difficult to find a clearly isolated band to measure or a neutral point to set as a reference. Very often bands will overlap leading to cross-sensitivity effects. These can frequently be negated by careful choice of the reference or extra filters can be used to measure the interfering species and subtract them from the output. This approach can be extended to a point were so many analysis wavelengths are used that the instrument virtually becomes a spectrometer and statistical methods are necessary to define the sample matrix.

Used in this way spectroscopy does not yield high sensitivities. With sample cells of a practicable length, one can think in terms of measuring gases down to the 100 ppm level. The method is more applicable to major component than to trace analysis. In the liquid phase, of course, more sensitivity is available. Typical instruments should be able to measure the level of water in an organic solvent down to about 1 ppm.

Although it is easy to fault such simple instruments on their spectroscopic limitations, in reality they can and do perform a surprising number of measurements. Basic filter photometers are by far the largest class of optical analyser. They find application in a vast number of areas and are manufactured in large numbers. The reason for this is quite simply that most analyses are not complicated, simple instruments can be used, they are a product occupying a niche.

In practice this basic design needs some refinement to realise its full potential. The source used will depend on the wavelength range used. In the NIR a tungsten halogen lamp will be satisfactory. To move farther into the IR a ceramic coated open filament will be more appropriate. In both cases it is advantageous to include some form of source temperature control. The performance is very much controlled by the filters themselves. These will be almost invariably multilayer interference devices. Energy should be collimated through them to preserve their profile; sharply focused beams will cause changes in centre frequency and will distort the shape of the transmission curve. A collimated system imposes some limitations on beam diameter and can limit the throughput of the system, but nevertheless this must be accepted as the price of performance. Ideally filters should be temperature controlled. Interference filters show definite temperature coefficients comparable with the performance limits of these instruments. In some applications, for instance when working on the edges of bands, unacceptable temperature sensitivity may be encountered. The simple remedy is to enclose the filter wheel in a heated enclosure and to control its temperature to say 30–40°C. Temperatures above this should be avoided: experience shows that filters can behave unpredictably if they are used above 45°C. Some form of filter indexing is necessary in order to identify the position of the filters and to provide a phase lock for the detector electronics. This can take the form of a photodetector reading index marks off the disc periphery.

The source and detector units can each be enclosed in separate casings. They can be linked by the module containing the sample cell. This achieves the necessary segregation of the analyser from the potential damaging sample. Often this casing will be made to some recognised standard for operation in hazardous areas. In any case, common sense would dictate that they be kept in good condition by inert gas purging. Signal processing will usually be done by a remotely mounted electronics unit, again generally built into a hardened enclosure.

6.2.2.3 Gas filtering systems. A very widely used technique is to arrange for the material being analysed to act as its own filter and detector. This method has many advantages, the most important of these is that high sensitivity can be obtained. Simple reliable instruments can be produced which have found many uses in simple applications.

The low sensitivity of filter-based instruments has been mentioned already. The cause of this is that these systems attempt to measure small changes in a large signal, always a difficult thing to do. The energy seen by the detector is the full source energy modulated by the absorption due to the sample. The dominating factor is noise and drift on this large baseline energy. Gas filter systems can overcome this problem by balancing out this high signal. As with all IR analysers there are an almost endless range of configurations, the one illustrated in Figure 6.4 will serve to illustrate the salient features. Energy from the source is divided into two beams by mirrors. One beam carries the sample cell with the

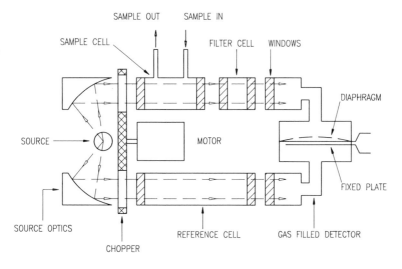

Figure 6.4 Typical gas filtering instrument.

process sample continuously flowing through it. The second beam acts as a reference and usually will be left empty or perhaps fitted with a matching nitrogen filled cell. The two beams deliver their energy into a microphone detector, the so called Luft detector. This consists of two chambers separated by a diaphragm that forms one plate of a capacitor microphone; the other being a fixed metal grid. The detector is filled at low pressure with the gas to be measured. Energy entering the device causes this gas to heat and expand. If more energy is received on either side the diaphragm is deflected, changing the capacity of the microphone and thus producing an electrical signal.

In the layout shown the source radiation is chopped mechanically. Radiation is passed alternately down the sample and reference arms. If the sample cell does not contain the gas to be detected, the energy in the two beams is equal, or at least can be made so by attenuating one of the beams. The presence of the measured gas causes energy to be absorbed in the sample cell so that less reaches the detector. As the chopper interrupts the beams the capacitor microphone vibrates at an amplitude depending on the energy imbalance. This can be detected electrically and used to generate a signal that can be turned into an output. The arrangement is inherently sensitive. A signal is being measured against a low background level. Selectivity is also inherent. The gas in the detector is the same as that being measured so that the detector is exactly matched to the sample.

This arrangement will serve for simple applications. It does however have many shortcomings such as interference from other components in the sample. This can be dealt with to some degree by using additional filtering. Cells containing a high concentration of the interfering gas can be placed in each beam This will suppress energy in the interfering region and reduce or possibly eliminate the problem. Optical filters can also be used to the same effect.

The response of the device is arbitrary. Sensitivity is controlled by the detector fill and the mechanical properties of the microphone. These are things that are difficult to reproduce, each instrument has to be individually calibrated. Regular calibration checks will be required to guard against changes in the detectors as gas leaks or is absorbed.

Obviously the range of analyses possible with this type of analyser is limited. The gas fill in the detector must be maintained; inert gases such as carbon dioxide are quite easy to accommodate, more reactive gases are always going to be a problem. Clearly, the arrangement is unsuited to anything but the simplest gas phase measurements. With the usual ingenuity shown by analysts, surprisingly difficult tasks can be performed but these are the exception. Gas filter instruments are simple sensitive instruments for doing simple sensitive jobs, in this role they perform a huge number of measurements very satisfactorily.

6.2.2.4 Process Fourier transform spectrometers. Dispersive IR spectrometers have been used in the laboratory from the very beginning of the techniques use. Such systems usually used a diffraction grating to generate spectra and are capable of high performance. The method has not, however, found great application for process analysis being more suitable for use in the laboratory environment. Their use has been restricted by their insensitivity and the simple fact that they are too slow. Early in the 1970s FTIR was developed and rapidly displaced grating instruments from the laboratory bench. To begin with, FTIR instruments were fragile instruments highly sensitive to vibration and acoustic interference, the very conditions so commonly found in the process analyser world. In recent years, very tough and reliable FTIR systems have been developed by several manufacturers. These are finding their way in increasing numbers into production analysis and the technique will clearly be a commonly encountered one in the future.

To understand why this technique is so important it is necessary to explain how it works. The key component is what is known as an amplitude division interferometer. An interferometer is a device that divides a light beam into separate paths, arranges for a difference in these paths and then recombines the energy. Two coherent light beams when combined will add together algebraically; constructive or destructive interference will occur. Since this is a function of path difference and wavelength, the information necessary to construct a spectrogram is contained in the resulting interferogram. The basic technique is illustrated in Figure 6.5. This shows Michelson's arrangement, one of the commonest forms of interferometer and the one found in most FTIR instruments. There are many different forms of interferometer, any good physics text book will show a bewildering variety of them, however the basic Michelson layout demonstrates all the features of the technique. Collimated light from a suitable source is projected onto a beamsplitter. This transmits part of the energy and reflects the rest forming two separate beams or arms. Two mirrors are used to reflect energy back onto the beam splitter; one of these is fixed and the other can be moved along the axis of the arm. The optical length of one arm relative to the other can thus be altered. The beams are recombined on the beam splitter and sent off to be focused onto a detector.

If we consider what happens when the source energy is perfectly monochromatic the devices operation should be clear. As the moving mirror travels along the arm, the combined energy on the detector will vary. This is caused by light waves interfering with each other, light waves in phase add together and those completely out of phase negate each other. The detector output will therefore be a sine wave of signal strength versus mirror position. If the source produces a broad IR spectrum then the detector will see the result of an infinite series of these

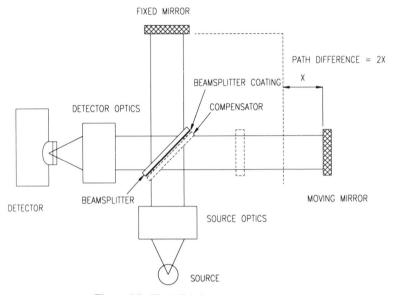

Figure 6.5 The Michelson interferometer.

sine waves superimposed on each other. Each will have a different period but each will have one point in the mirror motion where the path difference between the arms is zero. The combined signal, the interferogram, will look rather like the one illustrated (Figure 6.6). The central maximum is coincident with the zero path difference condition with the pattern being largely symmetrical about this point. Since the exact shape of this interferogram depends only on the spectrum of the input energy, it will be understood that this can be transformed into a spectrogram. The mathematical technique used to do this is the Fourier transformation, hence the naming of the method. To turn this device into a spectrometer it is only necessary to introduce a sample into one of the four arms, usually this is placed between the detector and the beam splitter. The resolution of the system is largely controlled by the length swept out by the moving mirror. The greater this is then the higher the resolution will be. This implies that the fine detail of any spectrum is contained in the wings of the interferogram, the central burst merely containing the lowest resolution detail of all, the baseline. On most instruments the resolution can be varied. A typical small process instrument will resolve down to 1 wavenumber and can be operated up to 128 to suit the measurement being made. As the mirror speed is fairly constant and the mirror sweep determines resolution, scan time is linked directly to resolution.

What may not be immediately apparent is just why this approach makes a superior spectrometer. There are several fundamental differences

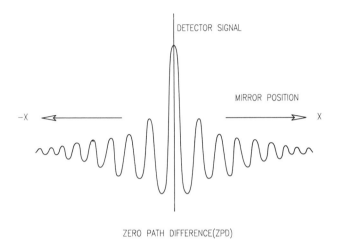

DETECTOR SIGNAL

MIRROR POSITION

−X ⟵ ⟶ X

ZERO PATH DIFFERENCE(ZPD)

Figure 6.6 An interferogram.

between an FT system and a grating instrument. The interferometer does not use slits, all the energy of the transmitted beam is available at the detector; the signal to noise performance is therefore that much better. All wavelengths are processed simultaneously instead of sequentially as in the case of a scanning grating; this gives much better stability. The normal arrangement is to sense the relative position of the mirrors by running a beam of laser light coaxially with the source beam. This is diverted onto a secondary detector and the interference fringes produced are used to measure the mirror positions. Since the frequency of the laser is accurately known, the wavelength scale of the instrument can be determined with great accuracy. The combination of high spectroscopic precision and superior photometric performance leads to the greatest single advantage of the method. Spectra are so reproducible that they can be processed numerically using powerful statistical techniques. Such an approach was never really possible with grating MIR instruments. These rely on mechanical methods to determine wavelength and with the variations found over long scan times the precision and stability were just not present. This combination of a better spectrometer with its link to modern small computers is what makes FTIR such a powerful method.

FTIR systems are relatively fast. Typically an interferogram is taken at around once per second. Many interferograms can be accumulated and co-added to improve signal to noise. The systems are so stable that this is possible without loss of spectroscopic performance. It will be understood that noise will be halved each time the number of co-added spectra is increased by a factor of four. The number of scans taken can be as many as a hundred or so depending on the response times being looked for. There is a trade-off between speed, resolution and noise. Reducing the resolution also directly reduces the scan time allowing more spectra to be accumulated and so reducing the noise level. The knowledge of the response time required by the measurement and the spectroscopic resolution needed to make the analysis, lets us choose the number of scans to be co-added.

The simple Michelson layout needs some refinement to make it work properly. One addition is the compensating plate (shown in Figure 6.5). This is to ensure that each beam travels through the same lengths of optical material; it balances each arm of the system. The interferometer measures very small differences, fractions of a micron; it has to be very rigid and precise. As can be imagined, it is very difficult to move a mirror with the great precision required and this is the problem that limited the use of FTIR in its early days. Any unwanted mirror motion, whether caused by vibrations or even by sound, affect the spectra produced. Recently most of these problems have been overcome and a range of good rugged systems is now available. Figure 6.7 shows the arrangement of one such system. This is a version of the Michelson configuration used by Bomen Inc. of Canada. It will be seen that the basic form is still retained although with some important alterations. Simple mirrors are replaced by corner reflectors, unlike mirrors these are inherently unaffected by tilt, the beam returns on a precisely parallel path. They also reject transverse motion. Both mirrors are moved by the simple device of mounting them on a yoke that oscillates around a flex pivot. This geometry halves the mirror motion needed as this is now shared between

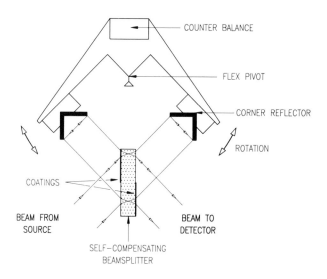

Figure 6.7 Modified Michelson interferometer (Bomen Inc.).

two moving surfaces. Another refinement is the use of a self-compensating beam splitter that eliminates the need for a separate compensator plate. Vibration sensitivity of the interferometer is kept to a minimum by this construction. The yoke is very carefully balanced and great pains are taken to ensure the dimensional stability of the assembly. The result is a robust and reliable interferometer ideal for process use. Process FTIR analysers are becoming commercially available and are finding increasing use.

6.2.2.5 Grating instruments.   Grating spectrometers are still the most common form of instrument used in the visible and NIR regions. Here the advantages gained by the FT approach are not so apparent as in the MIR. A typical grating instrument is illustrated in Figure 6.8. and as before, this refers to no particular instrument, the general principles only are shown.

Energy from the source, usually a high-temperature filament lamp, is focused onto the entrance slits of the grating monochromator. The grating itself can be a concave holographic element arranged to rock about an axis at relatively high speed (several times a second is possible). The spectrum formed by the grating is directed onto the detector through the exit slit. The rocking grating is driven by a stepper motor, its exact position is sensed by a precision shaft encoder. A separate chopper would usually be fitted between the grating and the detector to reduce the effects of background light and to improve signal to noise.

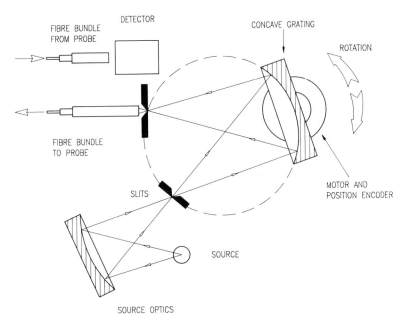

Figure 6.8   Typical grating instrument.

The system can be used for process analysis in several ways. The most common is to interpose a fibre optic link and an optical probe between the exit slit and the detector. Various commercial instruments based on these principles are available. They all differ in detail, some may monochromate light after it has passed through the sample and some before. Arrangements have to be made to isolate the appropriate order spectrum from the grating, this can be done with motor driven slits and various filter arrangements.

Generally, grating instruments are simple and robust. Their successful use for NIR analysis comes as much from general improvements in technology rather than anything special about the technique. The performance determining factors are things like stepper motor and shaft encoder accuracy and the available power of small computers.

6.2.2.6 Other whole spectrum methods. There are many different ways of making spectroscopic analysers and space has precluded mentioning all of them. Two in particular are however worth a brief outline description. These are the diode array spectrometer and the acousto-optical tuneable filter (AOTF). Both of these function without any moving parts, they are all solid state with all of the advantages of reliability and robustness that this implies. They are limited in their wavelength range but nonetheless each occupies its own particular niche. The diode array system is, at least in principle, very simple. Energy from a source it taken through the sample and into a grating monochromator in the usual way. In this case, however, the grating is fixed, it does not scan. The spectrum produced by the grating is laid onto a linear array of detectors. The output of these is taken directly to the system electronics to be processed into a spectrogram. The resolution and range of the device are controlled by how many and how closely spaced the detector elements are. In theory, this technique could be employed anywhere in the spectrum. In reality it is limited by the state of detector technology. In the visible and very short wave NIR (up to 1.1 µm), charge-coupled silicon detectors can be used to give excellent results. To move further into the NIR, more exotic devices are used but at present the method is distinctly limited in its wavelength range. Process diode array instruments are currently available for operation in the UV, visible and shortwave NIR regions.

The acousto-optical tuneable filter is an all solid state monochromator capable of extremely fast scanning. It has been around since the early 1960s but technical difficulties have limited its use until recently. Certainly this is something that still has to realise its full potential and it may be many years before such instruments are widely used. At present there are systems available operating in the NIR. These seem to give good results and unrivalled speed of operation. The AOTF consists of a specially cut crystal of, for example, tellurium dioxide. This is bonded onto a piezoelectric acoustic transducer that is driven by an oscillator at high frequency, typically 30–200 MHz. The acoustic waves induce refractive index changes throughout the crystal. The result of the interaction of transmitted light and the crystal produces three diverging beams, one of non-scattered energy and two of orthogonally polarised monochromatic light. Discrete wavelengths can be selected electrically without the need to scan through the intervening spectrum. Wavelength selection is therefore fast and this coupled to the small size of the unit makes it an interesting monochromator. The crystals used are generally limited to a maximum wavelength of around 5 μm although to the best of our knowledge, commercial instruments are presently restricted to the NIR up to 1.7 μm.

6.2.3 Optical accessories and sampling

Modern instruments are generally reliable and most analytical methods have been well researched and can be expected to work. Difficulties tend to arise at the interface between the analyser and the sample. Perfectly good measurements can be ruined by bad sampling and by bad sample presentation. The author recalls keeping a record of the problems of IR analysers in service. The conclusion reached was that around 80% of reported faults were false alarms and that of the remainder, 80% were sample system problems!

Many options are available. Frequently the approach is to extract a sample from the process and send it to the analyser, where it may be measured in some form of sample cell; this is usually called the 'on-line' approach. An alternative is to take the analysis to the process and to measure directly in the process line; the 'in-line' method. Both methods have their good and bad points and, as ever, it is a case of picking the right one for the job in hand. The choice will always be a compromise. Faced with the choice it is always tempting to go for the in-line solution. An NIR instrument coupled by fibre-optics to a transflectance probe, inserted directly into a line, seems to be a wonderful arrangement. It does however have its own problems. Can the probe be positioned in such a way as to 'see' a representative sample? How can the calibration be checked? How can the probe be removed for cleaning? Extractive sample systems can be expensive and may sometimes be unreliable. Again the questions are: Is a representative sample being taken? Is it always getting to the analyser? Is the sampling process changing the sample?

We can safely say that one of the hardest tasks in process analysis lies in the presentation of the sample to the instrument. This would apply equally to at-line measurements where the use of appropriate optical accessories may be key to the design of user friendly instruments.

6.2.3.1 Sample cells. The usual method of presenting a sample to an optical analyser is to use a transmission sample cell. The function of this is to contain a fixed length of sample in the analysing beam. This fixed path length is, of course, one of the factors controlling sensitivity. In its simplest form, a cell is merely a tube, with a transparent window at each end, through which the process material is passed. In practice, careful attention has to be given to the design of such a device. Obvious points are that the materials of construction must resist the sample and that the sample must be maintained at the correct conditions of temperature and pressure for the sample. A critical choice will be the selection of the correct window material. This must be transparent over the wavelength range to be used and must be unaffected by the sample. In the visible and NIR regions there is little difficulty here. Materials such as the optical glasses and quartz combine mechanical strength with corrosion resistance and good optical properties. Sapphire is an excellent, if rather expensive, choice for the more severe applications. It is incredibly robust and is inert as well as having excellent thermal properties such as a low sensitivity to thermal shock. In the MIR region optical materials can be more of a problem. Generally those available are comparatively weak and many are hygroscopic. Calcium fluoride and zinc selenide make good general purpose windows for many applications.

In the NIR sample path lengths are usually comparatively long. The region is not often used for gas analysis due to the weakness of the overtone absorption bands. For liquid phase work this weakness is a positive attraction of the region. Sample paths may be from a millimetre to several centimetres. Cells of this length are practical propositions for process work. They are long enough to allow unrestricted sample flow and are unlikely to become blocked by debris in the sample. The cells are large enough to be dimensionally stable and are usually easy to design. In the MIR, cells can be more problematical. For gas analysis things are quite simple. Plain tube cells in the range of 1–50 cm cover most applications. Liquid cells on the other hand can become unacceptably short. The high extinction coefficients of liquid samples generally call for cells of between a few tens of micrometres to half a millimetre in length. This is often too short to be practicable. Very narrow gaps are prone to block and it is difficult to be sure of good sample flow through them. Short cells are susceptible to dimension changes and hence can cause large variations in instrument calibration. There are solutions to this problem. An obvious step is to avoid the largest absorption bands and to concentrate on smaller features. If carried to extremes this can negate some of the reasons for working in this region and one may as well give up and look for a near-infrared solution to the problem. Another possible way out is to use an attenuated total reflectance technique (ATR). This method will be described later.

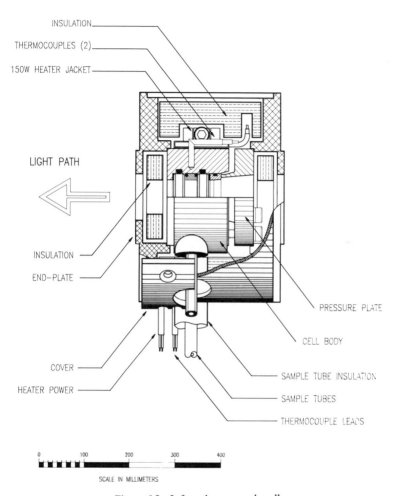

INSULATION

THERMOCOUPLES (2)

150W HEATER JACKET

LIGHT PATH

INSULATION

END−PLATE

PRESSURE PLATE

CELL BODY

COVER

HEATER POWER

SAMPLE TUBE INSULATION

SAMPLE TUBES

THERMOCOUPLE LEADS

0   100   200   300   400

SCALE IN MILLIMETERS

Figure 6.9 Infrared gas sample cell.

The sample cell shown in Figure 6.9 will illustrate several points. This was designed to carry a gaseous sample of mainly hydrofluoric acid at a temperature of 200°C and a working pressure of 16 bar gauge. The windows selected were zinc selenide and were 25 mm diameter and 6 mm thick. The sealing O rings were from Kalrez and all wetted parts were machined from Hastelloy C. An annular 150 VA electrical heater was wrapped around the body and two thermocouples were used to provide temperature control and over temperature shut down. The whole cell was insulated with a layer of glass fibre and enclosed in a light alloy casing. The cell was supported, in its housing, on glass loaded PTFE bushes carried on 'Tufnol' end pieces. In this case, the cell was found to have a limited but useful lifetime of about 30 weeks before the windows became so corroded as to become unusable. This was acceptable and in any case, no better solution appeared to be available.

6.2.3.2 Sample probes (optrodes).   One very practical alternative to the conventional sample cell is to use transflectance insertion probes. These are mounted directly in the process line and can be coupled to an instrument by optical fibres or tube light guides. These have become common accessories for NIR instruments. The device shown in Figure 6.10 shows one possible layout. The sample is defined by the gap between

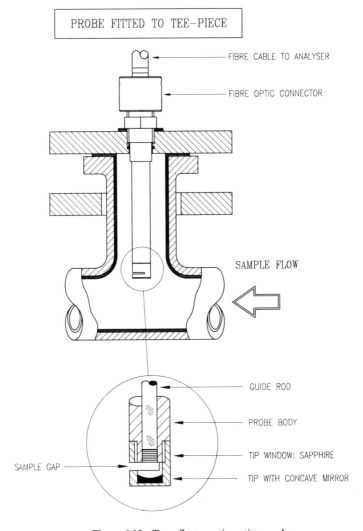

Figure 6.10   Transflectance insertion probe.

the probe tip and the return mirror. The light beam thus passes through the sample twice. Light is led to the probe from the analyser source, and passes down the central guiding rod and through the sample. A separate set of fibres, interleaved with those from the source, collects the returned energy and takes it to the monochromator for analysis. The advantages of the method are those usually associated with in-line systems. No sample system is required and the sample is examined directly under process conditions. Disadvantages are that it is difficult to check or to calibrate such a system without removing it. Removing the probe for cleaning or replacement is equally problematic. Some means has to be found of isolating the section of pipe-work containing the probe. This can be done by fitting it in a side arm with isolation valves and provision for emptying and flush. Another solution it to insert the probe through a pair of ball valves.

A very practical solution is to use a retractable system of the type developed for pH probes. The sketch in Figure 6.11 shows one arrangement. The probe can be retracted from the process line by a pneumatic

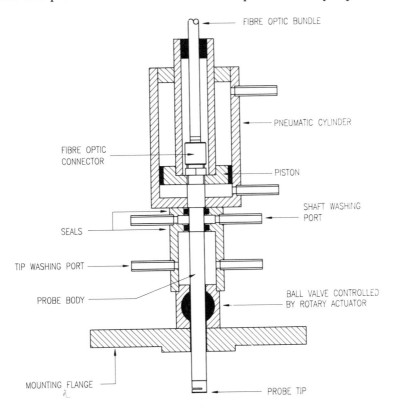

Figure 6.11 Retractable probe system.

cylinder. When retracted, O ring seals isolate the probe from the process. The probe tip is withdrawn into a sealed space that can be flushed to clean it or filled with standard mixtures for calibration. It will be noted that a second flushable space is provided to give added protection. This prevents sample being carried out of the line on the probe shaft as well as acting as a second sealing system for safety purposes.

Fibre coupled probes have proved very successful in the NIR. They are however of limited use in the MIR region. Fibres for this part of the spectrum are very limited at present, they are expensive, of relatively low transmission and consequently, use is restricted to short runs only. It is debatable whether fibre will ever be entirely satisfactory conduits in this region. If they are small enough to be sufficiently flexible then their diameter will be too small for the high throughput necessary in this region.

6.2.3.3 Attenuated total internal reflectance (ATR).   The concept of the simple cell, where an optical path of known length is defined by window, serves for many applications. In some cases it has severe limitations. When working with liquid samples in the MIR, very short paths are usual. These may be so small as to be unusable, small gaps (below 0.25 mm) are prone to fouling and it can be difficult to establish a flow of sample through them. One possible solution to this problem is the ATR technique. This functions by using the absorptions that occur when total internal reflection occurs at the interface between two materials. If a transparent material of high refractive index is placed in contact with the sample, light passing through this at angles above the critical will be internally reflected off the interface. In the process of reflection, the beam will actually penetrate slightly into the sample medium. The depth of this penetration will depend on the incident angle and upon the ratio of the indices of the two materials. Typically it is around one wavelength, e.g. a few micrometres. Figure 6.12 illustrates the process. It can be seen that this technique gives us a practical method of generating small sample path lengths. Different optical arrangements can be used to vary this. If the light beam is arranged to reflect down the inside of a crystal rod, at each reflection a path length of around one wavelength is produced. Such a rod might generate twenty reflections yielding a total path of the order of 100 μm. The method is good in that usable pathlengths can be produced in highly absorbing samples where other approaches would be quite unworkable.

Despite its obvious potential, ATR has not found quite as much application for process work as one might expect. In part this is due to the fragile nature of the interface between the ATR crystal and the sample. Any coating or contamination of this surface is likely to have serious effects on the results produced. In a process environment, such effects are all too likely to happen. Viscous samples can also create

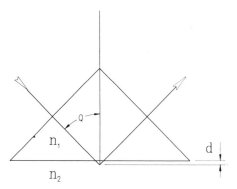

Figure 6.12 Principle of ATR spectroscopy. An element of refractive index n_1 in contact with a sample of lower refractive index n_2. Radiation penetrates the sample by a distance d, generating a spectrum.

problems. The thin boundary layer can remain stuck to the crystal resulting in long response times. Despite these drawbacks, the method is a valuable one and certainly should see far more use in the future. There are several excellent systems available. In particular, one is produced where an ATR probe is interfaced to an FTIR spectrometer by tube light guides. Two pass or multipass crystals are available in a variety of materials. The system can be used directly in-line or with an extractive sample system. ATR spectroscopy is also most suitable for 'at-line' analysis. Horizontal ATR accessories, configured as part of dedicated ATR instruments, can offer robust, cost-effective solutions to measurement problems in batch processing.

6.2.3.4 Fibre-optics and light guides. Many instruments, particular those operating in the NIR and the visible regions, can be interfaced with the process by fibre-optic links. This can in fact be one of the main advantages in using such instruments. Analysis can be carried out remote from the sample point with safety. The presence of flammable or toxic materials can be dealt with without recourse to heavy explosion proof enclosures and elaborate safety interlocks. Multiplexing arrangements can be used allowing one instrument to interrogate several sample points in sequence. For Fourier-transform-based systems, which tend to have large image sizes, several fibres can be run from an interferometer in parallel. Each fibre can be led to the sample and returned to individual detectors. The arrangement eliminates the need for optical multiplexing, instead carrying out the function with electronics. Measurements are effectively simultaneous, rather than sequential, with obviously faster cycle times. In either case each separate sample point can be making entirely different measurements giving clear economies in installation and maintenance costs. Generally speaking, the most valuable gain from fibres is that it becomes much easier to make in-line measurement. This opens up a whole range of measurements that were previously impossible. Going in-line gives obvious gains in terms of eliminating sample systems and all their problems. It also permits measurements or corrosive and toxic materials that were previously rated as intrinsically difficult to sample.

Two basic fibre configurations are used. Bundles of fibres are used for runs up to a few tens of metres. They give high throughput and generally are to be preferred for precision measurements. Single fibres must be used for long runs, indeed up to several kilometres may be possible in certain ideal cases. In this case, fibres tend to be larger than those used for telecommunications, 600–1000 μm core sizes being usual. The difficulties with fibres are their low energy throughput and their spectral limitations. In general, it is hard to input energy from the extended sources used in spectroscopy. This is not too severe a problem in the NIR but is very much so in the MIR region. It is questionable whether or not fibres will ever be of any great use in the latter region. All fibres have their own internal spectra. The common glass and silica types normally encountered are limited to a range from the visible out to around 3 μm. Special materials such as chalcogenide and fluoride glasses are available but these are only suitable for the shortest of runs. Even so, short runs can be useful, just a metre of fibre can be all it takes to get a measurement in-line.

An alternative to fibres, for the MIR, is to use light pipes. A large diamter (>6mm) tube with a polished inner surface makes a very effective light guide. Straight sections of almost any length can be joined with articulated mirrored joints to give enough flexibility for many purposes. Commercially produced guides of this type are available and can offer a good way of getting MIR measurements 'in-line'.

6.2.4 Applications

Most analysers currently in use on industrial processes are relatively simple instruments involving measurement at just a few wavelengths. Such systems are configured for operation anywhere in the UV–MIR region depending on the application. Reflectance applications are likely to use the Vis–NIR region. Despite such wide use, industrial applications are scant in recent literature, partly due to industrial secrecy but also as the application of such well established technology would not normally warrant publication. In most cases, calibration involves direct absorbance measurements. Here, to obtain quantitative data on specific chemical entities, it is necessary to identify fully resolved peaks associated with the analyte(s) of interest. Full characterisation of the process matrix and its variability is key to the successful use of these systems. In some cases, the versatility of these instruments is enhanced through the use of direct multicomponent analysis such as MLR (see chapter 8). The Infrared Engineering MM55 General Purpose Gauge [1] for example, uses only five narrow band filters but is capable of relatively demanding measurements such as the moisture and fat content of milk powder. Simple instruments may also be used to identify abnormal occurrences without providing data on specific chemical species. The use of Procal Pulsi 300 instruments on the Zeneca Huddersfield Works (UK) effluent treatment plant is an example [2]: Here, five broad band filters cover the range from 230 to 705 nm. The absorbance of works effluent entering the treatment plant is continually monitored in each of the five spectral regions. Deviations outside of specified limits are alarmed leading to effluent diversion to storage tanks. Extensive validation exercises have established that environmentally significant quantities of major on-site chemicals would be detectable using this approach.

These traditional analysers are generally quite robust being based on well established technology. They continue to provide cost-effective solutions to many industrial problems. Where one or two key analytes need to be monitored in a relatively simple and stable process matrix, these analysers cannot be beaten. Such systems offer little, however, when quantitative, multicomponent data from complex process streams is required (this type of data is increasingly required, especially in the pharmaceutical and fine chemical industries). Spectroscopy may be applied to such demanding applications but this usually requires the power of whole spectrum instruments and multivariate calibration.

Applied to spectroscopy, multivariate calibration (see chapter 8) involves taking the spectra of several samples (carefully selected to maximise the process variability represented) of known composition (reference data set). Advanced statistical routines are used to establish correlation between the spectra and reference data. If sound calibration models are produced, then accurate prediction of analyte values can be obtained from the spectrum of an unknown sample. This approach is very powerful and can appear to be almost 'magical' in that quantitative data can be extracted from a spectrum even where expert traditional interpretation would clearly fail to provide the same data. This 'magical' aspect can cause difficulty in identifying suitable applications. Some people refuse to believe that the technology can work while others are keen to apply it to every situation! In assessing application viability, the following should be considered.

Impact of analytes on whole matrix spectrum. The greater the impact of the analytes of interest on the spectrum, the better. Calibration models based on substantial spectral variance will tend to be of low dimensionality (low number of PCR or PLS factors—see chapter 8), can be based on relatively few calibration samples and are likely to be robust against minor process and instrument changes. It must be noted, however, that many robust calibrations are based on only subtle analyte related spectral variance.

System signal to noise. High signal to noise allows the extraction of useful information from 'low order' PCR or PLS factors. Where the spectral variance associated with the analyte of interest is small compared to the overall spectral variance, these low order factors are needed in the calibration model. In such cases, high signal to noise is key to feasibility. High signal to noise is one of the major attractions of NIR spectroscopy. Models of high dimensionality are routinely used in the prediction of analyte values even where analyte related spectral variance is minute.

Accuracy requirements. Clearly, the more stringent the accuracy requirements, the more demanding is the application! In attempting to achieve high accuracy, it is important to understand that errors from the reference data will be compounded with errors introduced through the spectroscopy and calibration.

Value of the measurement. This important factor should not be overlooked. The more demanding the application in terms of the above criteria, the higher the cost of successful implementation. A demanding application may require hundreds or even thousands of samples in the calibration set. This requires a substantial investment in the reference data before a major spectroscopy/calibration project can even commence. In addition, ongoing resource for the maintenance of calibrations will be required (periodic re-modelling may be necessary to compensate for minor process and instrument changes).

Many applications of whole spectrum instruments to both at-line and on-line analysis can be found in the literature. FT MIR ATR spectroscopy is becoming an important technique. Its application to fermentation analysis was investigated by Alberti *et al.* [3] Calibration models were developed for glucose, ethanol and glycerol concentrations. The multivariate models proved far superior to the simple Beer–Lambert models developed for comparison. Cahn and Compton [4] applied both PCR and PLSR (see chapter 8) to FTIR ATR data obtained for binary and ternary mixtures of xylene isomers. In this case a flowthrough ATR cell was used. Sound models were developed with prediction errors of around 1%. FTIR ATR with PLSR has also been used successfully for the prediction of non-transparent polymer compositions [5]. Doyle and Jennings [6] demonstrated the application of an ATR immersion probe for the on-line monitoring of the well understood esterification of acetic acid with methanol. The spctra of acetic acid and methyl acetate are very similar but ratioing two frequencies was shown to describe the reaction progress.

FT MIR ATR spectroscopy is attractive in that it exploits fundamental vibrations. Variation in the composition of a process stream will often lead to substantial changes in the MIR spectrum. Signal to noise, however, is not especially high; as a surface technique, ATR crystals are prone to fouling and the effective pathlength is sensitive to temperature variation. Where apparently useful calibration models exploit only very subtle spectral variance, it may prove difficult to achieve robust performance in the plant environment. FT MIR spectroscopy is particularly well suited to the monitoring of gaseous process streams and associated environmental releases. Here, fundamental vibrations are used together with sampling approaches that maintain reasonably high signal to noise. As an example, Cronin [7] has reported the development of a stack emissions system for DuPont's titanium dioxide plants. The simultaneous analysis of carbon monoxide, carbon dioxide, sulphur dioxide, hydrogen chloride and carbonyl sulphide was achieved.

Excellent signal to noise can be achieved in NIR spectroscopy but here the spectral changes resulting from compositional change are smaller than in MIR spectroscopy. NIR spectroscopy is now well established as a process control tool. Much of the early quantitative NIR work was applied to agricultural analysis using reflectance techniques [8, 9]. More recently it has been applied extensively to chemical process monitoring. Callis and co-workers have exploited the shortwave NIR region to monitor gasoline octane number [10], caustic and caustic brine measurement [11] and ethanol in fermentation process [12]. In each case, the quantitative analysis of the grossly overlapping spectra would not have been possible without the multivariate calibration routines which have been applied. Work on an epimerisation process [13] has also demonstrated the power of PLSR to resolve NIR data with minimal spectral variance. The proceedings of a recent European conference on NIR spectroscopy [14] confirmed its widespread use in a diverse range of industries (e.g. pharmaceuticals, alcoholic beverages, foods, dairy products, fine chemicals, commodity chemicals, animal feedstuffs, tobacco products).

6.2.5 Future developments in optical spectroscopy

One the most interesting and significant recent developments has been in the field of Raman spectroscopy. Instruments suitable for process use are just becoming available and offer considerable potential for the future. Raman differs from normal spectroscopy in that the interactions involved are inelastic. In the usual Rayleigh scattering (elastic) process, light is absorbed and re-emitted at the same wavelength as the scattering centre is energised and then recoils directly to the ground state. In the Raman process, energy returns to its ground state in steps, energy is re-emitted at wavelengths offset from the existing frequency.

If a beam of monochromatic light from a source, typically a laser, is directed at a sample, the spectrum of scattered light will be dominated by the laser frequency (the Rayleigh line). On either side of this band, small secondary bands will be seen (the so-called Stokes and anti-Stokes Raman lines). The frequency shifts of these secondary lines can be disassembled into the Raman spectrum. This is related to the various vibrational and rotational modes within the sample molecule, providing similar and complimentary data to the mid-infrared spectrum. The effect is weak in the extreme. Typically, the energies of the Raman lines are around seven or eight orders of magnitude lower than that of the Rayleigh line. This weakness of the Raman signal has limited its use. The scattering can easily become dominated by other processes such as fluorescence and it can be difficult to separate the signal from the massive Rayleigh line. For these reasons Raman spectroscopy has been confined to the research laboratory where it has been a valuable tool, albeit a difficult and expensive one.

Technology does of course improve and the development of very sophisticated holographic filters and gratings is leading to practical process instruments. The advantages to be gained from this type of spectroscopy are considerable and apply particularly to the process situation. Raman spectra contain information comparable to the MIR with the advantage of using visible or very shortwave NIR light. Conventional glass optics can be used with all the precision and sophistication that signifies. Raman spectroscopy offers the flexibility and information content of the MIR combined with the practicalities, such as confocal probes and optical fibres, that one associates with the NIR. It is perhaps too early to make confident predictions of the development of the technology. It does, however, seem very possible that this will be the major process analysis story of the late 1990s in the same way that NIR dominated the late 1980s.

6.3 Process mass spectrometry

6.3.1 Principle of operation

A mass spectrometer is capable of characterising a sample by producing and measuring a range of characteristic fragments in a highly controlled manner. The sample is fragmented or cracked as a result of an induced ionisation step and it is the relative characteristics of these fragments which can be used to identify the parent compounds. Results are

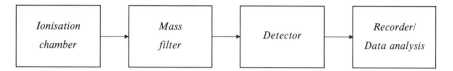

Figure 6.13 Mass spectrometer block diagram.

compared with known results which are derived from the analysis of standard mixtures. The results and raw data can be stored for subsequent collation or reprocessing. The stored data can be automatically compiled into reports or logs which can be user customised for a range of purposes. A block diagram of a mass spectrometer is shown in Figure 6.13.

A series of functions are performed by the mass spectrometer and these are usually computer controlled; they are:

(1) Ionisation of sample
(2) Separation of charged fragments according to mass
(3) Detection of fragments
(4) Data acquisition
(5) Sample identification

Advances in electronics and microprocessor control technology have enabled the functions detailed above to be sequenced rapidly. Multicomponent analysis can be achieved in seconds making mass spectrometry an ideal candidate for many environmental monitoring and process control measurement applications.

Resolving power. The resolving power of a mass spectrometer is a measure of its ability to separate and selectively detect ions which have only a slight difference in mass. Adjacent peaks in a spectrum are resolved if the height of the connecting valley above the baseline is less than 10% of the height of the smaller peak above the baseline. If we term the mass of one component as M and ΔM is the difference in mass between the adjacent peaks, then the resolving power is said to be $M/\Delta M$.

Ionisation. A controlled quantity of the sample is introduced to the mass spectrometer through an inlet which can take the form of a membrane, porous frit or a capillary. The molecules then enter the ionisation chamber, where they are ionised by collision with electrons which have been generated by a heated filament. The ions that are produced are accelerated out of the ionisation chamber and are focused into a narrow beam before they enter the analyser section.

This acceleration and focusing is achieved by the influence of electric fields which are produced by the use of high electrical potentials. During the ionisation process the collection of ions formed often include smaller fragment ions which are the result of larger unstable ions breaking up into smaller, more stable units. Large molecules are known to fragment in a highly reproducible manner, and it is this predictable behaviour which makes mass spectrometry such a powerful qualitative technique.

Mass filter. The mass filter is a section of the analyser which processes the collection of charged ions and separates them and quantifies them on the basis of their mass to charge ratio (M/Z). In most cases the ion charge is a single positive charge and therefore the separation can be directly referenced to mass. There are a number of techniques used for separating the groups of ions formed in a mass spectrometer. Process instruments are available in forms which utilise the three most common of these. The first is classed as time of flight, in this method an electronic shutter allows a brief pulse of the ions to enter a drift region. The ions migrate across the drift region under the influence of an electric field and the drift time is proportional to the ions mass to charge ratio. The second technique utilises a magnetic sector and here a steady beam of the charged species enter a magnetic field perpendicular to which is the direction of ion motion. The field strength can be varied to allow a narrow cut of the charged species to emerge from the sector unaffected. By scanning the magnets through a range of field strengths the magnetic sector can resolve the ion mixture into a series of groups of ions having similar mass to charge ratios. The third technique utilises a quadrupole mass filter, here as the name suggests, four rods are arranged in parallel so that the ends of each rod would occupy one of the four corners of a square. A combination of DC and RF voltages are applied to the electrodes, and this generates a three-dimensional electric field. The ions produced in the ionisation chamber are focused to a linear trajectory and enter the end of the open quadrupole assembly. The motion of the ions along the central axis of the quadrupole is extremely complicated. The ions oscillate around this central axis with rapidly increasing amplitudes and most ions strike the electrodes and are neutralised. However, for particular values of DC and RF voltages the trajectory of an ion with a certain mass to charge ratio will be stable along the central axis and these ions will emerge from the quadrupole unaffected. Thus, if the DC and RF components are varied different mass to charge ratios can be allowed to reach the detector in sequence. The quadrupole type mass filter has certain advantages over the magnetic sector mass filter for process use; it is more robust, smaller and cheaper and has a faster scanning capability.

Detector. Process mass spectrometers usually have a Faraday cup detector or an electron multiplier or in some cases a combination of both. In the case of the electron multiplier ions strike a collector plate at the first of a series of dynodes. This impact causes the displacement of one or more electrons which initiate a cascade effect as they move through the series of dynodes in the electron multiplier. A measurable current can be generated, which is a measure of the abundance of ions which impact the first dynode collector plate.

6.3.2 Applications

Some process applications are listed below.

6.3.2.1 Fermentation control. The process mass spectrometer can be used to monitor and maintain the ratio of CO_2 to O_2. Other volatile compounds produced during the fermentation process can also be measured. This information can be used to determine fermenter nutrient feed-rate requirements and to infer the overall condition of the fermenter.

6.3.2.2 Steel production. In many iron and steel plants, waste gases from the three main processes, iron production, steel production and coke making are recycled as fuels to minimise the expenditure on commercial fuels. Process mass spectrometers are used to continuously determine the calorific value of each stream by measuring their composition. In this application there is a requirement to measure several gas streams of different and fluctuating composition in a very short cycle time. Process mass spectrometry has ousted the use of on-line calorimetry for this demanding duty.

6.3.2.3 Acrylonitrile, ammonia, and methanol synthesis. Around the world, on many chemical plants synthesising ammonia, methanol, and acrylonitrile a single mass spectrometer equipped with a multistream inlet has replaced several other on-line analysers. A range of the inlet and outlet streams on the process are analysed in sequence and high frequency process control data is produced.

6.3.2.4 Environmental and hygiene monitoring. Plant area monitoring for selective measurement of volatile organic compounds from mixtures is usually achieved using a multipoint, sequential analyser system. These systems are used to reduce the risk and incidence of personnel exposure to unacceptable levels of volatile chemical compounds on the plant. They do not and cannot guarantee peoples safety because a limited number of points on the plant are sampled, and each point is visited once per analysis cycle. The process mass spectrometer is generally faster than other types of analyser for this duty and thus enables either a reduced cycle time or more points to be visited without time penalty. This means that for this particular application it offers a key advantage. Other safety, health and environmental applications include vent stack monitoring, landfill gas monitoring and vessel vent monitoring.

The above list of some of the more established applications is extremely limited as process mass spectrometry is an emerging technique in comparison to some of its competitors. The specificity, high speed, sensitivity and reducing cost of the technique will make the technique commonplace in many industrial processes during the forthcoming years.

References

1. Commercial literature on the Infrared Engineering MM55 General Purpose Gauge (ref 76/8633-01) Issue 5). Issued by Infrared Engineering, The Causeway, Maldon, Essex CM9, 7XD, UK.
2. Wheeler, V. private communication and Zenecca internal documentation.
3. J. C., Alberti, J. A. Phillips, D. J. Fink, and F. M. Wacasz, *Biotechnol. Bioengng* **15** (1985) 689.
4. F. Cahn and S. Crompton, *Appl. Spectrosc.* **42** (1988) 865.
5. J. Toft, O. V. Kvalheim, T. J. Karstang, A. A. Christy, K. Kveleland, and A. Henriksen, *Appl. Spectrosc.* **46** (1992) 1002.
6. W. Doyle and N. A. Jennings, *Spectrosc. Int.* **2** (1991) 48.
7. J. T. Cronin, *Spectroscopy* **7**(5) (1992) 33.
8. J. R. Piggott, *Statistical Procedures in Food Research.* Elsevier, Amsterdam, The Netherlands, 1986.
9. T. Naes, and H. Martens, *J. Chemom.* **8**(2) (1988) 155.
10. J. J. Kelly, C. H. Barlow, T. M. Jinguji, and J. B. Callis, *Anal. Chem.* **61** (1989) 313.
11. M. K. Phelan, C. H. Barlow, J. J. Kelly, T. M. Jinguji, and J. B. Callis, *Anal. Chem.* **61** (1989) 1419.
12. A. G. Cavinato, D. M. Mayes, Z. Ge and J. B. Callis, *Anal. Chem.* **62** (1990) 1977.
13. N. C. Crabb and F. McLennan, *Process Control Quality* **3** (1992) 229.
14. *2nd European Symposium on NIR Spectroscopy*, Kolding, Denmark, 1993.

7 Electrochemical methods

M. L. HITCHMAN and L. E. A. BERLOUIS

7.1 Introduction

Electrochemical methods of detection are used for process analysis in a number of important areas, such as in the control of industrial processes, the monitoring of effluents and toxic gases in the environment and also in medicine [1]. The nature of the electrochemical parameter measured (i.e. a potential or a flow of charge) means that the signal can be easily processed by modern digital electronics for data collection and analysis so that any subsequent remedial feedback can be rapidly effected. In addition, some sensors will detect the presence of one particular species in the presence of other species found in the same solution, even at very reduced concentrations and the signal can be proportional to the concentration or activity of that species over several decades, down to, in some cases, the nanomolar level.

However, on-line electrochemical sensors for process applications often have much more stringent requirements than for those usually employed in the laboratory. In particular, one of the most important factors is the stability, especially long-term stability, of the sensor electrode in the test environment. In many practical process analytical situations, there are species which can reduce the electrocatalytic activity and this results in variations in sensor response. Much effort, not to say ingenuity, has gone into trying to overcome these difficulties and we shall be discussing some of these in the sections which follow. Before doing so though, we outline the principles of the two main types of electrochemistry which are used for process analysers, namely potentiometry and amperometry. Examples of applications of the principles are then given, not in an encyclopaedic or comprehensive way but rather in an illustrative manner, with particular emphasis on the scientific base on which sensors are built. For whatever the process, the scientific foundations will remain unaltered and are generally applicable whether one is monitoring a reaction involving a boiler, a biosynthesis, a burner, a body, a brewery or whatever.

7.2 Potentiometry

7.2.1 Principles of measurement

7.2.1.1 Introduction. An electrode in contact with a solution will acquire a net positive or negative charge with respect to that solution. This is due to the interaction between the electrode and species in the solution. As a consequence of this interaction, a difference in the electrical potential between the electrode and the solution results and this can be measured as an electrode potential, E. An equilibrium value will be reached when the rates of transfer of charge to and from the electrode become equal. If the concentration of the potential-determining species (the analyte) in the solution changes, its chemical potential will correspondingly alter and charge transfer will occur across the interface until a new equilibrium is reached, giving rise to a new value for E. By measurement of this interfacial potential E it is possible to determine the concentration of the analyte.

The magnitude of the interfacial potential should ideally depend only on the concentration and charge of the determining species in the solution. However, other components of the solution can also have a marked effect on the response of the potential of the electrode. For example, the presence of complexing agents in the solution will reduce the 'free' concentration of the analyte. This reduction will depend on the relative amounts of the complexant and analyte present and the strength of the ligand–analyte bond. Species with similar properties to the analyte (e.g. as occurs in the measurement with an ion selective electrode of K^+ concentration in the presence of Na^+ ions — see section 7.2.2.4) can also interfere with the measurements. In addition, the total number of ions present in the solution, as represented by the ionic strength of the solution, also plays a role in the determination of the analyte concentration from the electrode potential measured.

In the following section the theory behind the measurement of the concentration of an ionic species in a solution by the potentiometric method is first outlined. Then various factors which can affect this measurement are considered and steps which can be taken to overcome the problems are discussed. Finally, some applications of potentiometry to process analysis are illustrated.

7.2.1.2 Theory. The dependence of electrode potential on the concentration of a particular ion in the solution is described by the analytical version of the Nernst equation, given by

$$E = E^0 \pm \frac{RT}{nF} \ln C \qquad (7.1)$$

where E^0 is the standard potential of the species to be detected, F is the Faraday constant ($96487\,A\ s\ mol^{-1}$), n is the number of electrons transferred in the reaction (this is usually, but not always, the charge on the ionic species concerned), and C is the concentration of the species in the solution. The sign is positive for cations and negative for anions. It should be noted that the electrode potential is not a function of electrode area and this means that not only can a wide range of sizes of electrode be used, but also that results from different electrode systems can be readily compared. The standard potential is the potential of an electrochemical cell comprising the *indicator* or *working electrode* in contact with the species at unit activity when it is connected to the standard hydrogen electrode (SHE). Strictly speaking, the potential response, E, is related to the activity of the species in the solution and this is, in turn, related to the concentration by

$$a = C \times f \tag{7.2}$$

where a is the activity of the species concerned and f is the activity coefficient of the ionic species. The activity coefficient represents the degree of interaction of the ion with all the other ions in the solution. The greater the number of ions in the solution, the greater this interaction but as the solution becomes more dilute the interaction reduces and $a \simeq C$ as $f \to 1$. The activity coefficient is therefore a function of the ionic composition of the solution which is commonly represented by the ionic strength I, where

$$I = \frac{1}{2} \sum_i C_i z_i^2 \tag{7.3}$$

and z_i is the charge on the ion. The expressions relating the activity coefficient to the ionic strength of the solution are based on the Debye–Hückel limiting law [2]. There are several forms for this expression, and one of the forms, equation (7.4), works reasonably well up to an ionic strength of $10^{-3}\ mol\,dm^{-3}$.

$$-\log(f_i) = A z_i^2 I^{1/2} \tag{7.4}$$

The constant A includes the density, the dielectric constant and the temperature of the solvent.

At higher ionic strengths, due to the increased interaction between all the ionic species in the solution, modifications to equation (7.4) have to be carried out in order that the activity coefficient will still approach unity as the concentration (and so the ionic strength) tends to zero. The modifications are, however, largely empirical. In making any potentiometric measurements, including those for process analysis, clearly errors can occur as a result of ionic strength variations. In order to overcome this problem of ionic strength errors, it is common practice that prior to the potentiometric measurement of the concentration of the analyte, a calibration curve is constructed using concentration standards of the analyte to which is added a known volume of an ionic strength adjustment buffer (ISAB). The ISAB is also added to the unknown solution before measurement. Since all the solutions will then be of the same ionic strength, f for the analyte will have the same value in all the solutions. Hence, on combining equations (7.1) and (7.2), we obtain:

$$E = \left(E^0 \pm \frac{RT}{nF} \ln f \right) \pm \frac{RT}{nF} \ln C \qquad (7.5)$$

and as f is now constant, this equation has a similar form to equation (7.1) and the concentration of the analyte can then be read directly from a calibration graph. The most commonly used ISABs are KCl, $NaNO_3$, NaOH and acetate buffer solutions, the choice depending on the ion to be analysed and the medium of the measurement. In practice, the bracketed term in equation (7.5) can also include other constants of measurement, such as instrumentation potentials and liquid junction potentials. This need to add an ISAB to a test solution can obviously be a limitation for on-line process analytical applications.

7.2.2 Methods of measurement

7.2.2.1 Introduction. In order to measure the interfacial potential difference at an indicator electrode another electrode has to be introduced, either directly or via a salt bridge, into the solution to complete the electrical circuit. This second electrode is called a *reference electrode*. The most common ones employed in practice are the saturated (KCl) calomel electrode (SCE) and the Ag/AgCl electrode, with 1 M of saturated KCl. The potential of these two electrodes can both be expressed by

$$E_{Ref} = E^0 - \frac{RT}{nF} \ln C_{Cl^-} \qquad (7.6)$$

where E^0 for the SCE and the Ag/AgCl (saturated KCl) are 0.2682 V and 0.2223 V, respectively, vs SHE. The potential of the electrochemical cell so formed, E_{Cell}, is measured using a high impedance voltmeter from which the potential, E_{Ind}, of the indicator electrode is obtained as

$$E_{Ind} = E_{Cell} + E_{Ref} \qquad (7.7)$$

The necessary criterion for the reference electrode must clearly be that its potential is reproducible and stable in the environment in which the measurement is carried out. Reference electrodes should be ideally unpolarisable, meaning that there is little resistance to electron transfer across their interfaces. Thus, virtually no potential difference is required to maintain equilibrium at this interface and so any potential change observed in an electrochemical cell comprising such a reference electrode and an indicator electrode can be attributed solely to the latter. Either a SCE or a Ag/AgCl electrode are commonly used in process analysis, although where Cl^- contamination is a problem then a Hg/Hg_2SO_4 reference electrode in contact with SO_4^{2-} ions is often employed (*cf.* section 7.2.3.2).

7.2.2.2 Limits of detection. The Nernst equation (equation (7.1)) predicts a logarithmic dependence of the potential on the analyte concentration. Below a certain concentration, however, the logarithmic dependence is no longer seen and the electrode eventually fails to respond to the presence of the analyte in the solution. The reason this occurs is because the electrode responds to another ion in solution which is at a fixed concentration higher than that of the analyte [3]. For example, in the determination of Ag^+ ions in solution a solid state membrane electrode comprising either a silver halide or silver sulphide may be employed. In the case where the membrane is AgCl, the equilibrium set up is:

$$AgCl_{(S)} \rightleftharpoons Ag^+_{(Aq)} + Cl^-_{(Aq)} \qquad (R7.1)$$

where the subscripts S and Aq refer to the solid and aqueous phases, respectively. Thus, if the analyte in question consists of silver ions, there will also be present in the solution a certain amount of that species from the dissolution of the membrane. As the concentration of the analyte in the test solution decreases and approaches that of the dissolved ion concentration (C_M) from the membrane, the plot of E vs $\log C$ is no longer linear and becomes curved and then independent of C, as shown in Figure 7.1. This is because the potential responds to ($C + C_M$) and when $C < C_M$, the electrode response will be determined by C_M which is constant. The limit where this occurs can be experimentally determined by extrapolating the unchanging potential region of Figure 7.1 to meet the extension of the Nernstian response in the calibration E vs $\log C$ plot.

The limit of detection can also be statistically derived by carrying out measurements in blank solutions; i.e. solutions that do not contain the analyte. If σ_B is the standard deviation for the potential readings in the blank solution, a subsequent potential reading obtained by the indicator in a solution containing the analyte can be quoted to within the 95% confidence limit if it differs from that of the blank solution by an amount greater than $4.652\sigma_B$ [4]. Thus, this allows the detection limit of an indicator electrode to be set for the particular analyte concerned. In practice for process analysis the limit of detection is typically *c.* 10^{-6} M, although with a glass electrode, which is highly selective, measurements can be made up to *c.* pH 12; i.e. down to a $[H^+] \simeq 10^{-12}$ M (see section 7.2.3.3.1).

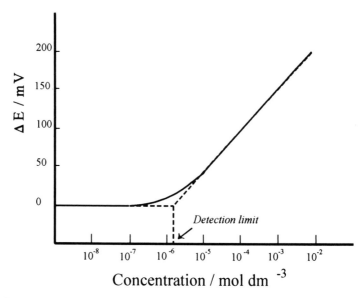

Figure 7.1 Determination of the detection limit for an ion selective electrode.

7.2.2.3 Types of sensor electrodes. The electrical potential difference at sensor electrodes can arise due to the difference in chemical activity of the analyte species in the solution and that of the solid electrode (i.e. an *ion selective electrode—ISE*) or a sensor electrode may consist of an inert metal (e.g. platinum) in a solution where the analyte exists in two different stable oxidation states (i.e. a redox electrode). An example of the latter would be where the Fe^{2+}/Fe^{3+} redox couple is present in solution. The inert metal effectively acts either as an electron sink or reservoir for the analyte species, depending on the relative activity of the two forms of the analyte in the solution. Here, the potential of the inert electrode is given by:

$$E = E^0 + \frac{RT}{nF} \ln \frac{C_{Ox}}{C_{Red}} \tag{7.8}$$

where C_{Ox} and C_{Red} are the concentrations of the oxidised and reduced forms of the analyte and E^o is the standard potential of the redox couple. Thus, if the concentration of one of the forms is known, the other can be obtained from the measurement of E against a reference electrode. This principle is used for oxidation–reduction potential measurements – see section 7.2.3.1.

As has been indicated above, not all sensing electrodes involve the direct transfer of electrons from the analyte to the electrode surface. Indeed, the vast majority involve the transport of an ionic species across a membrane which can be of the solid-state variety (e.g. the Ag/AgCl electrode already mentioned, and the glass pH electrode) or of the liquid ion-exchange type. In the latter, the membrane usually consists of an organic, water-immiscible phase positioned in between the test solution and an internal standard or reference solution. The potential difference across the membrane at 25°C is then given by:

$$E_M = \frac{0.059}{n} \log \frac{C_S}{C_R} \qquad (7.9)$$

where C_S and C_R are the concentrations of the analyte in the test solution and in the internal reference solution, respectively. Since C_R will be known, the analyte concentration can be readily obtained from the measurement of E, with the aid of a calibration plot or by standard addition. Different types of sensor electrodes are discussed further in section 7.2.3.3 under applications of potentiometric sensors.

7.2.2.4 Selectivity. A simple redox electrode shows no selectivity to any particular ionic redox species but an ion selective electrode does. However, it should be noted that the response of an ISE, of whatever type, towards a particular ion is selective rather than specific. This means that an ISE designed for the determination of, say, the potassium ion concentration in a solution can also respond to the presence of other ions, such as sodium, in the same solution, albeit usually to a lesser degree. This interference has to be taken into account, especially when the ISE has difficulty in differentiating between the two species concerned. There are various methods by which selectivity is conferred to the sensing electrode. Chemical modification of the electrode surface itself or the development of ion or molecular sensitive ligands are two of the methods currently utilised. In the latter case, the development of ionophores, such as macrocyclic polyether compounds (or crown ethers) which exhibit strong selectivity for cations such as K^+, Na^+, Ca^{2+} and Mg^{2+} have enabled the development of ISEs for these ions.

The response of an ISE in the presence of an interferent ion can still be described by a Nernstian-type relationship as long as both the analyte and the interferent ions individually follow the Nernst equation. In the case of a potassium ISE in the presence of sodium ions as the interferent, this relationship is of the following form [5]:

$$E = E' + \frac{2.303RT}{F} \log(C_K^+ + kC_{Na}^+) \qquad (7.10)$$

where k is the *selectivity coefficient* for this electrode to sodium. According to this expression (called the Nikolsky equation) a fraction k of the sodium ions behave as if they were potassium ions. Hence, the smaller the value of k, the greater will be the selectivity of the ISE to potassium when both potassium and sodium ions are present in the solution. For these pair of ions, the selectivity coefficient for a potassium ISE is as follows [6]:

$$k_{K^+/Na^+} = 2.6 \times 10^{-3}$$

indicating that the selectivity of potassium over sodium ions is 385:1.

For the Nikolsky equation one can consider three limiting cases with respect to the analyte and the interferent ion. The first is when the analyte is greatly in excess of the interferent ion, at which point the Nernst equation for the analyte is obeyed. In the second case, the interferent is greatly in excess and the ISE then follows the concentration of the interferent ion. The third case is when the both analyte and interferent are present in comparable concentrations, for which the full Nikolsky equation must be employed in order to determine the concentration of the analyte in the presence of the interferent. These limiting cases are further complicated by variations in the the magnitude of k for a given ion pair, as discussed below.

Selectivity coefficients are not constant but vary with the concentrations of both the analyte ion and the interferent ion. The value at a particular analyte concentration can be experimentally determined by increasing the concentration of the interferent ion until the potential alters from that of the pure analyte response. A graph can then be constructed, as shown in Figure 7.2, and the value of k is given by the ratio of the analyte concentration to that of the interferent when the

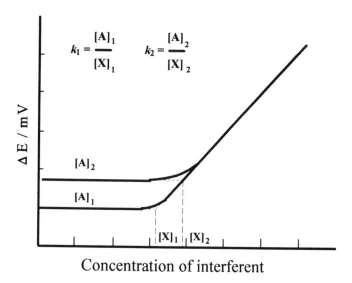

Concentration of interferent

Figure 7.2 Determination of the selectivity coefficient of an ISE at two different analyte concentrations in the presence of an interferent.

deviation in potential response occurs. By changing the concentration of the analyte, the value for k at the new concentration may similarly be obtained.

If the selectivity coefficient for a pair of ions (analyte/interferent) is known, it is possible therefore to determine whether the ISE can be employed in the presence of that interferent. For example, if the concentration of K^+ ions to be determined is $10^{-3}\,\mathrm{mol\,dm^{-3}}$, then, using the value for the selectivity coefficient k given above, the concentration of Na^+ ions in the solution that will lead to an error of $< 5\%$ in the concentration of K^+ ions is given by:

$$C_{Na^+} = \frac{10^{-3} \times 1.05}{(2.6 \times 10^{-3})} = 0.4\,\mathrm{mol\,dm^{-3}} \tag{7.11}$$

Thus, due to the small selectivity coefficient of the potassium selective electrode in K^+/Na^+, the concentration of K^+ ions can be determined to within a 95% confidence limit even in a solution containing sodium ions at $400\times$ the concentration of the K^+ ions. In other cases, though, the selectivity coefficient does not always favour the ion of primary interest. For example, in the measurement of chloride ion concentration using a solid state $Ag/AgCl$ electrode, typical selectivity coefficients, for OH^-, Br^-, I^- and CN^- are 1.25×10^{-1}, 3×10^2, 2×10^8 and 5×10^6, respectively. Thus, in the presence of all these ions, apart from OH^-, the response for chloride will be totally swamped by the interferent ions, even when their concentrations are very much lower than that of the chloride ions. The chloride ISE should not therefore be employed when these ions are present in the solution. For on-line, and even off-line, process analysis the presence of interferent ions can lead to difficulties in applying potentiometry.

7.2.3 Applications

7.2.3.1 Oxidation–reduction potential (ORP) measurements. As has been indicated earlier, one of the simplest electrochemical measurements to make is that of monitoring the potential of a solution at an inert electrode which in the presence of a redox couple can be related directly to the concentrations, or more strictly activities, of the oxidants and reductants. These type of measurements, which are known as oxidation–reduction potential (ORP) measurements, are probably used most extensively in water treatment processes [7, p. 6.123]. The basis for the measurement is clearly the Nernst equation (equation (7.8)) and all that is needed in practice is an inert indicator electrode (e.g. Pt, Rh or Au), an appropriate reference electrode (e.g. SCE, Hg/Hg_2SO_4 — although potentials are usually referred, for ease of comparison, with respect to the SHE), and a high impedance voltmeter. The procedure is equally straightforward, with the system being calibrated using, for example, a quinhydrone solution at, at least, two pH values in ISAB at a constant temperature, followed by measurement of the test solution in the presence of the same ISAB at the same temperature. A typical error at the 95% confidence level would be about 1%.

ORP measurements are widely relied on both to monitor and control cyanide oxidation and chromate reduction which are often required in waste treatment of plating and other metal treating processes before the effluent is discharged. The monitoring of freshwater streams by regulatory agencies to detect illicit dumping of redox species often employs ORP, and it is also sometimes used to monitor sewage influent, digester sludge, and in-plant chlorination for odour control. In the pulp and paper industry, ORP is used to control the input of chlorine and other oxidising agents in bleaching processes. In this context, a particularly interesting example of an ORP process analyser is the Draeger Chloralarm in which the electrodes and electrolyte are separated from the environment by a porous material through which the chlorine can diffuse. The electrolyte anion is bromide which is oxidised by the chlorine and the bromine/bromide ratio is monitored with a platinum electrode. The measured potential can be readily related to the environmental chlorine concentration. Control modules, with alarm facilities, can be set to measure over several ranges (e.g. 0–5, 10 and 50 ppm) with a precision of ±5% of full scale.

Clearly, ORP can be a simple and effective means of process analysis and control. However, as has been mentioned and is apparent from a consideration of the Nernst equation, it is non-selective and redox species other than the ones of interest can influence the observed potential. One means of overcoming this lack of selectivity to some extent is to use an indirect method of analysis rather than direct potentiometry.

7.2.3.2 Potentiometric titrations. The procedure here is exactly analogous to an ordinary titration except that the end point is detected not with a chemical indicator but by monitoring the variation of an electrode potential on addition of the titrant. The technique can be readily understood with an example, that of the determination of chloride in beer [8].

Chloride is one of the principal ions in beer and its concentration has a significant effect on flavour. The range is usually $100 < [Cl^-]$ mg $dm^{-3} < 1000$, with the highest levels being present in the strongest beers. For indirect potentiometric monitoring, the chloride in the beer is titrated with a standard silver nitrate solution in the presence of a silver electrode and a mercurous sulphate reference electrode. The latter is used in preference to, say, a SCE to overcome any possible contamination with chloride ions. A typical titration curve is shown in Figure 7.3.

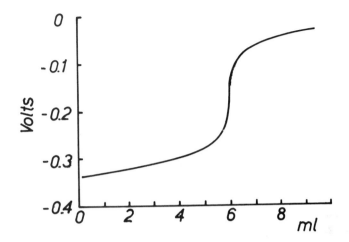

Figure 7.3   Titration curve for chloride ions in beer.

Initially the silver electrode responds to the Cl^- ion ($AgCl + e^- \rightleftharpoons Ag + Cl^-$) and after the end point it responds to the Ag^+ ion ($Ag^+ + e \rightleftharpoons Ag$). The sharp changeover occurs because of the significant difference between the standard potentials of the two couples: 0.22 V and 0.80 V, respectively.

A very high degree of precision can be achieved with the technique; e.g. an error (95% confidence level) of c. 0.2%. Using an autotitrator with an autosampler the method is suitable for routine process analysis, and a wide range of cations (e.g. Ag(I), Au(I), Cd(II), Cr(III), Cu(I), Ni(II), Pb(II), Sn(II), Sn(IV), Zn(II)) and anions (e.g. Cl^-, CN^-) can be readily determined [9]. By choosing a suitable titrant and indicator electrode it is possible to achieve a greater degree of selectivity than by simple ORP measurements. However, just as in traditional titrations, other species can cause interference and, in addition, the method does not readily lend itself to on-line analysis. In some instances of potentiometric titration, an ion selective electrode can be used as the indicator electrode (e.g. Cd(II) with a copper ISE), and in this case, a direct potentiometric measurement could also be made.

7.2.3.3 Direct potentiometry.

7.2.3.3.1 Ions.   As has been discussed earlier (section 7.2.2.3), an ion selective electrode relies not on charge transfer between an electronic and an ionic conductor but on charge transfer across two phases which are essentially ionic conductors. The potential difference at the interface in

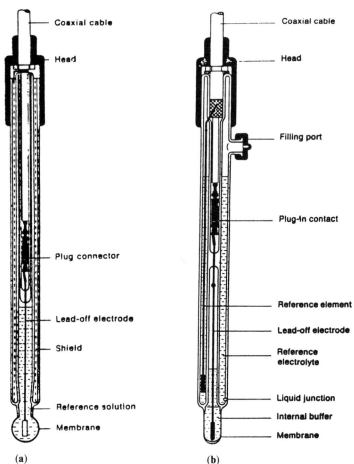

(a)   **(b)**

Figure 7.4 Schematic of internal designs of glass electrodes. (a) Simple electrode. (b) Combined electrode.

contact with the test solution arises because of the different mobilities of ions in the two phases leading to a concentration difference and hence a membrane or diffusion potential.

The best known and most widely used form of this type of electrode is the glass electrode. Typical designs are shown in Figure 7.4. The glass membrane consists of a matrix of negatively charged fixed sites into which protons, and a few small cations, can diffuse to give a space charge and an interfacial potential on contact with an aqueous solution. In the presence of water a hydrated layer, approximately 100 nm thick, is formed on the glass surface and sodium ions from this layer are exchanged for protons in solution:

$$H_3O^+(\text{solution}) + Na^+(\text{glass}) \rightleftharpoons H^+(\text{glass}) + Na^+(\text{solution}) + H_2O \quad (R7.2)$$

At pH < 11 this equilibrium shifts to the right until practically all exchangeable sodium ions in the hydrated layer are replaced by protons. It can be shown [10, Ch. 2] that the interfacial potential responds to proton activity according to the Nernst equation (equation (7.1)). At pH > 11, sodium ions are increasingly accepted into the hydrated ion exchange layer and the equilibrium shifts to the left. The membrane potential now no longer depends solely on proton activity. By using glasses with a high lithium content, the electrode can be made more specific for H^+ at higher pH values and the measurement range can be extended to a pH of 12 or higher. Alternatively, by using an aluminium-rich glass, the Na^+ response can be enhanced and a glass electrode then functions as a sodium ISE.

Measurements with glass electrodes are made in the usual potentiometric fashion with an external reference electrode and a voltmeter (Figure 7.5). The external reference electrode is often a SCE, with, more recently, silver–silver chloride electrodes being used. Because of the wide use of glass electrodes many of them have the reference electrode incorporated in the body of the glass electrode with a liquid junction to allow ionic contact via the test solution (cf. Figure 7.4(b)). The high resistance of the glass membrane means that it is essential that the measuring voltmeter must have a high input impedance, typically $> 10^{12} \, \Omega$, otherwise the current flow in the circuit can lead to significant errors in the measured voltage. Another source of voltage error is the asymmetry potential arising from the inner and outer surfaces of the glass membrane responding differently to identical solutions. This is usually corrected by a voltage adjustment during calibration with standard buffer solutions so that at pH 7 and 25°C the voltage output is zero. Temperature compensation to allow for the Nernstian temperature effect (cf. equation (7.1)) is also usually based on a common isothermal point at pH 7, and is achieved with the aid of a resistive network coupled with a thermistor. It should be noted that this temperature compensation does not take into

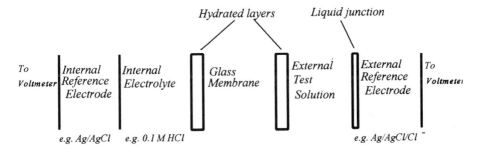

Figure 7.5 Schematic of circuit arrangement for pH measurement with a glass electrode.

account the effects of temperature on the solution characteristics; e.g. the degree of electrolytic dissociation.

Some very basic practical problems can also be encountered in using glass electrodes. Fouling of the glass membrane can cause significant errors in any pH measurement and often periodic cleaning is required; cleaning methods can range from simple manual wiping, through mechanical brushing, flow impingement, and chemical cleaning, to ultrasonic cleaning. Another problem commonly experienced with the electrodes is physical damage (e.g. cracking) to the glass membranes. The reference electrode can also experience fouling, particularly of the liquid junction, and even poisoning (e.g. by H_2S).

Notwithstanding these practical problems and the need for various corrections that have to be made, with careful and frequent calibration (to allow for changes in the glass membrane structure with time) precision and accuracy of better than ±0.02 pH units are routinely achievable. Also, with modern digital circuitry, the various compensations can be made automatically. These developments, together with (i) the fact that a wide range of forms of glass electrode (e.g. needle probes, flat probes, flow through electrodes, robust guarded membranes, ultrasonic deposit removing transducers) are available, and (ii) that no electron exchange is involved in the interfacial potential (measurements can thus be made in the presence of redox couples and common electrode poisons such as As, CN^- and organic species) have led to glass electrodes being used in more forms of process analysis than all other chemically based continuous process analysers combined [11, Ch. 13]. Applications range from industrial and waste water treatment, through pulp and paper, synthetic rubber, pharmaceutical and fertiliser manufacture, to power generation plant, and flotation processes in the mining industry. Benefits arising from pH monitoring have included greater processing efficiency, improved product quality, data acquisitions for evidence of regulatory compliance, and conversion from batch to continuous processing. We can illustrate some of these points by briefly describing one application of a glass electrode to process analysis.

Figure 7.6 shows schematically the use of two different types of insertion pH probes in a reactor system. A typical example of such a reactor could be a batch or continuous operation mode deep tank fermentor [12] for microbiological transformations such as the aerobic oxidation of D-fructose in the synthesis of vitamin C. Bacteria usually have an optimum pH range for growth between 6.5 and 7.4 and so monitoring and control is essential. The Ingold InFit® probes illustrated in the figure are capable of operating over a pressure range 0–6 bar and a temperature range of −30 to +130°C. They are also steam sterilisable. The advantages mentioned above are apparent in this example of in situ process monitoring of pH.

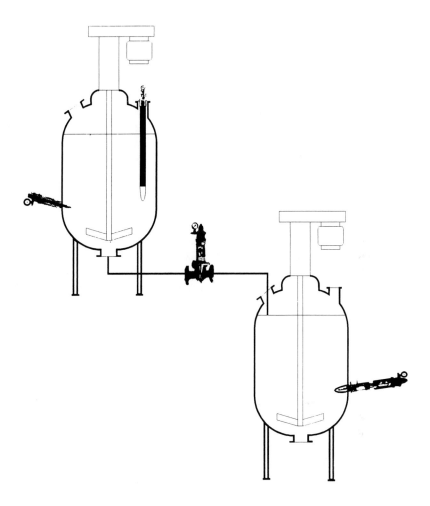

Figure 7.6 Schematic of the use of glass electrodes in a reactor system.

The glass electrode for pH measurement is thus a well established potentiometric sensor. A range of other solid state ISEs are also available. Variation of the glass composition can, as already indicated, allow Na^+ ions to be monitored (e.g. a glass with composition of 11% Na_2O, 18% Al_2O_3, 71% SiO_2 has a selectivity coefficient (equation (7.10)) of 2800 for Na^+ with respect to K^+ [10, Ch. 5], whilst other variations can provide selectivity for K^+, Li^+ and Ag^+, although H^+ ion response still usually dominates at low pH.

Membrane materials other than glass with fixed sites for ion exchange have been extensively described [10, 13, 14]. For example, silver halides are widely used for determination of Cl^-, Br^-, I^- and CN^-, metal sulphides (e.g. Ag_2S, CuS, CdS, PbS) for the determination of the corresponding cations, sulphide and other sulphur containing groups (e.g. thiocyanate, thiols, thiourea), and lanthanum fluoride for F^- measurement. In each case the principle of operation is similar to that of a glass electrode as is the mode of measurement (cf. Figure 7.5). The main differences arise from the lower resistance of the membrane material compared with a glass membrane and their generally lower selectivity compared with that of a glass electrode for protons. Nevertheless, they have been used extensively for a wide range of analyses, with many thousands of papers having been published on their applications. However, the great majority of these publications have been based on laboratory determinations. Problems of cross-interferences and long-term stability, which are more difficult to deal with in continuous on-line analysis than in discrete off-line analysis, have limited the widespread extension of such a simple direct method of analysis to on-site applications. In spite of this, though, as with the glass electrode, commercial systems have been produced for industrial monitoring, albeit often for end measurement in wet chemical analysers. Clevett [11, Ch. 13] lists some 10 or so manufacturers producing on-line specific ion monitors for anions such as halides sulphide and cyanide and cations such as Cu^{2+}, Cd^{2+}, Na^+, Pb^{2+}, Ca^{2+}, Mg^{2+} and Ag^+.

Other types of ISE rely not on inorganic solid state membranes but, as mentioned earlier, on organic molecules as the membrane material. Here the ion selective sites are mobile and so the ISEs based on these type of materials are known as liquid membrane electrodes. A good example is the K^+ ISE which uses valinomycin as the exchange medium. This is a 36-membered ring which has polar bonds directed into the centre while the outer non-polar envelope ensures lipophilicity of the aggregate. The dimensions of the cavity inside the ligand match closely the radius of the K^+ ion, ensuring the formation of a stable complex and a high selectivity. If valinomycin is incorporated into a high molecular weight matrix, such as PVC, the potassium ISE can be used for several months and this coupled with a good response time of less than one minute and a selectivity coefficient ($k_{K^+ Na^+}$) of better than 10^{-4} has allowed the routine determination of K^+ ions in a wide range of applications [13,15]. However, unlike solid state membrane ISEs, no significant non-laboratory based sensors have been reported. This is probably because of the lack of robustness and stability of the electrode.

Extensions of the principles of ISE to miniaturised electrode systems where the metal gate of a metal oxide field effect transistor (MOSFET) is replaced by a reference electrode, test solution and a chemically sensitive layer to produce a CHEMFET or, if it is specifically for ions, an ISFET have been widely reported on [16]. Still further extensions to include the incorporation of enzymes into the chemically sensitive layer (an ENFET) to provide sensitivity to biological substances have also been enthusiastically promoted [17]. However, all these types of electrode, whilst having been explored for process analytical and medical applications [18], have not yet really made it far out of the research laboratory to more practical situations.

7.2.3.3.2 Gases. A simple and elegant modification of the use of a glass electrode is as a probe for gases in the gaseous or liquid phase. This is achieved by covering the electrode with a thin film of a porous hydrophobic plastic such that the solution cannot penetrate into the pores. Between the inner surface of the plastic and the glass electrode there is a solution containing an inert electrolyte which also contacts the glass membrane and, through a liquid junction, the reference electrode. The gas being analysed (e.g. ammonia, carbon dioxide, chlorine) permeates the porous membrane, dissolves in the inert electrolyte solution, and establishes an equilibrium pH which, through a calibration curve, can be related to the concentration, or partial pressure, of the gas outside the membrane. One process analysis device based on this principle is a Sensidyne detector which allows the detection and measurement, with an alarm capability, of ammonia or hydrogen cyanide on up to twelve channels with a logarithmic range of 0–75 ppm and a precision of ±5% of full scale [11, Ch. 12]. Although a gas sensor of this type is very simple in principle, in practice difficulties can arise from, among other things, changes in membrane permeability (especially with temperature), slowness of response to step changes of external gas activity, and interference from other gaseous species. Despite these difficulties, however, the technique has interesting possibilities for process analysis applications.

A rather different type of sensor for direct potentiometric analysis of a gas is that based on the high temperature properties of zirconia. This material at temperatures above 600°C has an appreciable conductivity for oxygen ions over a range of oxygen pressures from above one atmosphere to about 10^{-20} atm. When it is incorporated into a cell of the type

Zirconia
$p'O_2$, Pt | ZrO_2 | Pt, $p''O_2$

then under equilibrium conditions the open circuit voltage (E) of the cell indicates the ratio of oxygen partial pressures at the electrodes according to the equation.

$$E = \frac{RT}{4F} \ln \frac{p''_{O_2}}{p'_{O_2}} \qquad (7.12)$$

where R, T and F have their usual meanings. The oxide conducting zirconia is acting as a solid state electrolyte equivalent to an ionic liquid electrolyte in more conventional electrode systems. The reason why a more traditional type cell with a liquid electrolyte cannot be used in the same manner for monitoring oxygen is that the O_2/OH^- couple in aqueous solutions is very irreversible (i.e. the interfacial resistance to electron transfer on Pt and other noble metals is high) and so it is difficult to achieve a true and reproducible equilibrium and, hence, a reliable value of the electrode potential. The O_2/O_2^- couple at elevated temperatures, on the other hand, is reversible so equilibrium at the electrode/electrolyte interface is readily established. Then if p''_{O_2} is considered to be a time independent and known reference (e.g. air with $p'_{O_2} = 0.2$ atm) the zirconia cell can indicate the unknown or variable p'_{O_2} with high accuracy.

Figure 7.7 illustrates schematically a zirconia electrolyte system. A pair of short (*c.* 1 cm) porous Pt electrodes (III and IV) are placed downstream and form the open circuit cell for the continuous monitoring of the oxygen partial pressure in the test gas. A pair of longer (*c.* 10 cm) porous Pt electrodes (I and II) can serve as an upstream 'pumping' cell. This pumping cell can be used either to generate known concentrations of oxygen coulometrically, in order to calibrate the system, or to pump residual oxygen out of an inert carrier gas; for the latter purpose it is possible under favourable conditions to reduce the residual oxygen level to *c.* 10^{-20} atm. The constriction mixer is to ensure good gas mixing just prior to monitoring and the thermocouple allows measurement of the monitoring temperature (*cf.* equation (7.12)).

An example of the application of a potentiometric zirconia oxygen sensor is in furnace firing for ethylene production by cracking of ethane

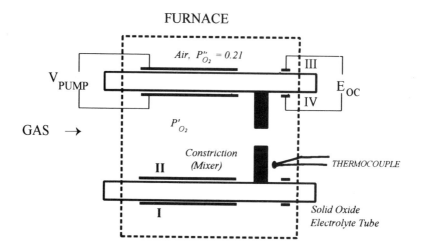

Figure 7.7 Schematic of a zirconia oxygen sensor.

at 800°C. A typical plant has a number of furnaces (e.g. 5–10) each using about 50 MW of energy. Clearly there is a strong incentive to optimise energy usage. One way of wasting energy is to use too much oxidant (i.e. O_2) since this has to be heated up and if it is unused, it is then released into the atmosphere. Monitoring of oxygen in the flue gas effluent with feedback control can provide cost efficient operation of the furnaces. Typical commercial examples of zirconia cells used in this context are produced by Westinghouse in varying probe lengths (0.45–3.7 m) for different stack diameters. An interesting feature of these cells is the gas diffuser upstream from the electrodes to filter out particulates which might macroscopically contaminate the electrodes.

7.3 Amperometry

7.3.1 Principles of measurement

Whereas the sensitivity of a potentiometric sensor is very much limited by the Nernstian response of the system under test (i.e. $60n$ mV for a 10-fold change in analyte concentration), amperometric devices do not suffer from this and exhibit a much greater sensitivity to the change in analyte concentration. For example, a redox reaction involving a two-electron transfer will only result in a 29 mV change for a 10-fold change in the analyte concentration at 25°C, half that for a single electron transfer reaction. But with amperometric devices, the current resulting from such an oxidation and for such a concentration change will be about twenty times that for a single electron transfer reaction.

The current flowing through an electrolysis cell is directly related to the amount of reaction taking place at the electrode surfaces. We can apply Faraday's Law which simply relates the amount of charge passed to the number of moles (or mass W) of a substance of molecular mass M liberated or reacted at an electrode during electrolysis. This can be written in the following form:

$$W = \frac{Q \times M}{n \times F} = \frac{J \times t \times M}{S \times n \times F} \tag{7.13}$$

where Q is the charge per unit electrode area (anode or cathode) consumed in the process. As the equation shows, the charge density Q is equivalent to a current density j (the ratio of the total current J to the electrode area S) flowing in the external circuit over a period of time t. Thus, assuming there is no other reacting species in the solution, the current density will be directly proportional to W, and, hence, to the analyte concentration. Equation (7.13) is also used as the basis for coulometric process analysis (section 7.3.4.5).

The current flowing through an electrolysis cell can also be related to the energetic driving force, the electrode potential, E. If we assume that the analyte is not directly adsorbed onto the electrode surface, the reacting species must move from the bulk of the test solution to the electrode surface where the electron transfer reaction occurs. The former process is a mass transport phenomenon and the latter is determined by kinetics of the electron transfer reaction at the electrode surface. The various regions of an electroactive species undergoing reduction or oxidation at an electrode surface are shown in the form of a J–E plot, as illustrated in Figure 7.8. At low overpotentials (where the overpotential η is defined as the difference between the applied potential and the Nernstian equilibrium potential), the current is determined by the kinetics of the electron transfer step and increases exponentially with E. With further increases in E, both mass transfer and kinetics control the rate of the electrode reaction and eventually, at high enough values of E, the reaction becomes totally mass transport controlled.

Mass transport in a solution is governed by three factors: migration, diffusion and convection. The effect of migration of the reacting species (the movement of charged species under an electric field gradient) is normally suppressed by the addition of an inert electrolyte at a concentration greatly in excess of the reacting species. Naturally, if the analyte is not charged its migration becomes insignificant. Diffusion arises from a spatial concentration difference in a solution with species moving from

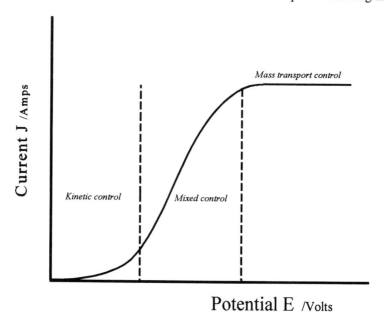

Figure 7.8 Steady-state current–voltage curve for an electrochemical reaction.

high to low concentration regimes until concentration uniformity is achieved. Convection is more difficult to define precisely as this depends on the relative movement of the solution with respect to the electrode surface as well as on the geometry of the cell used for the analysis. In practice, reproducible convection is only achieved when a well defined cell and elecrode geometry is employed, such as a that of a rotating disc electrode (RDE) [19] or a wall-jet electrode [20] (cf. section 7.3.2.3).

As is apparent from Figure 7.8, the current at an electrode surface can be controlled by the kinetics of the electron transfer step, by mass transport of the analyte to the electrode surface, or by a combination of these two processes. As can also be seen from Figure 7.8, the largest current is obtained from the mass transport controlled region; this is known as the *limiting current*. However, in order to achieve this, not only must convective mass transport be reproducible, but also higher overpotentials must be applied to the interface. A consequence of the latter requirement is that secondary reactions (e.g. solvent decomposition) may then start occurring at a significant rate. The current from this secondary reaction will thus have to be decoupled from that of the primary reaction of analyte reduction/oxidation. The degree of precision with which the measurement of the analyte concentration can be carried out will therefore depend on how effective this separation of primary and secondary reactions is. Usually a limiting current plateau greater than c. 100 mV long is sufficient to allow it to be used for process analytical applications.

7.3.2 Methods of measurement

7.3.2.1 Steady-state measurements. In this technique, the electrode of interest (the *working electrode*) is held at a constant potential and the steady-state current is recorded. The cell used for this purpose usually consists of the working electrode made of a suitably catalytic and stable material for the reaction concerned, an *auxiliary electrode* to complete the electrical circuitry and a third electrode, the *reference electrode*, which, via a potentiostat, precisely controls the potential applied to the working electrode–solution interface (Figure 7.9). For a kinetically controlled reaction, the current will be determined by a limiting form of the *Butler–Volmer* equation [21, Ch. 3] known as the *Tafel equation*. An example is

$$j = = j_0 \exp \frac{(1 - \alpha)nF\eta}{RT} \tag{7.14}$$

where the overpotential η is positive. In this equation j_0 is the *exchange current density* and it is defined as

$$j_0 = nFk^0 C_{Ox}^{(1 - \alpha)} C_{Red}^{\alpha} \tag{7.15}$$

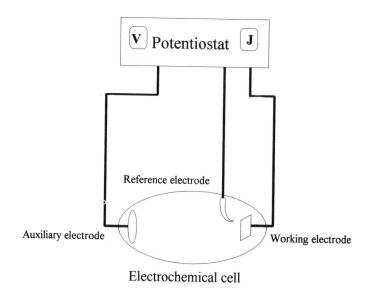

Figure 7.9   Block diagram of cell design and instrumentation for amperometric measurements.

where k^0 is the standard rate constant for the reaction and α is the cathodic charge transfer coefficient. The latter is the fraction of the energy supplied which goes into reducing the activation barrier for the cathodic reaction. In principle, since equation (7.15) contains concentration terms then equation (7.14), relating current and potential can be used for analytical purposes. In practice, since the relationship is for heterogeneously controlled kinetics the value of j depends very much on careful control of the electrode potential and on surface catalytic activity which can be dramatically affected by impurities. Therefore this region of current–voltage curve is not very useful for process analysis.

For control by mass transport, on the other hand, the current, known as the *limiting current*, j_L is given by:

$$j_L = nFk_L C\,' \qquad (7.16)$$

where k_L is the mass transport coefficient for the analyte in that particular cell and $C\,'$ is the concentration of the analyte in the bulk of the solution. This situation arises when sufficient energy (i.e. overpotential) is supplied to the interface so that all the analyte reaching the surface is completely oxidised or reduced. Further reaction can then only be maintained by the transport of analyte from the bulk solution to the interface. There will therefore exist a concentration gradient of the analyte which extends into the solution over a distance known as the *boundary layer*. When there is no forced convection, mass transport through this region will be given by Fick's First Law of Diffusion and $C`$ can be readily related to j_L in theory as well as in practice. When convection is present, however, the relationship of the concentration of the analyte to j_L is usually more complicated and cannot always be described theoretically. Nevertheless, an unknown concentration can still be readily obtained from a calibration plot of j_L vs C. It is this region of the current–voltage curve which is most useful for steady-state measurements in process analysis.

7.3.2.2 Non-steady-state measurements. A very wide range of non-steady-state electrochemical measurements have been developed for analytical purposes [21, Chs 5 and 6], but from the point of process analysis only one technique is of real interest. This is that based on chronoamperometry. In this technique, the potential is instantaneously changed from a region of no net reaction to one where a net reaction occurs, usually under diffusion control; i.e. the plateau region of Figure 7.8. The system response is characterised by a decaying current transient which is composed of two effects, namely, a current due to the electrical double layer charging and the Faradaic current due to the electrochemical reaction. The Faradaic current density is described by the Cottrell equation

$$j = \frac{nFD^{1/2}C`}{\pi^{1/2}t^{1/2}} \tag{7.17}$$

from which it can be seen that the current decays according to $t^{-1/2}$. D is the diffusion coefficient of the electroactive species. On the other hand, the charging current for the capacitor of the double layer decays exponentially with time according to:

$$j = \frac{E_{\text{Step}}}{R_S} \exp\left(-\frac{t}{R_S C_{\text{DL}}}\right) \tag{7.18}$$

Using typical values for the double layer capacitance, C_{DL}, of $40\ \mu F\,cm^{-2}$ and a solution resistance R_S of 50Ω, we can deduce that for a potential step, E_{Step} of 0.5 V, the charging of the double layer will be 95% complete after a time $t = 9 \times 10^{-3}$s. The Faradaic component of the current can therefore be readily obtained from the current measurement at any time beyond that. For optimum sensitivity, the current is sampled *c.* 50–60 ms after the application of the potential step. The contributions of these two effects are shown in Figure 7.10.

One major advantage of this type of non-steady-state technique is that the concentration of the analyte is measured at a time where the rate of mass transfer to the electrode surface is still very high. Consequently, the current will also be high and the sensitivity to the analyte concentration will therefore be greater than for the steady-state methods. Other advantages will become apparent in the discussion in section 7.3.4.

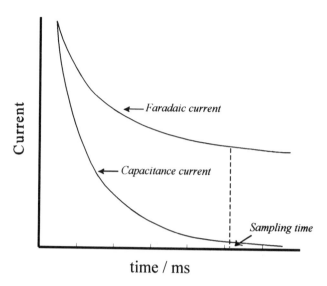

Figure 7.10 Double layer charging and Faradaic current as a result of a potential step to an electrode–solution interface.

7.3.2.3 Hydrodynamic voltammetry. This technique can be viewed as a particular extension of a general steady-state measurement but in a flowing stream of solution since it involves amperometric measurements carried out at constant potential under well-defined hydrodynamic conditions. The purpose of this well-defined hydrodynamic condition is to produce an equally well-defined boundary layer so that the mass transport coefficient is reproducible and equation (7.16) can be applied. The best known example of this type of electrode system is the rotating disc electrode. The disc electrode is spun at a typical rotation speed of 20 Hz and the solution close to the electrode surface is thrown out centrifugally. This solution is replaced from below the disc at a constant rate, and it can be shown [19, Ch. 2] that the limiting current is proportional to the square root of the rotation speed.

An example of a more practical device of this type of electrode is that based on the wall jet system. In this set-up, the test solution is introduced via a nozzle (with a diameter *c.* 0.1 mm) and hits the disc electrode surface normally, where it scatters. The mass transfer conditions here are well characterised and the limiting current arising for the oxidation or reduction of the analyte at the electrode surface can be expressed as

$$j = 0.898\, nFC^` D^{2/3}\, v^{-5/12} d^{1/2}\, V_{M}^{3/4} \tag{7.19}$$

where v is the kinematic viscosity of the solution, d is the diameter of the inlet and V_M is the volumetric flowrate. For analysis the important point is that there is a direct linear correlation between j and C'. The advantages of the wall jet system over the rotating disc system is that there are no moving parts, there is a small dead space, and a relatively high fraction ($c.$ 7%) of the reactant that passes through the jet is destroyed. Advances in design have allowed rapid sampling rates to be achieved under computer control and for alternative sampling of the carrier stream and sample to be carried out, followed by subtraction of the signals to give detection limits at the nanomolar level [22]. The method is equally applicable to the determination of both inorganic and organic compounds undergoing either oxidation or reduction [23]. An example is given in section 7.3.4.4.

7.3.3 Coulometry

This technique stems directly from equation (7.13) and is effectively an extension of amperometric measurements in that the total charge passed during the reaction is measured. Experimentally, this is carried out by integrating, through the use of operational amplifiers or digital circuitry, the current flowing through the electrochemical cell. Since the measuring system is unable to differentiate between the double layer current and the Faradaic current, this imposes three important criteria on the use of charge measuring devices or coulometers in the determination of the analyte concentration: (i) the reaction must be of known stoichiometry; (ii) the analyte must undergo one single reaction only or should have no side reactions where the stoichiometry is different to (i); and (iii) the reaction must proceed with close to 100% efficiency. If these conditions are satisfied, then the analyte concentration can readily be deduced from the application of equation (7.13). Two important points to note are, first, that all the terms in this equation are either known or readily measured and so no calibration is needed, and, second, that none of the terms has any temperature dependence. Process analysis applications are considered in section 7.3.4.5.

7.3.4 Applications

7.3.4.1 Introduction. The range of electrochemical techniques based on current measurements is, as can be concluded from the section 7.3.2, rather wider than those based on potential measurements. However, the range of applications of amperometry is just as limited, if not more so, as the range of applications of potentiometry. The reason for this is probably common to both techniques and has already been discussed in the context of potentiometry; namely, the poisoning of the indicator or working electrode in all but the cleanest test environments. Most electrochemical sensors, be they potentiometric or amperometric, rely on electron transfer to or from a heterogeneous catalyst and any loss of activity of that catalyst as a result of contamination will affect the electrochemical response. It is noteworthy in this context that, as has been mentioned earlier, the most widely used potentiometric sensor in process anlaysis is one which does not involve electron transfer but rather ion exchange; i.e. a glass electrode for pH measurement. The application of amperometry to process analysis is also dominated by one type of measurement in which electrode poisoning is minimised. This is the use of sensors for gas analysers where the electrode system is protected by a gas permeable membrane from the test environment. This type of sensor is described in the next section.

7.3.4.2 Membrane-covered gas sensors

7.3.4.2.1 Oxygen. Figure 7.11 shows a schematic diagram of a membrane-covered amperometric oxygen detector [24, Ch.4]. The detector

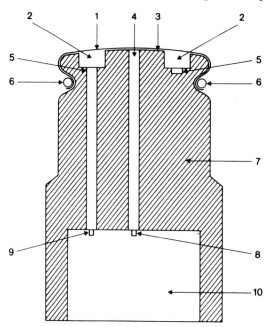

(1) = Membrane; (2) = Electrolvte reservoir; (3) = Land between cathode

Figure 7.11 Schematic diagram of a membrane-covered amperometric oxygen detector. (1) = Membrane; (2) = electrolyte reservoir; (3) = land between cathode and electrolyte reservoir containing thin film of electrolyte solution; (4) = cathode; (5) = anode; (6) = O-ring; (7) = body of detector in insulating material; (8) = connection to cathode; (9) = connection to anode; (10) = space filled with insulating material.

consists of a plastic body into which are incorporated two electrodes. The central electrode is an inert noble metal, such as gold or platinum, which acts as a cathode at which oxygen is reduced.

$$O_2 + 2H_2O + 4e^- \rightarrow 4OH^- \quad \text{(R7.3)}$$

The outer electrode, the anode, is usually either silver or lead. With silver, a voltage has to be applied to the two electrodes (typically $c. -0.8$ V to the cathode with respect to the anode) corresponding to the limiting current regime (cf. Figure 7.8) and the cell operates in a steady-state amperometric mode. This type of oxygen sensor is also often known as a polarographic detector. The electrolyte in the cell usually contains chloride ions and so the anodic reaction is

$$Ag + Cl^- \rightarrow AgCl + e^- \quad \text{(R7.4)}$$

If lead is used as the anode with an alkaline electrolyte the anodic reaction is

$$Pb + 2OH^- \rightarrow PbO + H_2O + 2e^- \quad \text{(R7.5)}$$

The overall cell reaction in this case

$$2Pb + O_2 \rightarrow 2PbO \quad \text{(R7.6)}$$

has a negative free energy change and so the oxygen reduction is driven thermodynamically without the need for an external applied voltage. In this method the sensor is operating in what is known as a galvanic mode. This type of sensor is also known as a fuel cell detector. Much has been made in the literature of the differences between an amperometric and a galvanic sensor, but really the only significant difference arises from the operation of the two cells at low oxygen concentrations as is discussed further below. The electrolyte contained in the cell depends, as just mentioned, on the anode material. However, since the reduction of oxygen generates OH^- ions (reaction (R7.3)) the pH of the electrolyte tends to rise, especially in the narrow electrolyte film above the cathode. Thus, for an amperometric sensor, OH^- ions are often added as well as Cl^- ions to adjust the pH to $c.$ 13 [25]. The membrane which separates the electrodes and electrolyte solution from the test environment is usually 10–50 μm thick, and it serves to protect the electrodes from gross contamination by species in the environment and also maintains reproducible conditions for electrolysis. It has a high permeability to gases in general and oxygen in particular (typical materials for the membrane are PTFE, FEP and silicone rubber), and it is held in position so that it is taut over the cathode with a suitable fixing (e.g. an 'O' ring). In this way a thin layer of electrolyte (5–10 μm thick) is sandwiched between the membrane and the cathode.

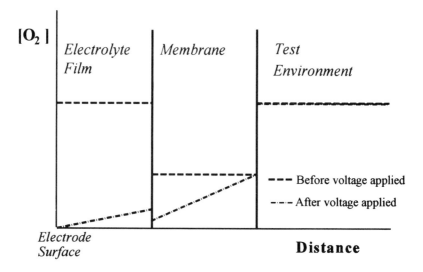

Figure 7.12 Concentration profiles for a membrane-covered amperometric oxygen detector.

The principle of measurement is one of steady-state amperometry, but because of the fixed diffusion layer thickness it could also be regarded as analogous to a type of hydrodynamic voltammetry. Referring to the schematic diagram in Figure 7.11, and also the plot of concentration as a function of distance shown in Figure 7.12, oxygen from the test environment enters the membrane, dissolves in it and diffuses through into the film of electrolyte over the cathode. As already mentioned, the cathode either has a negative voltage applied to it or else it is thermo-dynamically negative with respect to the anode. Either way, the energy of the cathode surface is sufficiently great to reduce all of the oxygen reaching it and this results in a current flow which is proportional to the concentration, or more strictly the partial pressure [26], of the oxygen in the test environment. It may be readily shown [24, Ch.4] that the current (J) is under mass transport control and can be described by the following equation:

$$J = nFSP_m C_e/b \qquad (7.20)$$

where n is the number of electrons involved in oxygen reduction, F is Faraday's constant, S is the cathode area, P_m is the permeability coefficient of the membrane for oxygen, C_e is the oxygen concentration in the test solution, and b is the membrane thickness. This deceptively simple relationship has been the subject of considerable study [24,27], and many complications can arise in applying it to oxygen measurement. Nevertheless, membrane-covered probes can, with care, provide reliable and reproducible results in a wide variety of situations. From the point of view of process analysis these include (i) corrosion control for enhanced oil recovery and for boiler feed waters; (ii) monitoring the atmosphere in contact with oxygen sensitive chemicals during manufacture, storage and transportation; (iii) evaluating the explosion risks in volatile and combustible environments [11, Ch. 7]; (iv) contributing to savings in aeration power consumption in activated sludge treatment plants [28]; (v) alarms for personnel working in hazardous environments; [29] and (vi) control of colour and taste of food and drink products [30]. As with the use of potentiometric analysers, we choose just one application to illustrate the use of a membrane-covered amperometric sensor for oxygen evaluation. In this instance it is for a brewery [31,32].

Oxygen monitoring is of importance to a brewer for two main reasons. First, many substances in beer are readily oxidised, particularly during the high temperature boiling, pasteurisation and sterilisation stages of production. Oxidation occurring during these process steps can lead to detrimental changes in taste and clarity of the final product. On the other hand, high oxygen levels are necessary in the early stages of fermentation to promote yeast growth. The use of in-line and off-line dissolved oxygen sensors for breweries has been investigated [31,32]. An Orbisphere Laboratories amperometric system with a gold cathode gave reliable and reproducible results with a more than adequate sensitivity, typically better than 1 µg of oxygen per kg of solution. Furthermore, the sensor did not show any significant effects arising from pressurised carbon dioxide. Again the usefulness of an electrochemical process analyser is apparent. Clevett [11, Ch. 7] gives a résumé of the operating characteristics of a range of commercially available on-line dissolved oxygen sensors.

It has been pointed out above that membrane-covered amperometric gas sensors are examples of both steady-state amperometry and a type of hydrodynamic voltammetry. They can also illustrate the third method of measurement discussed in section 7.3.2, namely non-steady-state measurements. The Cottrell equation discussed in section 7.3.2.2 and as given in equation (7.17) can be considered as applying to a membrane-covered oxygen sensor for $t < c.\,0.1$ s and if it is assumed that the distribution coefficients at the test solution/membrane and electrolyte layer/membrane interfaces are the same [24, Ch. 4]. For any electrode system, one of the advantages of measuring the non-steady-state current as described by equation (7.17) is that a high current at short times gives increased sensitivity. For an oxygen sensor further advantages accrue because at short times the current is independent of the membrane. Thus, any changes in the properties of the membrane and any deposits on the membrane will have no effect on the measurement. In addition, the temperature coefficient of the detector is now no longer dependent on the permeability coefficient of oxygen in the membrane but on its diffusion coefficient in the electrolyte, and this is a much better behaved parameter. The temperature correction, for example, can be done automatically and with more confidence. Also, since the diffusion layer has not spread to the outer surface of the membrane while a measurement is being made, there is no influence of solution flow on the reading obtained. Stirring of the test solution is thus not necessary. Finally, a measurement can be made very rapidly and so, the overall life of the detector should be increased. The membrane and the problems associated with it have always been a major preoccupation for those working with and trying to understand membrane-covered oxygen sensors. Simply by making measurements at short times, all concern about the membrane can be removed, other than the fact that it must still be there to protect the electrodes against gross pollution. There are, though, some disadvantages to non-steady-state measurements which have inhibited the development of the technique to practical applications. The major drawbacks are that current transients are generally slower than expected on the basis of equation (7.17) and there is an even slower return to equilibrium after the application of the voltage pulse and a return to open circuit. These effects have been investigated and some improvements have been made in detector response and reproducibility [33], and one could expect to see greater use made of chronoamperometry for the monitoring of oxygen and other gases with membrane-covered detectors.

Before passing on to other gas sensors we return to the point made earlier about there being only one significant difference between amperometric and galvanic oxygen sensors. This primarily arises in the context of the background current; i.e. corresponding to zero oxygen concentration. For the amperometric sensor the negative voltage to the cathode can also reduce water:

$$2H_2O + 2e^- \rightleftharpoons H_2 + 2OH^- \qquad E^0 = -0.83 \text{ V vs SHE} \qquad (R7.7)$$

If, as is likely, the partial pressure of hydrogen (p_{H_2}) is less than 1 atm, then the equilibrium potential (E_{eq}) for this reaction will be more positive than E^0 and an applied potential of c. -0.8 V vs SCE (or c. -0.6 V vs NHE) can be more negative than E_{eq} so that water will then be reduced at the cathode. This gives rise to the background current often observed in amperometric sensors at low partial pressures of oxygen. For a galvanic sensor in the absence of oxygen the overall reaction will be a combination of (R7.5) and (R7.7).

$$Pb + H_2O \rightarrow PbO + H_2 \qquad (R7.8)$$

which, under standard conditions, will have a value of E^0_{Cell} of -0.26 V and thermodynamically the reaction cannot take place since the free energy change is positive. If, however, $p_{H_2} < 1$ atm, the equilibrium cell potential will move in a positive direction and when the conditions are such that $E_{Cell} > 0$, then water reduction will take place at the cathode and a background current again results. The relative size of the background currents in amperometric and galvanic sensors will depend on a variety of factors, but for identical conditions of pH, p_{H_2} and cathode material, the determining factor will be, in the case of the amperometric sensor, how far negative the applied potential is in relation to the equilibrium potential for (R7.7). For the galvanic sensor, it will be how positive the equilibrium cell potential is. In general, galvanic sensors give lower background currents than amperometric sensors, but they will not necessarily be zero, as is implied by many authors.

7.3.4.2.2 Other gases. A similar principle to that used for oxygen measurement with membrane-covered amperometric sensors has been applied for monitoring a number of other gases. In some cases the process involves, as for oxygen, electrochemical reduction (e.g. Cl_2, NO_2, H_2) but there are also a number of anodic processes (e.g. CO, SO_2, HCN, NO). In the latter case where the sensor is based on galvanic rather than amperometric operations, it is essentially functioning as a fuel cell. Examples are CO and H_2S measurement by anodic oxidation at a working electrode with oxygen reduction at the counter cathode; e.g.

Working anode $CO + H_2O \rightarrow CO_2 + 2H^+ + 2e^-$ (R7.9)
Counter cathode $1/2O_2 + H_2O + 2e^- \rightarrow 2OH^-$ (R7.3)
Overall reaction $CO + 1/2O_2 \rightarrow CO_2$ (R7.10)

For both galvanic and amperometric devices, the construction of the electrochemical cell can be very similar to that shown in Figure 7.11 for an amperometric oxygen sensor. However, because often there is not a reversible couple at this counter electrode (*cf.* reaction (R7.3) above) an additional reference electrode has to be introduced into this cell (Figure 7.13). Nevertheless, the mode of operation of the gas sensor is essentially as described earlier. Some commercial gas sensors, though, make use of a gas porous membrane instead of a gas permeable membrane (*cf.* Figure 7.13). The latter type of membrane, as mentioned above, requires the gas to dissolve in the plastic and to diffuse through into the inner thin electrolyte film. A gas porous membrane, on the other hand, whilst also being made of a polymeric material such as PTFE, consists of an open network of micropores such as is commonly used in fuel cells [34–36]. Gas entering the membrane therefore does not have to dissolve in the membrane material but can transfer from the test environment to the inside of the electrochemical cell in the gaseous phase through the pores. For a gaseous test environment, provided the average pore diameter is greater than the mean free path of the gas molecules, then diffusive transport in the membrane is Fickian and will be very similar to that

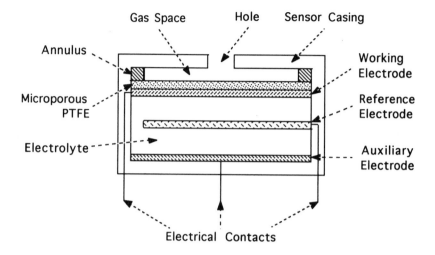

Figure 7.13 Schematic diagram of a porous membrane-covered amperometric sensor.

which applies in the environment close to the external surface of the membrane. In the case where the molecular mean free path is comparable or greater than the pore size, then Knudsen diffusion becomes important and transport is no longer strictly comparable to that in the free gas phase [37]. For Fickian diffusion in a porous membrane, one might therefore expect that the overall mass transport characteristics would be simpler than the case of diffusive transport in a solid membrane. However, there are additional complications. One complication is that in common with fuel cell electrodes the electroactive catalyst is often in intimate contact with the porous membrane. For example, a fuel cell electrode described by Niedrach and Alford [38] consisted of a mixture of platinum black and PTFE and this is essentially the same form as the combined gas membrane and electrode used in many gas sensors [39]. In addition, the Niedrach and Alford electrode and gas sensor electrodes form part of a layered structure with the combined electrode/membrane being compressed together with another porous sheet on top of it. The resulting overall structure then consists of a system of macro and micro pores of varying lengths and ill-defined diameters [36]. The charac-terisation of such a porous media is difficult and is complicated even further since porous capillaries will not be straight but will follow a tortuous path. These will, furthermore, not necessarily interconnect with other pores, and may also be interspersed with larger voids.

An additional complicating feature of a porous gas membrane arises from the penetration of the cell electrolyte into pores of the catalyst membrane. The top sheet of porous PTFE mentioned above, prevents complete seepage of the liquid electrolyte through the combined membrane into the test environment, partly because of the hydrophobic nature of the PTFE and partly because of the small diameter pores generally found in the outer membrane. For the inner part of the membrane, though, the presence of electrocatalyst particles means that here the membrane is less hydrophobic and pore sizes are larger than in the outer part so that electrolyte solution permeates the porous medium. However, this permeation is such that the pores are not flooded with electrolyte. Rather, away from the inner surface of the catalyst membrane, the pores remain filled with gas and there is only a thin layer of electrolyte covering the catalyst [40]. This results in much higher currents than if the pores were flooded with electrolyte since in that case the reactive gas would have to diffuse through a substantial thickness of electrolyte before it could react at a catalyst site. Nevertheless, the presence of the electrolyte film over a catalyst site means that in this type of membrane sensor the mass transfer of electroactive gas molecules is complicated by transport at a three-phase boundary between gas, liquid and solid. The overall electrochemical process, therefore, in a porous membrane-covered gas sensor is somewhat different from that of a permeable membrane-covered gas sensor and this is reflected in a number of operating characteristics. For example, the current output of a gas porous membrane sensor will be much greater than that from a gas permeable membrane sensor with the same geometric area exposed to the gas in both cases. This is because of, amongst other factors, the much larger electrocatalytic area in the former case. Also because of the essentially different mechanism involved in getting the gas from the external environment to the working electrode, the temperature and pressure characteristics of both type of sensor are rather different [41]. Some features of sensor operation such as calibration and cross-sensitivity remain much the same for the two sensor types. Little has been published on comparisons between these two different types of sensor, but an awareness of the different *modus operandi* can provide valuable insights into what may otherwise appear to be some very disparate specifications from different manufacturers and suppliers of gas sensors.

A major application of amperometric gas membrane covered sensors in process industries is for monitoring ambient air, especially for regulatory control and stack emissions. Examples of commercially available analysers for SO_2 and Cl_2 have been described by Clevett [11, Ch. 12]. An example of a range of sensors for toxic gases such as CO, H_2S, SO_2, Cl_2 and NO_2 is the EXOTOX system produced by Neotronics Ltd [29]. In addition to instantaneous monitoring of the gases, the sensors will monitor both the long-term exposure limit (TWA) and short-term exposure limit (STEL). If any of the measurements exceed the recommended national level both audible and visual alarms are initiated.

Examples of applications are for protection of workers exposed to CO at fossil fuel fired power stations, road toll booths, steel works or within mines, and where H_2S is a danger, for instance, in sewage works, oil production and processing plants.

7.3.4.3 Gas sensors without membranes. So far all of the amperometric gas sensors we have described have been based on membrane-covered devices. A case where the membrane is dispensed with is in a high temperature combustion analyser based on the zirconia cell described earlier in section 7.2.3.3.2. The construction of the cell for amperometric application is essentially the same as that for potentiometry, but now oxygen is reduced at the test gas side under diffusion controlled conditions and a cathodic current proportional to oxygen partial pressure is produced. If the oxygen level falls and combustible gases come into contact with the working electrode an anodic current flows and oxygen diffuses from the reference side of the cell to the test side to combine with the combustible gases. In this way, the source is able to monitor combustion conditions on either side of the stochiometric point. Here, the condition is that there is no net current, but with either excess oxygen or excess combustibles, a cathodic or anodic net current, respectively, flows.

Other examples of membraneless amperometric gas sensors are residual chlorine analysers [11, Ch. 13]. Several methods of operation are used, but one which is particularly interesting is that of Fisher and Porter since here the gold working electrode is rotated at 1550 rpm to provide a well defined hydrodynamic boundary layer. In most of these process analysers the chlorine is not monitored directly but rather indirectly by liberating I_2 or Br_2 from I^- or Br^- solutions, as in the case of the potentiometric chlorine alarm mentioned earlier (section 7.2.2.4). The more reversible I_2/I^- and Br_2/Br^- couples are easier to handle electrochemically since they are less susceptible to loss of electrode activity by poisoning. If electrode poisoning is found to be a problem with particular test samples various methods of in situ electrode cleaning have been suggested and applied. An example is that of the Wallace and Tiernan residual chlorine analysers which have grit added to the cell and this is carried by the flow of the sample solution and bombards the electrodes to prevent build up of contamination [11, Ch. 13; 7, p.6.103]. Another example from the same company is the use of hydrochloric acid to clean the measuring electrode, while the Capital Controls Co. use the continuous action of PVC balls agitated by a motor driven rotator to keep the cell electrodes clean.

7.3.4.4 Biosensors. Process analysis is often thought of solely in the context of industrial systems, but, as has been indicated in the discussion of potentiometry, an increasingly important area of on-line monitoring is for medical and biochemical applications. Unlike the use of potentiometric sensors in this area, though, amperometric sensors have not been used for main group cations but rather for biological molecules. This is not always easy, however, because usually conventional metallic electrode surfaces do not show suitable electrochemical kinetics with complex biological molecules, such as proteins. In order to get rapid and reversible electrochemistry of biomolecules it is often necessary to modify electrode surfaces with a mediator for electron transfer. Perhaps the most successful example of a sensitive, selective and stable modified electrode is the glucose electrode [17, 42] which incorporates ferrocene as a mediator between the electrode and an immobilised layer of the enzyme glucose oxidase. The enzyme oxidises glucose in the test sample and is itself reduced. The ferrocence returns the enzyme to an oxidising state and is then reactivated by electron transfer from the underlying metal electrode. This type of sensor does not consume any reagents and so can function for long periods. In a practical version of this type of electrode a non-steady-state measurement is made with the electrode potential being stepped from open circuit, or an equilibrium value, to a value where the rate of the electrochemical process is determined by the diffusive mass transport. This gives a current that varies inversely with the square root of time, corresponding to the increase of thickness of the diffusion layer as the electroactive species is concerned (*cf.* section 7.3.2.2). The current at a given time after the start of the transient is directly proportional to the concentration of reactant (equation (7.17)). A linear calibration can be obtained with a glucose electrode for blood samples with a glucose content over the concentration range 2–30 mM of glucose, the range relevant to diabetics. A commercial product based on this principle is available for diabetics to monitor their own blood glucose levels and allow a determination of the size of insulin injection required. This is, effectively, a process analyser with limited feedback control. A closed-loop artificial pancreas consisting of an automatic continuous monitor of blood glucose level and an automatic injector of insulin coupled with a feedback system has great potential for prevention of diabetic complications [43] and is the next stage of glucose sensor process analysis development. In fact, this type of closed-loop artificial pancreas is already being clinically tested in laboratories and hospitals.

Another form of glucose sensor combines the use of glucose oxidase with a membrane-covered amperometric oxygen detector. Here the enzyme reaction is

$$\text{Glucose} + \text{O}_2 \xrightarrow[\text{oxidase}]{\text{Glucose}} \text{H}_2\text{O}_2 + \text{Gluconic acid} \quad \text{(R7.11)}$$

The decrease of oxygen reduction current as measured by the oxygen sensor, and arising from the consumption of oxygen in the glucose oxidation, is an indication of the glucose concentration. Clearly, this indirect type of sensor is very different from that based on the direct ferrocene mediated device. As mentioned, the role of the ferrocene was to regenerate the enzyme, but that is not done here and so the sensor can only be used once, which has the advantage of avoiding the risk of contamination if multipatient use is intended. Another significant difference between the oxygen sensor and ferrocene-based systems is that the former is usually operated in a steady-state mode, whereas the latter is based on chronoamperometry. There would, however, in the use of a glucose sensor, be an advantage in making a non-steady-state measurement with an oxygen detector. This is especially so if the time of measurement is restricted to the period either before the diffusion layer at the electrode has reached the membrane surface or before it has moved a significant distance into the membrane as discussed in section 7.3.4.2.1.

A rather different form of biosensor is based on a wall jet electrode. The principle of this electrode was described in section 7.3.2.3 and Figure 7.14 illustrates a wall jet ring disc electrode [44]. In this system the central disc electrode is used to generate electrochemically a simple reactant, some of which reacts with the bioreactant in the incoming solution and some of which is transported radially to the outer concentric ring electrode by the impinging jet of solution. The ring electrode is set at the potential corresponding to the transport limited reduction or oxidation

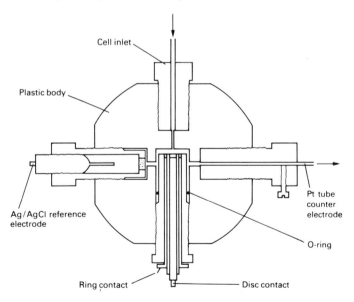

Figure 7.14 Schematic diagram of a wall jet ring disc electrode.

of the disc generated reactant. Since the technique does not involve direct electrochemistry of the bioreactant in solution, the electrochemical generation at the disc and the monitoring at the ring are often of a simple reversible couple and, as a result, electrode modification is not usually needed. A good example of the application of the technique is in the monitoring and determination of proteins and amino acids eluting from a chromatography column with electrogenerated bromine [45]. The overall electrochemistry can be represented by

$$\text{Disc electrode } Br^- \rightarrow \tfrac{1}{2} Br_2 + e^- \qquad (R7.12)$$
$$\text{Solution } nBr_2 + \text{Protein} \rightarrow \text{products} \qquad (R7.13)$$
$$\text{Ring electrode } \tfrac{1}{2} Br_2 + e^- \rightarrow Br^- \qquad (R7.14)$$

For proteins the solution pH should be 9.2 so that the bromine is present as OBr^-, and at this pH one molecule of protein can consume several hundred molecules of OBr^-—the bromine number for the protein. Thus there is a large amplification compared with a direct determination by reduction or oxidation of a protein on an electrode. Figure 7.15(a) shows a typical ring disc titration curve with N_0, the collection efficiency of the ring in the absence of any reaction in the solution. Figure 7.15(b) gives some typical results obtained on successive addition of aliquots of the protein. More than 20 proteins have been determined in this way and if a second determination is made at pH 5, where bromination is much slower and the bromine number is lower than at pH 9.2, a ratio of bromine numbers can be used to identify the proteins.

7.3.4.5 Coulometric sensors. Just as with amperometric sensors, one of the most widely investigated applications of coulometric sensors has been to oxygen monitoring. This has been reviewed and discussed in some detail [24, Ch. 7]. We commented earlier (section 7.3.3) that for a coulometric efficiency of 100% no calibration of a coulometric device is necessary and this feature is one which distinguishes a coulometric sensor from any other electrochemical monitor. In practice however, it is worthwhile to check that the efficiency is actually 100% from time to time. Another distinguishing feature of coulometric sensors is that to achieve a high coulometric efficiency the ratio of the electrode area to cell volume is very large, and this also leads to a high sensitivity.

The advantages of coulometric sensors and their simple mode of operation are often offset by practical problems, such as that of achieving a high coulometric efficiency in practice [24, Ch. 7]. This has perhaps limited their range of applications. Coulometric sensors are used, though, for gaseous oxygen analysis in petrochemical plant streams and they have also been applied to dissolved oxygen analysis in boiler waters. Other gases can be analysed coulometrically as well. These are usually oxidising gases such as F_2, Cl_2, Br_2, and O_3 which are capable of liberating I_2 by

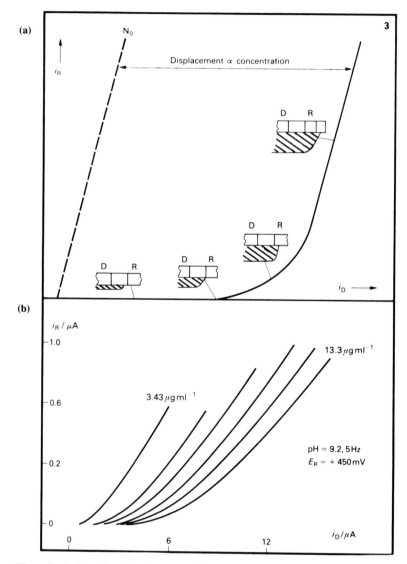

Figure 7.15 Ring disc titration curves. For explanation of parts (a) and (b) see text.

reaction with a KI solution. The free iodine reacts with electrochemically generated hydrogen gas and the charge needed to regenerate the hydrogen consumed is measured and related to the quantity of oxidising gas in the original sample.

Another example of a coulometric sensor is for moisture analysis. Again, the principle is based on Faraday's Law with, assuming 100% current efficiency for water electrolysis, the current being directly proportional to the mass of the water in the sample. In this technique the cell consists of two noble metal wires (e.g. Pt or Rh) which are wound in bifilar form on a cylindrical quartz or PTFE former and the space between them is coated with phosphorus pentoxide. The dry insulating oxide absorbs moisture which passes over it and is then electrolysed by the applied voltage to give, as mentioned, a direct measure (through the current magnitude) of the amount of water. There are two main problems with this type of analyser. The first arises from the fact that the hydrogen and oxygen produced by electrolysis are generated in close proximity to one another on electrodes in parallel. If, as is often the case, the spacing between the electrodes is small then catalytic recombination of the gases can occur regenerating water for further electrolysis leading to a high apparent measure of moisture content of the test gas. The second problem occurs when the sample moisture levels are high since this leads to a reduced lifetime for the electrolytic cell. Nevertheless, commercial analysers with ranges up to 10 000 ppm are available with repeatability being typically $\pm 5\%$ of full-scale reading.

7.4 Conclusions

For potentiometric methods of process analysis, glass electrodes for direct potentiometric measurement of pH, zirconia sensors for pO_2 monitoring and redox electrodes for monitoring of oxidation–reduction potentials are the only types of analysers which are routinely used for on-line process analysis. Clevett [11] gives useful summaries of commercially available systems. The wide ranging and extensive studies, including theoretical, practical and technical investigations, of ISEs other than glass electrodes for pH measurement have so far had only a small impact on industrial process analysis methodology. Their application to on-line biomedical analyses seems to be the most popular area currently being explored [17]. There are undoubtedly significant practical obstacles to be overcome before ISEs will be more extensively accepted with, as we have indicated, lack of physical and chemical robustness being two major problems. However, a glass electrode is itself not particularly robust and needs regular checking and calibration. Therefore, for other ISEs, particularly those based on solid state membranes, there is a significant possibility of them making a valuable contribution to process analysis and control. The scientific base is certainly already there. What is needed are further technological developments to bring the electrodes to a wider market place coupled with a commercial will to make them economically viable.

For amperometric methods of process analysis, there are also a limited range of types of sensors with membrane-covered gas sensors being the dominant type of process analyser. Again, Clevett gives a summary of

commercial systems [11, Ch.7]. The very wide range of other amperometric techniques [21] has had little impact on process analysis and as
mentioned in section 7.3.4.1, one major reason for this arises from
difficulties associated with electrode poisoning. As with potentiometric
sensors though, there is a good scientific base on which to build.

We conclude with an example of a very recent novel application of
three different sensors. Two are amperometric and one potentiometric
and they illustrate how a good scientific understanding can lead to
electrochemical analysers which are relatively simple and yet can provide
valuable insights into a complex process. The example concerns a
workstation for monitoring the fermentation of beer [46]. The instrument
has electrochemical sensors for measuring oxygen, glucose and carbon
dioxide. The oxygen sensor is essentially the same as that described
above (section 7.3.4.2.1). The glucose sensor is based on yet another
principle from those described earlier (section 7.3.4.4). In this case, the
electrode material is a conducting organic salt consisting of the cation
N-methylphenazinium (NMP^+) and the anion, tetracyanoquinodimeth-
anide ($TCNQ^-$) [47] with the usual enzyme of glucose oxidase (GOD/
FAD). As in the case of the oxygen sensor, the electrode is protected
from the wilder ravages of the test environment by a membrane, but for
this electrode it is a dialysis material which is used. The glucose in the
test solution diffuses through the membrane and is oxidised by the
enzyme which itself is reoxidised and regenerated on the NMP^+TCNQ^-
electrode.

$$\text{Glucose} + \text{GOD/FAD} \longrightarrow \text{Gluconolactone} + \text{GOD/FADH}_2 \qquad \text{(R7.15)}$$

$$\text{GOD/FADH}_2 \xrightarrow{NMP^+ TCNQ^-} \text{GOD/FAD} + 2H^+ + 2e^- \qquad \text{(R7.16)}$$

Finally, the CO_2 sensor is an ingenious extension of the design of an
oxygen sensor. It consists of a central iridium oxide electrode for
measuring pH [48], a concentric platinum electrode for generating OH^-
amperometrically, another concentric platinum ring that acts as a
counter electrode and a Ag/AgCl reference electrode at the outer edge of
the platinum ring [49]. The electrode is used by generating OH^- at
constant current for a fixed time (t_g). The current is then stopped and the
CO_2 diffusing through the membrane titrates the OH^-. The time (t_1) to
the end point, as monitored by the iridium oxide electrode, is measured
and the CO_2 concentration is proportional to the ratio t_g/t_1. The combination of sensors has been used to monitor the variation of the three
analytes in the fermentation of beer, both in the laboratory and in a 1700
dm^3 vessel in a pilot plant. These measurements are of great value as it
is found that for good fermentation, control of oxygen levels at the start
of the process is critical. Furthermore, the glucose concentration was
observed to pass through a maximum. The time to this maximum and

the initial rate of CO_2 production provide an early indication of yeast vitality.

In conclusion therefore, we can say that electrochemical sensors are playing and will undoubtedly continue to play an important role in process analysis. As we have mentioned, for both potentiometric and amperometric sensors, there is a good scientific base. This, coupled with the relatively low costs of electrochemical systems when compared with, for example, spectrometric devices, strongly suggests that there could be a significant and growing market for electrochemical process analysers.

References

1. H. V. Venkatasetty, *Chem. Eng. Prog.* **88** (1992) 63.
2. P. Debye and E. Huckel, *Physikal. Z.* **24** (1923) 185.
3. M. L. Hitchman and F. W. M. Nyasulu, *Talanta* **40** (1993) 1449.
4. J. B. Roos, *Analyst* **87** (1962) 832.
5. M. Dole, *The Glass Electrode.* Wiley, New York, USA, 1941.
6. IUPAC, *Pure Appl. Chem.* **51** (1979) 1913.
7. J. N. Harman, *Process Instruments and Controls Handbook* (3rd edn), ed. D. M. Considine. McGraw Hill, New York, USA, 1985.
8. J. Grigsby, *J. Am. Soc. Brewing Chem.* **38** (1980) 96.
9. D. Midgley and K. Torrance, *Potentiometric Water Analysis* (2nd edn). Wiley, Chichester, UK, 1991.
10. J. Koryta, *Ion Selective Electrodes.* CUP, Cambridge, UK, 1975.
11. K. J. Clevett, *Process Analyzer Technology.* Wiley & Sons, New York, USA, 1986.
12. J. M. Coulson and J. F. Richardson, *Chemical Engineering* (Vol. 3). Pergamon, Oxford, UK, 1979, Ch. 5.
13. J. Vesely, D. Weiss and K. Stulik, *Analysis with Ion Selective Electrodes.* Ellis Horwood, Chichester, UK, 1978.
14. H. Freiser (ed.), *Ion Selective Electrodes in Analytical Chemistry.* Plenum, New York, USA, 1978.
15. W. E. Morf and W. Simon, In: *Ion Selective Electrodes in Analytical Chemistry*, ed. H. Freiser. Plenum, New York, USA, 1978, p. 211.
16. H. Wohltjon, *Anal. Chem.* **56** (1984) 87A.
17. A. P. F. Turner, I. Karube and G. S. Wilson, *Biosensors, Fundamentals and Applications.* OUP, Oxford, UK, 1989.
18. J. Haggin, *C & E News* **June 4** (1984) 7.
19. W. J. Albery and M. L. Hitchman, *Ring-Disc Electrodes.* Clarendon Press, Oxford, UK, 1971.
20. B. Fleet and C. J. Little, *Chromatogr. Sci.* **12** (1947) 747.
21. A. J. Bard and L. R. Faulkner, *Electrochemical Methods.* Wiley, New York, USA. 1980.
22. J. Wang and H. D. Dewald, *Anal. Chem.* **56** (1984) 156.
23. H. Gunasingham and B. Fleet, In: *Electroanalytical Chemistry* (Vol. 16), ed. A. J. Bard. Marcel Dekker, New York, USA, 1989.
24. M. L. Hitchman, *Measurement of Dissolved Oxygen.* Wiley, New York, USA, 1978.
25. J. M. Hale and M. L. Hitchman, *J. Electroanal. Chem.* **107** (1980) 281.
26. M. L. Hitchman and E. Gnaiger, In: *Polarographic Oxygen Sensors*, ed. E. Gnaiger and H. Forstner. Springer-Verlag, Berlin, Germany, 1983.
27. V. Linek, V. Vacek, J. Sinkule and P. Benes, *Measurement of Oxygen by Membrane-Covered Probes.* Ellis Horwood, Chichester, UK, 1988.
28. C. Excell and I. Rowland, *Water and Waste Treatment,* **July** (1991).
29. Neotronics Ltd, *Portable Multi-Gas Monitors* (Application Notes 058–0050, 058–0071). Takeley, 1993.

30. Orbisphere Laboratories, O_2, O_3, CO_2 or N_2 Analysis in Beer, Wine, Soft Drinks and Mineral Waters (Beverage Training Manual BTM 9403). Geneva, Switzerland, 1991.
31. J. M. Hale, Technical Note TN 179409, Orbisphere Laboratories, Geneva, Switzerland, 1991.
32. R. Mitchell, J. Hobson, N. Turner and J. M. Hale, Am. Soc. Brew. Chem. J. **41** (1983) 68.
33. M. L. Hitchman, results to be published.
34. S. Srinivasan and E. Gileadi, In: Handbook of Fuel Cell Technology, ed. C. Berger. Prentice Hall, New Jersey, USA, 1968, p 219.
35. M. W. Breiter, Electrochemical Processes in Fuel Cells. Springer-Verlag, Berlin, Germany, 1969, p. 238.
36. H. A. Liebhafsky and E. J. Cairns, Fuel Cells and Fuel Batteries. Wiley, New York, USA, 1969, Ch. 7, p. 245.
37. T. K. Sherwood, R. L. Pigford and C. R. Wilke, Mass Transfer. McGraw Hill, New York, USA, 1975, Ch. 2, p. 8.
38. L. W. Niedrach and H. R. Alford, J. Electrochem. Soc. **112** (1965) 117.
39. J. P. Hoare, The Electrochemistry of Oxygen. Interscience, New York, USA, 1968, p. 188.
40. L. G. Austin, In: Handbook of Fuel Cell Technology, ed. C. Berger. Prentice Hall, New Jersey, USA. 1968, p. 1.
41. M. L. Hitchman, N. J. Cade, T. K. Gibbs and N. J. M. Hedley, results to be published.
42. M. L. Hitchman and H. A. O. Hill, Chem. Brit. **22** (1986) 1118.
43. K. Ito, S. Ikeda, K. Asai, H. Naruse, K. Ohkura, H. Ichihashi, H. Kamei and T. Kondo, In: Fundamentals and Applications of Chemical Sensors, ed. D. Schuetzle and R. Hammerle. ACS Symposium Series 309, Washington, DC, USA, 1986, Ch. 23.
44. W. J. Albery and C. M. A. Brett, J. Electroanal, Chem. **148** (1983) 201.
45. W. J. Albery, L. R. Svanberg and P. Wood, J. Electroanal. Chem. **162** (1984) 29.
46. W. J. Albery, M. S. Appleton, T. R. D. Pragnell, M. J. Pritchard, M. Uttamlal, L. E. Fieldgate, D. R. Lawrence and F. R. Sharpe, J. Appl. Electrochem. **24** (1994) 521.
47. W. J. Albery, P. N. Bartlett and D. H. Craston, J. Electroanal. Chem. **194** (1985) 223.
48. M. L. Hitchman and S. Ramanthan, Analyst **113** (1988) 35.
49. W. J. Albery, M. Uttamlal, M. S. Appleton, N. J. Freeman, B. B. Kebbekus and M. D. Neville, J. Appl. Electrochem. **24** (1994) 14.

8 Process chemometrics
B. M. WISE and B. R. KOWALSKI

8.1 Introduction

Chemometrics is often thought of as a subdiscipline of chemistry, and in particular, analytical chemistry. Recently, 'chemometric techniques' have been applied to problems that are most often thought of as being in the domain of the chemical engineer, i.e. chemical processes. These applications can be roughly divided between those directed at maintenance of the process instruments, e.g. calibration and calibration transfer, and those that are concerned with the maintenance of the process itself, e.g. statistical process control and dynamic modeling. In this chapter we will provide a tutorial on commonly used chemometric methods and will illustrate the methods with some example problems in chemical process monitoring, process instrument calibration and dynamic process modeling.

8.2 Definition

Before moving on to chemometric techniques, it is important that we define chemometrics. This has been a matter of debate in the technical community for some time; [1,2] however, the authors choose the following definition: *Chemometrics is the science of relating measurements made on a chemical system to the state of the system via application of mathematical or statistical methods.* It is clear from the definition that chemometrics is data based. The goal of many chemometrics techniques is the production of an empirical model, derived from data, that allows one to estimate one or more properties of a system from measurements. Chemical systems include chemical processes, the analysis of which form the basis of this text.

8.3 Process monitoring

It has been pointed out several times in the recent literature that chemical processes are becoming more heavily instrumented and the data is recorded more frequently [3, 4]. This is creating a data overload, and the result is that a good deal of the data is 'wasted,' i.e. no useful information is obtained from it. The problem is one of both compression and extraction. Generally, there is a great deal of correlated or redundant information in process measurements. This information must be compressed in a manner that retains the essential information and is more easily displayed than each of the process variables individually. Also, often essential information lies not in any individual process variable but in how the variables change with respect to one another, i.e. how they co-vary. In this case the information must be extracted from the data. Furthermore, in the presence of large amounts of noise, it would be desirable to take advantage of some sort of signal averaging.

8.3.1 Principal components analysis

Principal components analysis (PCA) [5] is a favorite tool of chemometricians for data compression and information extraction. PCA finds combinations of variables or *factors* that describe major trends in the data. Mathematically, PCA relies upon an eigenvector decomposition of the covariance or correlation matrix of the process variables. For a given data matrix X with m rows and n columns, with each variable being a column and each sample a row, the covariance matrix of X is defined as

$$\text{cov}(X) = \frac{X^T X}{m - 1} \tag{8.1}$$

provided that the columns of X have been 'mean centered,' i.e. adjusted to have a zero mean by subtracting off the original mean of each column. If the columns of X have been 'autoscaled,' i.e. adjusted to zero mean and unit variance by dividing each column by its standard deviation, equation (8.1) gives the correlation matrix of X. (Unless otherwise noted, it is assumed that data is either mean centered or autoscaled prior to analysis.) PCA decomposes the data matrix X as the sum of the outer product of vectors t_i and p_i plus a residual matrix E:

$$X = t_1 p_1^T + t_2 p_2^T + \ldots + t_k p_k^T + E \tag{8.2}$$

Here k must be less than or equal to the smaller dimension of X, i.e. $k < \min\{m, n\}$. The t_i vectors are known as *scores* and contain information on how the *samples* relate to each other. The p_i vectors are *eigenvectors* of the covariance matrix, i.e. for each p_i

$$\text{cov}(X) p_i = \lambda_i p_i \tag{8.3}$$

where λ_i is the *eigenvalue* associated with the eigenvector p_i. In PCA the p_i are known as *loadings* and contain information on how the *variables* relate to each other. The t_i form an orthogonal set ($t_i^T t_j = 0$ for $i \neq j$), while the p_i are orthonormal ($p_i^T p_j = 0$ for $i \neq j$, $p_i^T p_j = 1$ for $i = j$). Note that for X and any t_i, p_i pair

$$\mathbf{Xp}_i = \mathbf{t}_i \qquad\qquad (8.4)$$

i.e. the score vector $\mathbf{t}_i$ is the linear combination of the original $\mathbf{X}$ data defined by $\mathbf{p}_i$. The $\mathbf{t}_i$, $\mathbf{p}_i$ pairs are arranged in descending order according to the associated λ_i. The λ_i are a measure of the amount of *variance* described by the $\mathbf{t}_i$, $\mathbf{p}_i$ pair. In this context, we can think of variance as *information*. Because the $\mathbf{t}_i$, $\mathbf{p}_i$ pairs are in descending order of λ_i, the first pair capture the largest amount of information of any pair in the decomposition. In fact, it can be shown that the $\mathbf{t}_1$, $\mathbf{p}_1$ pair capture the greatest amount of variation in the data that it is possible to capture with a linear factor, and each subsequent pair captures the greatest possible amount of variance remaining at that step.

Generally it is found (and it is usually the objective) that the data can be adequately described using far fewer factors than original variables. Thus, the data overload experienced in chemical process monitoring can be solved by monitoring fewer scores (weighted sums of the original variables) than original variables, with no significant loss of information. It is also often found that PCA turns up combinations of variables that are useful descriptions, or even predictors, of particular process events. These combinations of variables are often more robust indicators of process conditions than individual variables due to the signal averaging aspects of PCA.

8.3.2 Application of PCA to a chemical process

As an example of this we now consider the use of PCA on the data produced by a high temperature process, a slurry-fed ceramic melter (SFCM) test system [6]. A schematic drawing of the process is shown in Figure 8.1. The process is used to convert waste from nuclear fuel reprocessing into a stable borosilicate glass. The SFCM consists of a ceramic lined steel box surrounding a pool of molten glass. Feed to the melter, a slurry consisting of reprocessing waste and glass-forming chemicals is introduced onto the pool of molten glass, which is heated by passing a large current through it. Water and other volatiles in the feed are driven off, producing a 'cold cap' of dried feed which eventually melts into the glass. Off-gas is routed to a treatment system. Molten glass is drawn off through a riser (not shown) and allowed to solidify for eventual long-term storage in a geologic repository.

Raw data from the SFCM process, consisting of temperatures in 16 locations (of 20 available) within the melter, is shown in Figure 8.2. The temperatures are measured at eight levels in each of two vertical thermowells that are inserted into the molten glass pool. (The variables are arranged in the data set so that variables 1 to 8 are from the bottom to the top of thermowell number 1, and variables 9 to 16 are from the

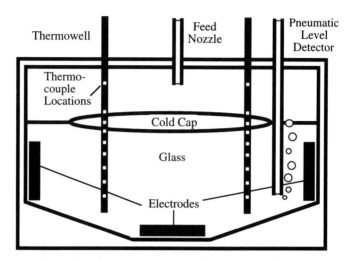

Figure 8.1   Schematic diagram of slurry-fed ceramic melter.

bottom to the top of thermowell number 2.) It is immediately apparent from Figure 8.2 that there is a great deal of correlation in the data. Many of the variables appear to follow a sawtooth pattern.

When PCA is performed on this data (after mean centering), it is found that three factors capture nearly 97% of the variance or information in

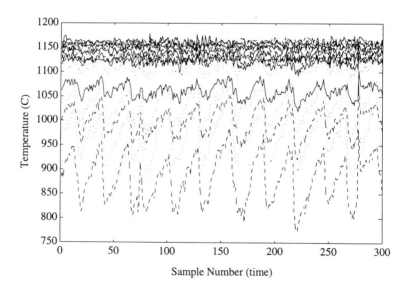

Figure 8.2   Data from a high temperature process.

the data set, as shown in Table 8.1. Thus, the 16 original variables could be replaced with three combinations of variables with very little loss of information.

Table 8.1 Variance captured by PCA model of SFCM data

| PC number | Percent variance captured | |
	This PC	Total
1	88.0711	88.0711
2	6.6974	94.7686
3	2.0442	96.8127
4	0.9122	97.7249
5	0.6693	98.3942
6	0.5503	98.9445
7	0.3614	99.3059
8	0.2268	99.5327

The first principal component scores for the SFCM data are shown in Figure 8.3. The scores capture the sawtooth nature of the process data, which is attributable to changes in the level in the molten glass pool, which is a controlled variable. Remember that the scores are the linear combination of the original temperature variables that describe the greatest variation in the process data. The loadings plot, Figure 8.4, shows the coefficients of the first eigenvector or principal component. Examination of the loadings indicates that the variables that contribute most strongly to the sawtooth trend are those near the surface of the molten glass pool, variables 6–8 and 14–16. The second

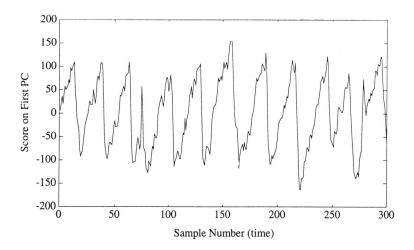

Figure 8.3 Scores on first principal component of SFCM data.

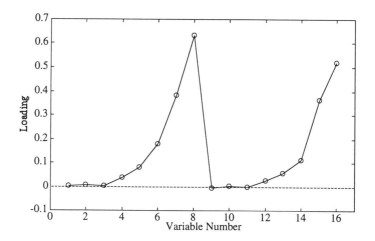

Figure 8.4   Loadings for first principal component of SFCM data.

and third PCs (not shown) capture variation that occurs between the two groups of measurement locations (which is not controlled) and variations of the overall process average temperature (which is controlled).

It is also possible to display the lack of model fit statistic, Q. Q is simply the sum of squares of each row (sample) of $\mathbf{E}$ (from equation (8.2)), i.e. for the ith sample in $\mathbf{X}$, $\mathbf{x}_i$

$$Q_i = \mathbf{e}_i\mathbf{e}_i^T = \mathbf{x}_i(\mathbf{I} - \mathbf{P}_k\mathbf{P}_k^T)\mathbf{x}_i^T \qquad (8.5)$$

where $\mathbf{e}_i$ is the ith row of $\mathbf{E}$, $\mathbf{P}_k$ is the matrix of the k loadings vectors retained in the PCA model (where each vector is a column of $\mathbf{P}_k$) and $\mathbf{I}$ is the identity matrix of appropriate size ($n \times n$). The Q statistic indicates how well each sample conforms to the PCA model. It is a measure of the amount of variation in each sample *not* captured by the k principal components retained in the model.

The sum of normalized squared scores, known as Hotelling's T^2 statistic, is a measure of the variation in each sample *within* the PCA model. T^2 is defined as

$$T_i^2 = \mathbf{t}_i\lambda^{-1}\mathbf{t}_i^T = \mathbf{x}_i\mathbf{P}\lambda^{-1}\mathbf{P}^T\mathbf{x}_i^T \qquad (8.6)$$

where $\mathbf{t}_i$ in this instance refers to the ith row of $\mathbf{T}_k$, the matrix of k scores vectors from the PCA model and λ^{-1} is the diagonal matrix containing the inverse of the eigenvalues associated with the k eigenvectors (principal components) retained in the model.

8.3.3 Multivariate statistical process control

Once developed, PCA models can be combined with tools and techniques from univariate statistical process control (SPC) to form multivariate statistical process control (MSPC) tools [4,6,7]. Control limits can be placed on the process scores, 'sum of scores' T^2, residual Q, or residuals of individual variables (single columns of **E**). There are multivariate analogs to Shewart, range and CUSUM charts. When used for MSPC purposes, PCA models have the additional advantage that the new scores variables produced, which are linear combinations of the original variables, are more normally distributed than the original variables themselves. This is a consequence of the central limit theorem, which can be stated as follows: *if the sample size is large, the theoretical sampling distribution of the mean can be approximated closely with a normal distribution* [8]. In our case, we are typically sampling a large number of variables when we form the PCA scores and thus we would expect the scores, which like a mean are a weighted sum, to be approximately normally distributed. A word of caution is advised here, however. In some instances, if score values are associated with a controlled variable (such as the level of the SFCM in the example above) we would not expect the scores to be normally distributed.

A theoretical connection exists between PCA and state-space models typically used by the process control community for describing the dynamics of a chemical process. Much literature is available concerning the state-space formalism, [9,10] however, it is useful to provide a brief introduction here. Consider a linear, time-invariant, discrete, state-space process model:

$$\mathbf{x}(k + 1) = \mathbf{\Phi}\mathbf{x}(k) + \mathbf{\Gamma}\mathbf{u}(k) + \boldsymbol{\nu}(k) \tag{8.7}$$

$$\mathbf{y}(k) = \mathbf{C}\mathbf{x}(k) + \mathbf{e}(k) \tag{8.8}$$

where $\mathbf{x}(k)$ is the $n \times 1$ state vector at sampling period k, $\mathbf{u}(k)$ is the $r \times 1$ vector of process inputs and $\mathbf{y}(k)$ is the $p \times 1$ vector of process measurements. The vector $\boldsymbol{\nu}(k)$ represents the state noise or disturbance input to the process; $\mathbf{e}(k)$ is measurement noise. The $\mathbf{\Phi}$, $\mathbf{\Gamma}$, and $\mathbf{C}$ matrices are assumed constant. Equation (8.7) shows that the process state at time $k + 1$ is a function of the state, process inputs and disturbances at time k. Equation (8.8) shows that the process measurements are related to the process states and measurement noise.

It has been shown that, for processes where there are more measurements than states, i.e. when there are more elements in **y** than **x**, variations in the process states appear primarily as variations in the PCA scores, while noise mainly affects the residuals [4], i.e. the difference between the original variables and their projection into the PCA model. This allows one to consider only the noise properties of the system when deriving limits on PCA residuals; the dynamics of the process need not be considered explicitly. The rapid increase in the availability of process analytical instrumentation has drastically increased the instances of processes having more measurements than (significant) states.

Once a PCA model of the system measurements (**Y**) has been obtained, confidence limits can be established for T^2, the overall residual Q and for the residuals on individual variables. Confidence limits can be calculated for Q, provided that all of the eigenvalues of the covariance matrix of **Y**, the λ_i, have been obtained: [7]

$$Q_\alpha = \Theta_1 \left[\frac{c_\alpha \sqrt{2\Theta_2 h_0^2}}{\Theta_1} + 1 + \frac{\Theta_2 h_0 (h_0 - 1)}{\Theta_1^2} \right]^{\frac{1}{h_0}} \tag{8.9}$$

where

$$\Theta_i = \sum_{j = k + 1}^{n} \lambda_j^i \qquad \text{for} \quad i = 1, 2, 3 \tag{8.10}$$

and

$$h_0 = 1 - \frac{2\Theta_1 \Theta_3}{3\Theta_2^2} \tag{8.11}$$

In equation (8.9), c_α is the standard normal deviate corresponding to the upper $(1 - \alpha)$ percentile. In equation (8.10), k is the number of principal components retained in the process model and n is the total number of principal components (equal to the smaller of the number of variables or samples in **Y**).

Statistical confidence limits for the values for T^2 can be calculated by means of the F-distribution as follows

$$T_{k, m, \alpha}^2 = \frac{k(m - 1)}{m - k} F_{k, m - k, \alpha} \tag{8.12}$$

Here m is the number of samples used to develop the PCA model and k is the number of principal component vectors retained.

Some discussion of the geometric interpretation of Q and T^2 is perhaps in order. As noted above, Q is a measure of the variation of the data outside of the PCA model. Imagine for a moment a process with three variables where those variables are normally restricted to lie on a plane that cuts through three-space. Such a system would be well described by a two PC model. Q is a measure of the distance off the plane, in fact $\sqrt{Q}$ is the Euclidean distance of the operating point from the plane formed by the two PC model. The Q limit defines a distance off the plane that is considered unusual for normal operating conditions. T^2, on the other hand, is a measure of the distance from the multivariate mean to the projection of the operating point onto the two PCs. The T^2 limit defines an ellipse on the plane within which the operating point normally projects.

The residual variance for each variable can be estimated for the PCA model. If it is assumed that the eigenvalues associated with the eigenvectors (PCs) not retained in the model are all the same (a common assumption), then the variance in the residual of the jth variable can be estimated from

$$\hat{s}_j^2 = \left(\sum_{i=1}^{n} \lambda_i - \sum_{i=1}^{k} \lambda_i \right) \left(1 - \sum_{i=1}^{k} p_{i,j}^2 \right) \tag{8.13}$$

which requires only the PCs and eigenvalues retained in the model. All of the eigenvalues are not required since the first term on the right-hand side in equation (8.13) can be replaced with the sum of the diagonal elements of the covariance matrix. Given the estimate of the variance of the residuals, the standard F test with the appropriate degrees of freedom can be used to determine if the system or its sensors have changed. This test is best performed using several samples. The test checks to see if

$$s_{j\text{new}}^2 / s_{j\text{old}}^2 > F_{v\text{new, } v\text{old, } \alpha} \tag{8.14}$$

where

$$v_{\text{new}} = m_{\text{new}} \tag{8.15}$$

$$v_{\text{old}} = m_{\text{old}} - k - 1 \tag{8.16}$$

where m_{new} and m_{old} are the number of samples in the test and training sets, respectively, and k is the number of PCs retained in the model. When the inequality of equation (8.16) holds, then a change has occurred in the system to a confidence level of $1 - \alpha$.

The mean residual should be zero for all the variables. The t-test can be used to detect a shift in the mean away from zero. In this case the hypothesis that the means are equal is to be tested. Thus the t-test reduces to

$$t_{v\text{ tot}} = \frac{(\overline{X}_{\text{old}} - \overline{X}_{\text{new}})(v_{\text{old}} + v_{\text{new}})^{0.5}}{(1/v_{\text{old}} + 1/v_{\text{new}})^{0.5}(v_{\text{old}}S_{\text{old}}^2 + v_{\text{new}}S_{\text{new}}^2)^{0.5}} \tag{8.17}$$

where the degrees of freedom are both one greater than for the case given above. For the purpose of setting limits, the variances can be assumed to be equal to the variance of the residuals of the calibration set as calculated by equation (8.13). Once the desired confidence level is chosen, it is possible to solve for the difference between the old and new means that is just significant.

Control limits can also be established for the scores, **T**. However, because the scores are related to the states of the process, they are typically quite correlated in time (autocorrelated) and therefore cannot

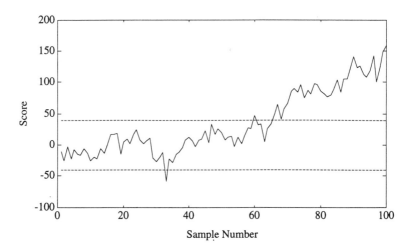

Figure 8.5 New sample score on second PC showing process upset starting at sample 50.

be assumed to be random and normally distributed. Thus, control limits on the scores must be established based either on judgment concerning desired process operating limits or using more sophisticated time series modeling techniques that are beyond the scope of this chapter.

As an example application of MSPC, imagine that a failure occurred in the SFCM system that caused the power input to one side of the melter to decrease. This would result in an abnormally high temperature on one side relative to the other and an abnormally high or low score on the second principal component. A simulated failure of this type is demonstrated in Figure 8.5, which shows scores on the second PC going outside the normal range. If a single thermocouple developed a bias, it would show up best in the model residual Q and in the residuals of the individual variables, the e_i. This is because the correlation of the sensor with the remaining sensors would have changed, i.e. variation outside of the model would have occurred. A failure of this sort is shown in Figure 8.6, where the residual Q becomes unusually large starting at sample 51. Note that in both of these examples of off-normal events it is not required that any of the *individual* variables go outside their normal range for the MSPC model to detect a change in the system. Instead, the MSPC model detects a change in the relationship *between* variables.

8.4 Multivariate calibration

In years past process measurements consisted primarily of traditional 'engineering' variables such as temperature, pressure, flow rate and

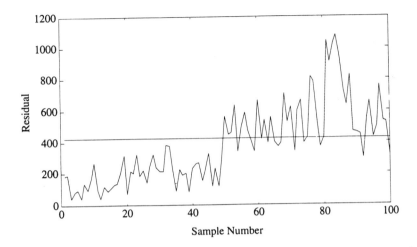

Figure 8.6 New sample residuals showing failure of thermocouple starting at sample 50.

electrical power. In distillation control, for example, although the aim was to achieve a specific concentration of product in the overhead stream, the column was controlled based upon pressure and temperature measurements. Due to the advent of on-line spectrometers, better estimates of chemical composition can be obtained. Even spectroscopy is indirect, however, and relies on the ability to develop a calibration model that relates the spectral intensities at different wavelengths to the chemical composition of the product. The development of calibration models for such systems is the focus of this section. Note, however, that the techniques to be discussed have much wider applicability than calibration of spectrometers.

8.4.1 Classical least squares

The classical least squares (CLS) model (also known as the **K**-matrix model) assumes that measurements are the weighted sum of linearly independent signals [11]. In spectroscopy, for example, the CLS model assumes that measured spectra are the sum of pure component spectra weighted by the concentration of the analytes. Thus, the model is:

$$\mathbf{x} = \mathbf{pS} \tag{8.18}$$

where $\mathbf{x}$ is the measured response vector, $\mathbf{S}$ is the matrix of pure component responses and $\mathbf{p}$ is the vector containing the weights, i.e. concentrations of the analytes. Generally, given the vector of measurements, $\mathbf{x}$, one would like to know $\mathbf{p}$, the degree to which each component contributes to the overall measurement. This can be determined from

$$p = xS^+ \tag{8.19}$$

where S^+ is the pseudo inverse of S, defined for CLS by:

$$S^+ = S^T(SS^T)^{-1} \tag{8.20}$$

The main disadvantages of CLS is that the pure responses S must be known *a priori*. This includes the responses of any minor components that may not be of interest themselves but may contribute to the measured signal. Furthermore, the pure component responses must be linearly independent. If they are not, $(SS^T)^{-1}$ is not defined. In the case where the pure responses are nearly collinear, S^+ may change a great deal given small changes in the estimates of the pure responses due to noise or other interferences. In this case, estimates of p obtained by equation (8.19) may not be very accurate.

If the pure component responses are not known, they can be determined using the K-matrix method provided that the concentrations of all spectroscopically active agents are known. Given a concentration matrix, C, containing the concentrations of active species, an estimate of S, usually called K, can be obtained from the measured responses, R, as follows:

$$K = (C^TC)^{-1}C^TR \tag{8.21}$$

Of course, this procedure also introduces error in the analysis, since the estimate K may also be prone to error because of colinearity problems and noise in the spectra.

8.4.2 Inverse least squares

It is possible to get around the problem of having to know S by using an inverse least squares (ILS) model. ILS assumes that a regression vector b can be used to determine a property of the system y from the measured variables x (a row vector such as a spectra). Thus, the ILS model is

$$xb = y \tag{8.22}$$

The regression vector b must be determined using a collection of measurements X and the known values of the property of interest, y. Thus, b is estimated from

$$b = X^+y \tag{8.23}$$

where X^+ is the pseudoinverse of X. There are many ways to determine a pseudoinverse, but perhaps the most obvious is multiple linear regression (MLR, also known as ordinary least squares). In this case, X^+ is defined by

$$X^+ = (X^TX)^{-1}X^T \tag{8.24}$$

Unfortunately, this approach often fails in practice because of the collinearity of **X**, e.g. some columns of **X** (variables) are linear combinations of other columns, or because **X** contains fewer samples than variables (fewer rows than columns). For example, the spectroscopy calibration problem is extremely ill-conditioned due to a high degree of correlation between absorbances at nearby wavelengths. It is also typical that there are fewer samples available than the number of wavelengths considered.

When equation (8.23) is used with systems that produce nearly collinear data, the solution for **b** is unstable, i.e. small perturbations in the original data, possibly due to noise or experimental error, cause the method to produce wildly different results. While the calibrations may fit the data, they are typically not useful for predicting the properties of new samples.

8.4.3 Principal components regression

Principal components regression (PCR) is one way to deal with the problem of ill-conditioned matrices [12]. Instead of regressing the system properties (e.g. concentrations) on the original measured variables (e.g. spectra), the properties are regressed on the principal component scores of the measured variables (which are orthogonal and, therefore, well conditioned). Thus, X^+ is estimated as

$$X^+ = P(T^TT)^{-1}T^T \tag{8.25}$$

As in PCA, the number of principal components to retain in the model must be determined. In our case, the purpose of the regression model is to predict the properties of interest for new samples. Thus, we would like to determine the number of PCs that optimizes the predictive ability of the model. This is typically done by cross-validation, a procedure where the available data is split between training and test sets. The prediction residual error on the test samples is determined as a function of the number of PCs retained in the regression model formed with the test data. The procedure is usually repeated several times, with each sample in the original data set being part of the test set at least once. The total prediction error over all the test sets as a function of the number of PCs is then used to determine the optimum number of PCs, i.e. the number of PCs which produces minimum prediction error. If all of the PCs are retained in the model, the result is identical to that for MLR (at least in the case of more samples than variables). In some sense, it can be seen that the PCR model 'converges' to the MLR model as PCs are added.

8.4.4 Application of PCR

As noted in the introduction to this section, the possible applications of 'calibration' methods go beyond the spectral calibration problem.

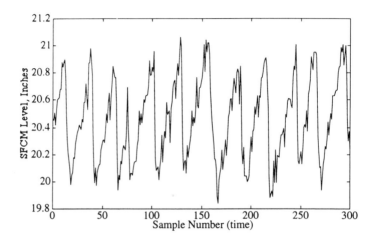

Figure 8.7 SFCM glass level versus time.

Therefore, as an example of the application of PCR, we will now develop a regression model that relates the temperatures measured in the SFCM (which take the place of spectra in the calibration problem) to the level of the molten glass (which takes the place of concentration). The glass level for the period of temperatures shown in Figure 8.2 is shown in Figure 8.7. As noted previously, the measured temperatures rise and fall with the glass level, which is a controlled variable. A model relating temperatures to level would serve as a backup to the level measuring instrument, which is a pneumatic device.

In our example, there are 300 samples available for the calibration. We must first decide how to split the available data for cross-validation to determine the optimum number of PCs to retain in the PCR model. When many training samples are available, a good rule of thumb is to use the square root of the number of samples for each test set and the number of test sets. Given 300 samples this would lead to 17.32 samples per test set and 17.32 test sets, a somewhat inconvenient number.

We must also decide how to select the test sets from the data. One possibility would to form each test set using every 10th point of the data, each subset starting from sample 1, 2, 3, etc. When dealing with data from a dynamic system, however, this is usually not a good choice because noise in the system is usually not completely random. Instead, noise is generally serially correlated, which means that the amount of variation due to noise on the nth point is correlated to the amount of noise on the $n + 1$th point. Thus, if the test sets were chosen using every 10th point, noise on the test data would be highly correlated with

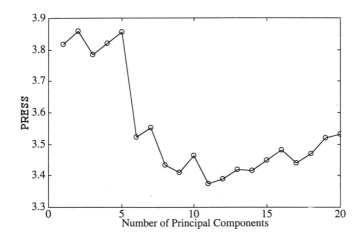

Figure 8.8 PRESS versus number of PCs in PCR model for SFCM example.

noise in the training data, and the result would be a model that fit much of the system noise. Instead, when dealing with time series data produced by a dynamic system, it is better to choose contiguous blocks of data for the test sets where the length of the block is longer than the settling time of the process. In our case, it is convenient to use 10 training sets of 30 samples each. Thus the first test set will be samples 1–30, the second will be samples 31–60 and so on.

Figure 8.8 shows the results of the cross-validation procedure on the temperature/level data. Note how the prediction error or PRESS (prediction residual error sum of squares) is a minimum at 11 PCs. Thus, we choose 11 PCs for construction of the final model using all 300 training samples. Note, however, that some PCs do not contribute to the prediction error in a positive way. When PCs 2, 4, 5, 7 and 10 are added, the PRESS actually gets worse. This suggests that these factors are not relevant for prediction of the level. In the next section we will look at a technique that will help us avoid these irrelevant factors.

8.4.5 Partial least squares

Partial least squares (PLS) regression [13, 14] is related to both PCR and MLR and can be thought of as occupying a middle ground between them. PCR finds factors that capture the greatest amount of variance in the predictor variables, e.g. spectra or the temperatures in the previous example. MLR seeks to find a single factor that best correlates predictor variables with predicted variables, e.g. concentrations or level. PLS attempts to find factors which do both, i.e. capture variance *and* achieve correlation. We commonly say that PLS attempts to maximize *covariance*.

There are several ways to calculate PLS model parameters (see, for example, the work of de Jong [15]), however, perhaps the most instructive method is known as NIPALS for non-iterative partial least squares. NIPALS calculates scores and loadings (similar to those used in PCR) and an additional set of vectors known as weights, $\mathbf{W}$. The addition of weights is required to maintain orthogonal scores. Unlike PCR and MLR, the NIPALS algorithm for PLS also works when there is more than one predicted variable, $\mathbf{Y}$, and therefore scores $\mathbf{U}$ and loadings $\mathbf{Q}$ are also calculated for the Y-block. A vector of 'inner-relationship' coefficients, $\mathbf{b}$, which relate the $\mathbf{X}$- and $\mathbf{Y}$-block scores, must also be calculated. Using NIPALS the scores, weights, loadings and inner-coefficients are calculated sequentially as shown below.

The PLS decomposition is started by selecting one column of $\mathbf{Y}$, $\mathbf{y}_j$ as the starting estimate for $\mathbf{u}_1$ (usually the column of $\mathbf{Y}$ with greatest variance if chosen). Of course in the case of univariate $\mathbf{y}$, $\mathbf{u}_1 = \mathbf{y}$. Starting in the $\mathbf{X}$ data block:

$$\mathbf{w}_1 = \frac{\mathbf{X}^T\mathbf{u}_1}{\|\mathbf{X}^T\mathbf{u}_1\|} \tag{8.26}$$

$$\mathbf{t}_1 = \mathbf{Xw}_1 \tag{8.27}$$

In the $\mathbf{y}$ data:

$$\mathbf{q}_1 = \frac{\mathbf{u}_1^T\mathbf{t}_1}{\|\mathbf{u}_1^T\mathbf{t}_1\|} \tag{8.28}$$

$$\mathbf{u}_1 = \mathbf{Yq}_1 \tag{8.29}$$

Check for convergence by comparing $\mathbf{t}_1$ in equation (8.27) with the one from the previous iteration. If they are equal within rounding error, proceed to equation (8.30). If they are not return to equation (8.26) and use the $\mathbf{u}_1$ from equation (8.29). If the Y-block is univariate, equations (8.28) and (8.29) can be omitted, set $\mathbf{q}_1 = 1$, and no iteration is required.

Calculate the $\mathbf{X}$ data block loadings and rescale the scores and weight accordingly:

$$\mathbf{p}_1 = \frac{\mathbf{X}_1^T\mathbf{t}_1}{\|\mathbf{t}_1^T t_1\|} \tag{8.30}$$

$$\mathbf{p}_{1new} = \frac{\mathbf{p}_{1old}}{\|\mathbf{p}_{1old}\|} \tag{8.31}$$

$$\mathbf{t}_{1new} = \mathbf{t}_{1old}\|\mathbf{p}_{1old}\| \tag{8.32}$$

$$\mathbf{w}_{1new} = \mathbf{w}_{1old}\|\mathbf{p}_{1old}\| \tag{8.33}$$

Find the regression coefficient b for the inner relation:

$$b_1 = \frac{\mathbf{u}_1^T \mathbf{t}_1}{\mathbf{t}_1^T \mathbf{t}_1} \tag{8.34}$$

After the scores and loadings have been calculated for the first factor (commonly called a latent variable [LV] in PLS). The $\mathbf{X}$-and $\mathbf{Y}$-block residuals are calculated as follows:

$$\mathbf{E}_1 = \mathbf{X} - \mathbf{t}_1 \mathbf{t}_1^T \tag{8.35}$$

$$\mathbf{F}_1 = \mathbf{Y} - b_1 \mathbf{u}_1 \mathbf{q}_1^T \tag{8.36}$$

The entire procedure is now repeated for the next latent variable starting from equation (8.26). $\mathbf{X}$ and $\mathbf{Y}$ are replaced with their residuals $\mathbf{E}_1$ and $\mathbf{F}_1$, respectively, and all subscripts are incremented by 1.

It can be shown that PLS forms the matrix inverse defined by:

$$\mathbf{X}^+ = \mathbf{W}(\mathbf{P}^T \mathbf{W})^{-1}(\mathbf{T}^T \mathbf{T})^{-1} \mathbf{T}^T \tag{8.37}$$

where the $\mathbf{W}$, $\mathbf{P}$, and $\mathbf{T}$ are as calculated above. Note that the scores and loadings calculated in PLS are *not* the same as those calculated in PCA and PCR. They can be thought of, however, as PCA scores and loadings that have been rotated to be more relevant for predicting y. As in PCR, the PLS model converges to the MLR solution if all latent variables are included.

If the PLS algorithm does not appear transparent, do not despair. The important thing to remember is that PLS attempts to find factors or LVs which are correlated with $\mathbf{Y}$ while describing a large amount of the variation in $\mathbf{X}$. This is in contrast to PCR, where the factors (in this case PCs) are selected solely on the amount of variation they explain in $\mathbf{X}$.

8.4.6 Application of PLS

PLS can be applied to our example problem of developing a model that relates the SFCM temperatures to the level. As before, we will divide the 300 samples available for training 10 times, leaving a contiguous block of 30 samples out each time. Figure 8.9 shows the PRESS versus number of LVs for the PLS model. Note how the PRESS goes through a minimum at five LVs, as opposed to the 11 PCs used in the PCR model. This is typical of the behavior of PLS relative to PCR. Because PLS attempts to find factors that describe variance (like PCR) *and* achieve correlation (like MLR), PLS models tend to 'converge' to the MLR solution faster than PCR models. For the same number of factors, they also tend to capture more of the relevant information in $\mathbf{X}$, which results in minimum PRESS at a smaller number of factors.

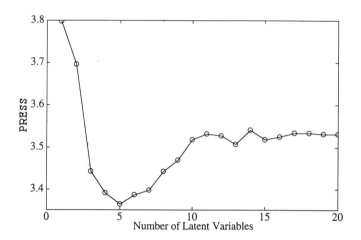

Figure 8.9 PRESS versus number of LVs in PLS model for SFCM example.

As mentioned above, there are often factors in PCR models which do not contribute positively to the predictive ability of the model, e.g. factors 2, 4, 5, 7 and 10 in our PCR example, where the PRESS goes up (see Figure 8.8). This happens in PCR because the factors (the PCs) are chosen without consideration of how they relate to the predicted variable. This happens much less frequently in PLS, because the factors (the latent variables) *are* chosen with regard to how correlated the factor scores are to the predicted variable. It is possible to form the PCR model leaving out the these factors, simply by deleting the appropriate columns from the **P** and **T** matrices in equation (8.25).

We can compare the regression vectors calculated by PLS and PCR with the regression vector calculated by MLR, as shown in Figure 8.10. It can be seen that the MLR regression coefficients vary considerably, switching from positive to negative and back several times. This 'ringing' in the coefficients is typical of models produced by MLR when the problem is ill-conditioned. Note how the PLS and PCR models are much smoother. In fact, the PLS and PCR models make more 'sense' from a physical standpoint. Because of the close proximity of the temperature sensors in this system, we would not expect the relationship between adjacent sensors and the level to be of opposite signs, as they are in the MLR model.

8.4.7 Ridge regression

Ridge regression (RR) is another technique for dealing with ill-conditioned data. Ridge regression gets its name because a constant is added

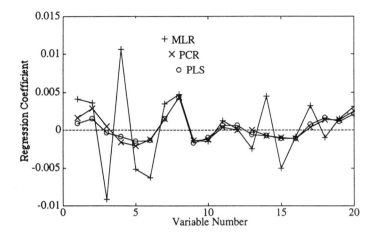

Figure 8.10 Regression coefficients for MLR, PCR and PLS models for SFCM example.

to the 'ridge' of the covariance matrix in the process of forming the pseudoinverse:

$$\mathbf{X}^+ = (\mathbf{X}^T\mathbf{X} + \mathbf{I}\Theta)^{-1}\mathbf{X}^T \tag{8.38}$$

The addition of the constant Θ to the ridge has the effect of stabilizing the values of the coefficients determined from the regression. It also has the effect of shrinking the coefficients. It has been shown [16] that the regression vector calculated from ridge regression is the least squares solution subject to the constraint that the vector is confined to a sphere centered around the origin. Thus, in some sense, RR assumes that the regression coefficients are more likely to be small (near zero) than large. Note that it is also possible to impose a non-spherical constraint on the regression vector by using a specific diagonal matrix instead of the identity matrix in equation (8.38). This is often useful when one knows that certain coefficients are more likely to be close to zero than others. This is the case with certain types of dynamic models, as will be shown below.

Ridge regression is also used quite often when one is concerned with the values of the regression coefficients themselves, rather than in prediction. If one can derive needed information from the coefficients themselves, RR can be an appropriate choice.

The trick in ridge regression is to determine the optimum value of Θ for developing a predictive model. A common statistical method was outlined by Hoerl *et al.* [17]. It is also possible to determine the optimum value through cross-validation.

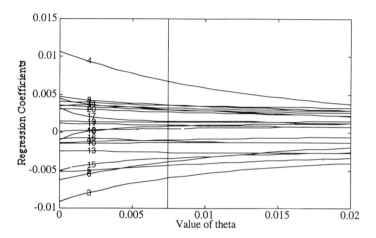

Figure 8.11 Regression coefficients for SFCM problem as a function of the ridge parameter Θ. Vertical line shows the best fit for Θ = 0.00743276.

8.4.7.1 Application of ridge regression. As an example of the use of ridge regression, we will now revisit the SFCM data used in the examples of PCR and PLS. The regression coefficients are shown as a function of the ridge parameter in Figure 8.11. Note how the values shrink as the value of Θ is increased. The value of the ridge parameter using the method of Hoerl and Kennard is 0.0074, while the value determined using the cross-validation procedure outlined above is 0.0076. The regression coefficients are compared with the those determined from MLR in Figure 8.12. Note how the regression coefficients have been shrunk in RR as compared to MLR.

8.4.8 Comparison of linear models on SFCM examples

Of course, the real reason for forming the regression models is to make predictions of the SFCM level given the temperatures in the event of a failure of the level instrument or as a check on the performance of the instrument. Thus the best test of the models is their ability to predict the level given new temperature measurements.

Before comparing the predictive ability of the models, it is useful to introduce several measures of a model's fit to the data and predictive power. In all of the measures considered, we are attempting to estimate the 'average' deviation of the model from the data. The root-mean-square error of calibration (RMSEC) tells us about the fit of the model to the calibration data. It is defined as:

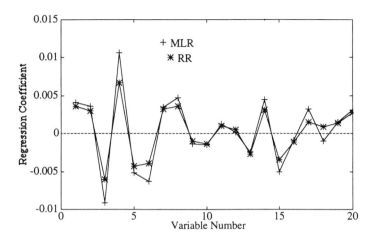

Figure 8.12 Comparison of regression coefficients for SFCM example from multiple linear regression and ridge regression.

$$\text{RMSEC} = \sqrt{\frac{\displaystyle\sum_{i=1}^{n} (\hat{y}_i - y_i)^2}{n}} \qquad (8.39)$$

where the $\hat{y}_i$ are the values of the predicted variable when all samples are included in the model formation and n is the number of calibration samples. RMSEC is a measure of how well the model fits the data.

This is in contrast to the root-mean-square error of cross-validation (RMSECV) which is a measure of a model's ability to predict new samples. The RMSECV is defined as in equation (8.39), except the y_i are predictions for samples *not* included in the model formulation. RMSECV is related to the PRESS value for the number of PCs or LVs included in the model, *i.e.*

$$\text{RMSECV}_k = \sqrt{\frac{\text{PRESS}_k}{n}} \qquad (8.40)$$

where PRESS_k is the sum of squares prediction error for the model which includes k factors. Of course, the exact value of RMSECV depends not only on k but on how the test sets were formed. It is also common to calculate PRESS, and thus RMSECV, for a leave-one-out cross-validation, *i.e.* where each sample is left out of the model formulation and predicted once.

It is also possible to calculate a root-mean-square error of prediction (RMSEP) when the model is applied to new data provided that the reference values for the new data are known. RMSEP is calculated exactly as in equation (8.39) except that the estimates y_i are based on a previously developed model, not one in which the samples to be 'predicted' are included in the model building.

The MLR, PCR, PLS, and RR models developed for the SFCM example can now be compared using the measures just introduced. This information is summarized in Table 8.2. Two new data sets of 100 samples each were used for the calculation of RMSEP (RMSEP – 1 and – 2) and the RMSEP for the combined data was also calculated. The actual level is compared with the prediction from PLS in Figure 8.13.

Table 8.2 Comparison of MLR, PCR, PLS and RR models used for SFCM example

	MLR	PCR	PLS	RR
RMSEC	0.0991	0.1012	0.1015	0.0996
RMSECV	0.1122	0.1069	0.1068	0.1117
RMSECV-LOO	0.1062	0.1050	0.1032	0.1067
RMSEP-1	0.1072	0.1077 (0.1004)	0.0994	0.1042
RMSEP-2	0.1824	0.1790 (0.1828)	0.1725	0.1800
RMSEP-1 & 2	0.1496	0.1477 (0.1474)	0.1424	0.1471

Note how the MLR model shows the smallest value for RMSEC, the error of calibration. This indicates that, as expected, the MLR model 'fits' the data best. Note, however, from the RMSECV that the MLR does not predict the best, even for samples within the original calibration data. Here the 'best' model is PLS. This is also true for the RMSECV for leave-one-out cross-validation (RMSECV-LOO). Note that the numbers for RMSECV-LOO are all lower than those for RMSECV. This is

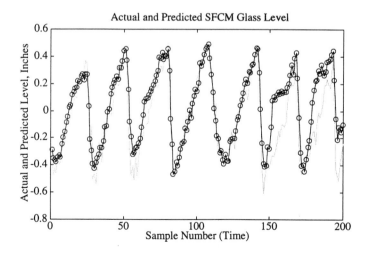

Figure 8.13 Actual (solid line) and predicted (circles) glass level for SFCM using PLS model.

expected, because the ability to predict improves as the number of samples included in the model formulation is increased.

For the prediction on new samples, the PLS model outperforms all other models. This is true for both of the test sets individually and the combined test set. In this example, conditions for the first test set were relatively quiescent in the SFCM system, while there were some process disturbances during the period of the second data set. Note how RMSEP goes up for all models on the second set of data. The values in parentheses are for the PCR model which used only the PCs that appeared to improve the PRESS, as indicated in the section on PCR above. Note that this model is somewhat better than the PCR model, with all PCs included.

It should be evident from the comparison of models that *fit* and *prediction* are entirely different aspects of a model's performance. If prediction is the goal, a model should be built with this criteria in mind. The calculation of PRESS from a cross-validation procedure is one way to measure a collection of models' predictive abilities.

8.4.9 Continuum regression

Recently, PCR, PLS, and MLR have all been unified under the single technique continuum regression (CR) [15, 18, 19]. CR is a continuously adjustable technique which encompasses PLS and includes PCR and MLR at opposite ends of the continuum, as shown in Figure 8.14. When using CR, cross-validation must be done to determine both the number of factors and the technique within the continuum that produces the optimum model (i.e. the model that predicts most accurately when used with new data).

In the version of CR considered here, the first step is to perform a singular value decomposition (SVD) on the independent variable block $\mathbf{X}(m \times n)$.

$$\mathbf{X} = \mathbf{USV}^{\mathrm{T}} \qquad (8.41)$$

where $\mathbf{U}$ $(m \times m)$ and $\mathbf{V}$ $(n \times n)$ are orthonormal matrices and $\mathbf{S}$ $(m \times n)$ is diagonal, *i.e.* zero everywhere except on the matrix diagonal. The

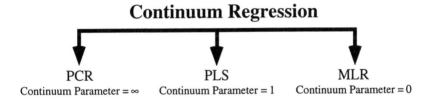

Figure 8.14 Relationship between PCR, PLS, MLR and CR.

elements of **S** are known as the *singular values* and are placed in **S** in descending order, *i.e.* $s_{11} > s_{22} > s_{33}$ and so on. The SVD is closely related to PCA. In fact, the eigenvectors or loadings **P** are identical to **V**, that is:

$$V = P \tag{8.42}$$

The relation between the scores **T** and the SVD is

$$US = T \tag{8.43}$$

In fact, the PCA model proposed in equation (8.2) can be written as

$$X = T_k P_k^T + E = U_k S_k V_k^T + E \tag{8.44}$$

The next step in CR is to form the matrix S^m by taking each of the singular values to the desired power, m. The new **X** block, defined as X^m, is then formed as follows:

$$X^m = US^m V^T \tag{8.45}$$

The PLS algorithm described earlier can now be used along with the matrix inverse defined by equation (8.37) to produce a regression vector, **r**, for any number of latent variables desired. That is

$$r = W(P^T W)^{-1} (T^T T)^{-1} T^T y \tag{8.46}$$

The regression vector obtained is not properly scaled because of the rescaled **X** block. Any regression vector, **r**, can be rescaled, however, by projecting it into the SVD basis set **V** and multiplying by the ratios of the singular values used in equation (8.45) to the original singular values calculated in equation (8.41). Thus:

$$r_{scl} = r V S_{scl} V^T \tag{8.47}$$

where

$$S_{scl} = S^m \!\int\! S \tag{8.48}$$

where the symbol '$\int$' indicates term by term division of the non-zero elements of the singular value matrices.

The application of PCR, PLS and CR to the calibration of on-line near-infrared (NTR) spectrometers, will serve as our example of calibration in process chemometrics. Figure 8.15 shows the NIR spectra of 30 pseudogasoline samples [20]. Cross-validation was used to find the optimal model within the CR parameter space for five analytes in the gasoline samples. The cross-validation prediction error sum of squares results are shown in Table 8.3 for one of the analytes. Note how the prediction error initially decreases as factors are added to a model. The error then goes through a minimum (bold data) and then increases,

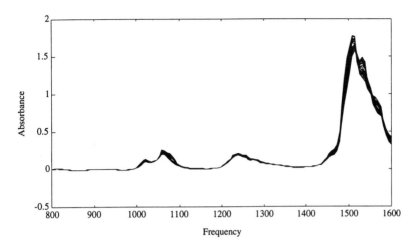

Figure 8.15   NIR spectra of gasoline samples.

finally reaching a stable value. As one approaches the MLR end of the technique, there are fewer LVs in the model with optimum PRESS. Note that the optimum CR model has almost 4% less prediction error than the optimum PLS model and over 14% less prediction error than the model which uses all of the factors which describe the data.

Table 8.3   Continuum regression prediction error results for NIR spectrometer calibration

Number of factors in continuum regression model	Continuum regression parameter						
	1.0 (PLS)	0.707	0.500	0.354	0.250	0.176	0.125
1	463.95	326.22	217.03	137.85	84.41	50.73	30.90
2	175.83	127.85	79.02	41.90	21.58	13.37	10.87
3	76.95	48.19	27.95	16.58	11.79	10.47	**10.38**
4	46.24	27.62	15.96	11.07	**10.10**	**10.30**	10.54
5	33.64	18.68	11.32	9.85	10.22	10.56	10.67
6	17.96	11.41	**9.36**	**9.82**	10.51	10.68	10.70
7	14.26	10.31	9.90	10.55	10.68	10.70	10.70
8	13.02	10.09	10.30	10.63	10.69	10.70	10.70
9	11.40	**10.05**	10.50	10.67	10.70	10.70	10.70
10	**9.71**	10.18	10.57	10.69	10.70	10.70	10.70
11	10.27	10.39	10.64	10.70	10.70	10.70	10.70
12	10.32	10.43	10.68	10.70	10.70	10.70	10.70
13	10.24	10.54	10.70	10.70	10.70	10.70	10.70
14	10.40	10.67	10.70	10.70	10.70	10.70	10.70
15	10.46	10.64	10.70	10.70	10.70	10.70	10.70

For explanation of bold data see text.

This example demonstrates why the authors believe that it is largely because of advances in chemometrics that sophisticated instruments such as NIR spectrometers are beginning to be common for on-line use instruments. Such instruments place large demands on calibration and fault detection algorithms. Prior to the advent of modern multivariate calibration techniques and (the required computer power) it was unrealistic to run such an instrument in the process environment.

8.5 Non-linear calibration methods

There are many non-linear methods currently being researched. These methods include non-linear adaptations of PCR and PLS, artificial neural networks (ANNs), projection pursuit regression (PPR), alternating conditional expectations (ACE), multivariate adaptive regression splines (MARS) and locally weighted regression (LWR), to name a few. Of these, we will consider only ANNs, due to their large number of application areas, and LWR, because of its conceptual simplicity and generally good performance.

8.5.1 Locally weighted regression

The basic idea behind LWR is that local regression models are produced using points that are near the sample to be predicted in the independent variable space. Furthermore, each calibration sample is weighted in the regression according to how close it is to the sample to be predicted. There are several ways of performing LWR; here we will outline the method developed by Næs *et al.* [21] where a reduced order description of the calibration data is achieved by forming a PCA model of the independent variables. (The result is something like locally weighted PCR.) Nearby points for the regression are selected based on their Euclidean distance based on the principal component scores adjusted to unit variance, which is also known as the Mahalanobis distance based on the PCA scores. The relationship between the scores t_i defined by equation (8.2) and the adjusted scores, t_{ia} is

$$t_{ia} = t_i/\text{std}(t_i) \tag{8.49}$$

where $\text{std}(t_i)$ is the standard deviation of the ith score vector. Thus, for a calibration set where k PCs were used to describe the data, the distance between the jth calibration set point and the lth point to be predicted, d_{jl} is defined as

$$d_{jl} = \sqrt{\sum_{m=1}^{k} (t_{jm} - t_{lm})^2} \tag{8.50}$$

where t_{jm} is the score on the mth principal component for the jth sample and t_{lm} is the score on the mth principal component for the lth sample.

A weighting function is defined using the Mahalanobis distance calculated in equation (8.50) according to

$$w_{j1} = \left(1 - \left(\frac{d_{j1}}{d_{max}}\right)^3\right)^3 \quad \text{for} \quad d_{j1} < d_{max} \tag{8.51}$$

and

$$w_{j1} = 0 \quad \text{for} \quad d_{j1} > d_{max} \tag{8.52}$$

where d_{max} is the distance of the furthest point considered in the regression (furthest away of the closest n points selected). The local regression model is then formed using the n points selected and the calculated weights.

$$\mathbf{b} = ([\mathbf{T}_a\mathbf{1}]^T\mathbf{h}[\mathbf{T}_a\mathbf{1}])^{-1}[\mathbf{T}_a\mathbf{1}]^T\mathbf{h}\mathbf{y} \tag{8.53}$$

where $[\mathbf{T}_a\mathbf{1}]$ is the matrix of adjusted scores augmented by a vector of ones of appropriate length (to accommodate an intercept term) and $\mathbf{h}$ is a diagonal matrix with the elements defined by

$$h_{jj} = (w_{j1})^2 \tag{8.54}$$

Note that the regression vector defined by equation (8.53) is in terms of the new sample scores and, therefore, can be used to predict the value of y_1 based on the scores of sample x_1.

As with PLS and PCR, LWR models must be cross-validated in order to achieve maximum predictive ability. Here the parameters to cross-validate against are the number of points to be considered local and the number of PCs to use in the distance measure and local model formation.

In essence, LWR works by 'linearizing' the process which produced the data around the point of interest, i.e. the point to be predicted. The approach works quite well provided that the data is not highly non-linear and there are points in the calibration set which are 'near' the new samples.

8.5.2 Artificial neural networks

Artificial neural networks (ANNs) consists of multiple layers of highly interconnected processing elements known as nodes [22–24]. There are many neural network structures, but the most suitable structure for modelling chemical process and developing non-linear calibrations is the multilayer feedforward configuration. Such a network is shown in Figure 8.16. The network consists of an input layer (on the left in the figure), a hidden layer, and an output layer (on the right). There can be more than one hidden layer, however, in most applications a single hidden layer is adequate. Each layer consists of a collection of nodes,

Figure 8.16 Multilayer feedforward artificial neural network structure.

indicated by the circles in the Figure 8.16. The configuration of a single node is shown in Figure 8.17. As shown in the figure, each node calculates the weighted sum of the inputs to the node and transforms that sum with some function. The most common node function is the sigmoid

$$o_j = \frac{1}{1 + e^{-\left(\frac{x}{\gamma}\right)}} \tag{8.55}$$

where

$$x = \sum_{i=1}^{I} w_{ij} o_i + \Theta_j \tag{8.56}$$

where Θ_j is a bias term, o_i is the output from the ith node of the previous layer, and w_{ij} represents the weight between node i and node j. In equation (8.55), γ is known as the gain and adjusts the slope of the sigmoid transfer function. The value of the bias term Θ_j, the gain γ, and

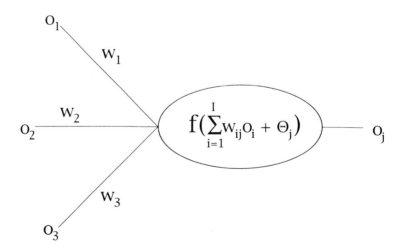

Figure 8.17 Artificial neural network node schematic.

the weights w_{ij} are adjusted using a numerical optimization method such as a gradient descent technique (or other more sophisticated approach) so that the network gives the desired output when presented with a set of inputs. This is commonly referred to as 'training' the network.

Many applications have been reported in the literature, and it is clear that when properly structured and parameterized, ANNs can outperform conventional techniques when applied to highly non-linear problems. This is confirmed by the study of Sekulic [25]. There are many problems with the application of ANNs, however. The amount of expertise required to develop a working ANN is quite high, and the time required (both personnel and computer) is generally large. This is due to the very large number of adjustable parameters in the network and in the optimization algorithms required to train them. In systems where linear techniques will suffice, there is little incentive to develop an ANN.

One way to reduce the size of the ANN parameterization problem is to preprocess the data using PCA or PLS. Several authors [26,27] have reported that hybrid PLS-ANN structures outperform both PLS and ANNs alone in highly non-linear applications.

8.6 Instrument standardization

Once a calibration for a process analytical instrument has been developed, it would be convenient if it could be transferred to other similar instruments. This would eliminate the need to run a large number of calibration samples and to develop a completely new calibration model. Direct transfer of calibrations is not usually satisfactory, however, because instruments are never identical. It also happens that instruments tend to drift, and this causes instruments to change their characteristics after some time. Instrument standardization methods were developed in order to deal with these similar problems.

This problem has been addressed in several different ways by various researchers, however, perhaps the most successful technique is the piecewise direct standardization (PDS) method [20, 28, 29] It works by forming 'local' linear models that relate the response of the lower resolution instrument over a range of frequencies to the response of the higher resolution instrument at a single frequency. This is shown schematically in Figure 8.18. This is in contrast to direct standardization (DS), which uses the entire spectra.

The general idea behind instrument standardization is that it is possible to develop a mathematical transform that maps the response of one instrument into another. This transform is developed using a small number of samples that are run on both instruments (or a single instrument at two different times).

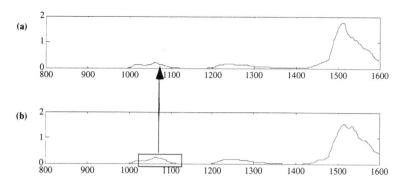

Figure 8.18 Relationship between spectra in development of PDS transform. (a) NIR spectrum from instrument 1; (b) NIR spectrum from instrument 2.

8.6.1 Sample subset selection

The first step in developing a standardization transform is to select samples that will be measured on both machines. Samples should be chosen based upon their multivariate leverage, which is a measure of their uniqueness in the calibration set. This is a simple procedure which starts by selecting the sample with the greatest deviation from the multivariate mean of the calibration samples. All other samples are then orthogonalized with respect to the first sample and the procedure is repeated. Given a calibration set from instrument 1, $\bar{\mathbf{R}}_1$ (m samples by n wavelengths,) that has been mean-centered, calculate the leverage matrix $\mathbf{H}$ as:

$$\mathbf{H} = \bar{\mathbf{R}}_1\bar{\mathbf{R}}_1^T \tag{8.57}$$

Note that $\mathbf{H}$ is m by m. The diagonal elements of $\mathbf{H}$ are the leverage of the samples, i.e. h_{ii} is the leverage of the ith sample. Select the sample with the highest leverage, $\mathbf{r}_{max}$. Set the elements of this spectra equal to zero, then orthogonalize each remaining spectra $\mathbf{r}_i$ in the data set by performing.

$$\mathbf{r}_{io} = \mathbf{r}_i - \mathbf{r}_{max}((\mathbf{r}_{max} - \mathbf{r}_i^T)/(\mathbf{r}_{max} - \mathbf{r}_{max}^T)) \tag{8.58}$$

for each spectra where $\mathbf{r}_{io}$ is the orthogonalized $\mathbf{r}_i$. Note that this procedure subtracts from $\mathbf{r}_i$ the portion of $\mathbf{r}_{max}$ that projects onto it. This procedure is then repeated as in equation (8.57) using the orthogonalized spectra in place of $\bar{\mathbf{R}}_1$ and the next highest leverage sample is found and so on.

Generally, it is found that three spectra are sufficient for building a transfer function between instruments, however, additional samples can be beneficial. Once the transfer samples are identified based on the leverage of their spectra, the samples must be measured on both spectrometers and the calculation of the standardization can commence.

8.6.2 Development of the standardization transform

It is assumed that the calibration model is formulated as

$$\mathbf{y} = \mathbf{R}_1\beta + \mathbf{1}b_1 \tag{8.59}$$

where $\mathbf{y}$ is the concentration vector of the analyte of interest, $\mathbf{R}_1$ is the response matrix, β is the regression vector, b_1 is the constant offset of the model and $\mathbf{1}$ is an $m \times 1$ vector of ones, where m is the number of samples. After mean-centering:

$$\bar{\mathbf{y}} = \bar{\mathbf{R}}_1\beta \tag{8.60}$$

where $\bar{\mathbf{y}}$ and $\bar{\mathbf{R}}_1$ are the mean-centered $\mathbf{y}$ and $\mathbf{R}_1$, respectively. The relationship between the instruments is then modeled as:

$$\mathbf{S}_1 = \mathbf{S}_2\mathbf{F}_b + \mathbf{1}\mathbf{b}_s^T \tag{8.61}$$

where $\mathbf{S}_1$ is the response of instrument 1 to the p transfer samples, $\mathbf{S}_2$ is the response of instrument 2 (or instrument 1 at a later time), $\mathbf{F}_b$ is the transformation matrix, $\mathbf{b}_s$ is the background correction which accommodates the additive background difference between the instruments and $\mathbf{1}$ is a p by 1 vector of ones, where p is the number of transfer samples (rows) in $\mathbf{S}_1$ and $\mathbf{S}_2$. Note that it is assumed at this point that $\mathbf{S}_1$ and $\mathbf{S}_2$ are *not* mean centered.

The transfer function matrix, $\mathbf{F}_b$, is then calculated to satisfy

$$\bar{\mathbf{S}}_1 = \bar{\mathbf{S}}_2\mathbf{F}_b \tag{8.62}$$

where $\bar{\mathbf{S}}_1$ and $\bar{\mathbf{S}}_2$ are the mean-centered $\mathbf{S}_1$ and $\mathbf{S}_2$. As noted previously, $\mathbf{F}_b$ can be calculated several ways, one of which is the PDS method as described above. In PDS, $\mathbf{F}_b$ has the diagonal structure shown in Figure 8.19, where each local model (generally determined from PLS or PCR) relates the response of a number of channels of instrument 2 to a single channel on instrument 1. One such model is shown in bold in the figure. Another method is the DS method [29]), which uses

$$\mathbf{F}_b = \bar{\mathbf{S}}_2^+\bar{\mathbf{S}}_1 \tag{8.63}$$

where $\bar{\mathbf{S}}_2^+$ is the pseudo inverse of $\bar{\mathbf{S}}_2$. This is formed through use of the SVD as follows. Take

$$\bar{\mathbf{S}}_2 = \mathbf{U}\mathbf{S}\mathbf{V}^T \tag{8.64}$$

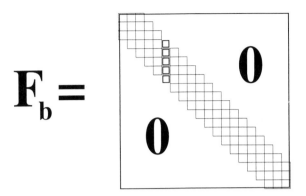

Figure 8.19 Structure of the PDS transformation matrix F_b.

then

$$\bar{\mathbf{S}}_2^+ = \mathbf{U}\mathbf{S}^{-1}\mathbf{V}^{\mathsf{T}} \tag{8.65}$$

where $\mathbf{S}^{-1}$ is $\mathbf{S}$ with all of the non-zero singular values replaced by their reciprocals.

The additive background term, $\mathbf{b}_s$, is calculated as

$$\mathbf{b}_s = \mathbf{s}_{1m} - \mathbf{F}_b^{\mathsf{T}}\mathbf{s}_{2m} \tag{8.66}$$

where $\mathbf{s}_{1m}$ and $\mathbf{s}_{2m}$ are the mean vectors of $\mathbf{S}_1$ and $\mathbf{S}_2$, respectively.

8.6.3 Prediction using standardized spectra

After calibration and formation of the standardization transform, the concentration estimate for any spectrum, $\mathbf{r}_{2un}$, measured on instrument 2 can be obtained using the calibration model developed on instrument 1 as

$$y_{2un} = (\mathbf{r}_{2un}\mathbf{F}_b + \mathbf{b}_s^{\mathsf{T}})\beta + b_1 \tag{8.67}$$

where b_1 is defined from equation (8.59).

8.6.3.1 Standardization example. As our example of standardization we will use the NIR spectra of pseudo gasoline mixtures used for the CR section. The spectra were measured on two instruments. A calibration was developed for each of the five analytes using PLS. The fit of the calibration samples to the calibration model is shown in Figure 8.20.

The calibration models were then applied to the raw spectra from instrument 2. The results are shown in Figure 8.21. A large amount of bias in the predictions is evident.

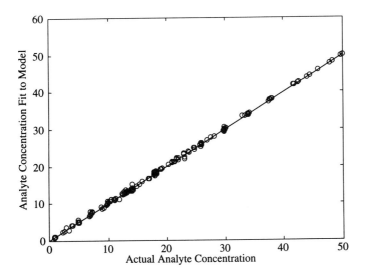

Figure 8.20 Actual vs fit concentrations for instrument 1.

Five samples were selected from instrument 1 based on their leverage. Instrument standardization transforms were then developed from the five samples using the DS method and the PDS method, both with additive background correction. In the case of PDS, a window of five channels

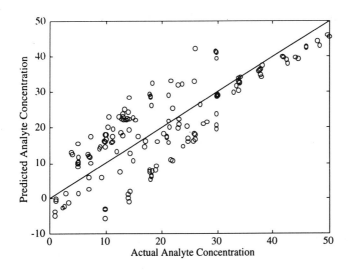

Figure 8.21 Predictions based on instrument 2 spectra and instrument 1 calibration.

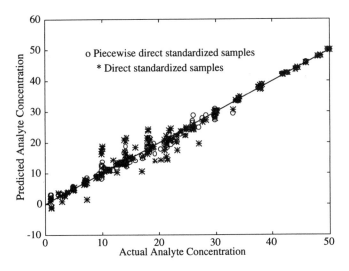

Figure 8.22 Predictions based on standardized instrument 2 spectra and instrument 1 calibration.

was used. The calibration models were then applied to the transformed instrument 2 spectra. The result is shown in Figure 8.22. Note that the predictions are now much closer. Also, the PDS method has outperformed the DS method.

8.7 Process modeling

There are several types of process modeling, but when control engineers speak of process modeling, they are usually referring to the development of a dynamic process model. Such a model relates the time history of the process inputs to the time history of the process outputs. It is often possible to obtain a dynamic system model directly from theoretical considerations (such as material and energy balances, heat and mass transfer considerations and known reaction kinetics), but there are many instances where the models must be identified from process data. In this section we focus on the latter case.

8.7.1 Dynamic process model forms

We have already introduced the state-space model form in a previous section. Here we introduce two somewhat simpler model forms that are often used for describing the dynamics of a single input/single output (SISO) chemical process. These models (or collections of them in the case of multiple input/multiple output (MIMO) processes) can be transformed into the equivalent state-space model form. The simplest of these forms is the finite impulse response (FIR) model. The premise of a FIR model is that the output of a process depends only on the history of the inputs, and that inputs sufficiently far in the past have no effect on the current output. Such a process is said to be asymptotically stable and most chemical processes meet this definition. The FIR model form is

$$y_t = b_1 u_{t-1} + b_2 u_{t-2k} + \ldots + b_i u_{t-ik} + \ldots + b_n u_{t-nk} \qquad (8.68)$$

where y_t is the process output at time t, k is the sampling interval (the time between points where the process input and output is measured, which must be at regular intervals), u_{t-ik} is the input to the process at time $(t - ik)$ and b_i is the coefficient which relates the effect of an input at time $(t - i)$ on the current output. Here nk is the effective settling time of the process, i.e. the time beyond which past inputs have little or no effect on the current output. The FIR model can be written much more simply in matrix notation as

$$y_t = \mathbf{u}_{t-k, t-nk}^{\mathrm{T}} \mathbf{b} \qquad (8.69)$$

where $\mathbf{u}_{t-1, t-nk}$ is the (column) vector of inputs from time $(t - k)$ to $(t - nk)$ and $\mathbf{b}$ is the vector of associated FIR coefficients. Typical chemical processes are sampled at intervals that are considerably shorter than the process settling time, thus, an adequate FIR description of a process will typically have many coefficients, 15–30 being common and 100 not unheard of.

Another common form is the autoregressive extensive variable (ARX) form. The ARX form assumes that the current value of the process output is a function of recent inputs to the process (the extensive variable) and recent outputs from the process (the autoregressive part). The ARX model form is

$$y_t = a_1 y_{t-1} + \ldots + a_m y_{t-mk} + b_1 u_{t-1} + \ldots + b_n u_{t-nk} \qquad (8.70)$$

where y_{t-ik} is the process output at time $(t - ik)$, a_i is the coefficient which relates the effect of the output at time $(t - i)$ on the current output, and the other symbols have the same meaning as above. Note that the ARX form considers n past values of the input but m past values of the output. Generally, for a model to be stable $m < n$. Again, we can write the ARX model in a simpler matrix notation as

$$y_t = y_{t-k, t-mk}^{\mathrm{T}} \mathbf{a} + \mathbf{u}_{t-k, t-nk}^{\mathrm{T}} \mathbf{b} \qquad (8.71)$$

where $\mathbf{y}_{t-1, t-mk}$ is the (column) vector of outputs from time $(t - k)$ to $(t - mk)$ and $\mathbf{a}$ is the vector of associated autoregressive coefficients.

The number of terms in an ARX model, known as the order of the model, has to do with the dynamics of the process being modeled. For instance, simple processes such as a constant volume, constant flow rate continuous stirred tank reactor (CSTR) where the input is the concentration of a non-reacting species in the feed stream and the output is the concentration of that species in the bulk of the tank is well modeled by a constant over a first-order input. Note that this is in contrast to an FIR model, which must have enough coefficients to span the settling time of the process. It is typical that ARX models have far fewer parameters than FIR models.

8.7.2 Dynamic model identification

As mentioned above, often it is not possible to obtain an adequate dyanmic process model directly from theoretical considerations. In these instances, a model must be identified from process data. There has been a great deal of work in the field of identifying models of the ARX structure [30–31]. This problem can actually be quite involved, particularly if the process is prone to disturbances which themselves have a significant settling time. In these cases, the process can be said to be corrupted by autocorrelated noise. Prediction error methods (PEM) have been developed to identity models of ARX and similar structure in the face of autocorrelated noise. These techniques perform an optimization on the ARX parameters until a model is achieved that gives optimum prediction when using only the known inputs to the process. In most instances, where one is allowed to perform a well designed experiment on the process inputs, PEM will yield a more than adequate process model in ARX form.

There are, however, some drawbacks to the ARX model form. The foremost of these is that, once the order of the model is decided, only certain types of process behavior can be described. Another drawback is that an optimization routine is required for identification of the ARX form in the presence of auto-correlated noise. Simple regression analysis will not result in an adequate model.

The FIR form, in contrast, can effective describe essentially all types of (stable) process behavior. The only parameter that must be determined *a priori* is the settling time of the process, which is usually reasonably well known. Furthermore, FIR models can be determined directly from regression analysis, although the problem is typically quite ill-conditioned. Of course, it is this type of problem that the PLS methods were developed to handle. In some instances where it is not possible to perform a well-designed identification experiment, the data will be particularly ill-conditioned and the identification techniques used must be robust enough to still achieve an adequate model.

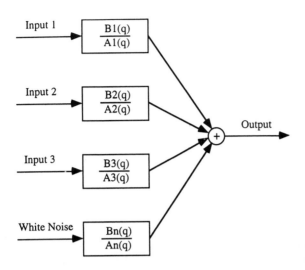

Figure 8.23 Schematic representation of linear additive process for identification tests.

8.7.3 Example identification problem

In this example, a multi-input single output (MISO) process model will be identified using CR and Ljung's prediction error method (PEM) [30]. The true process can be represented schematically as shown in Figure 8.23. The parameters in the model are given in Table 8.4 in shift operator form. The input signals and the output from the process are shown in Figure 8.24 (see for example [9]). As can be seen from Figure 8.24, there is a great deal of correlation in the input signals. Five hundred calibration and test samples were generated for the identification experiment.

Table 8.4 Model parameters for example identification problem

Bl(q)	0	0.2036	-0.0178	-0.0105	0.0007	-0.0000
Al(q)	1	-1.2683	0.5217	-0.0814	0.0040	-0.0000
B2(q)	0	0.2956	0.1158	-0.0639	0.0046	-0.0000
A2(q)	1	-0.7951	0.5692	-0.5071	0.0851	-0.0002
B3(q)	0	0.7796	0.9311			
A3(q)	1	0.1509	0.5598			
BN(q)	0.1367	0.1367				
AN(q)	1.0000	-0.7265				

FIR models were identified from this data using CR. There were 15 FIR coefficients used for each input and the models were identified simultaneously. Nine values of the continuum parameter, in logarithmically spaced intervals from 4 (near PCR) to 0.25 (near MLR), were tested. The best model, based on cross-validation against the test set, had 16 latent variables and a continuum parameter of 0.7071.

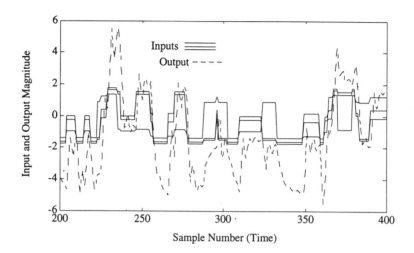

Figure 8.24 A segment of the identification data produced by the process shown in Figure 8.23.

ARX models of the process were identified using PEM. As mentioned above, the PEM routine performs an optimization on the parameters to minimize the prediction error. The true model orders were used in the ARX model estimation. Reduced order models were also identified but were not found to be significantly better than the model of correct order.

The actual and predicted values of the process output are shown in Figure 8.25. Note that the FIR model prediction (identified with CR) is nearly indistinguishable from the actual output. The prediction error for the ARX models (identified with PEM) is very large. Note that the predictions are effectively for infinitely far out into the future, rather than one step ahead. The ARX model would show much better performance for one-step-ahead prediction.

It is important to note that this example was contrived to be a case where the CR/FIR models should significantly outperform PEM/ARX models. As the inputs become less correlated, the performance of the two methods becomes quite similar. When it is possible to perform a well-designed identification experiment and the problem is properly formulated (i.e. the correct model order chosen), PEM will produce superior models. However, the CR/FIR form offers an attractive alternative in some particularly difficult cases.

8.7.4 Kalman filtering

It will be shown later that dynamic process models can be used directly in the control of processes. Process models may also be used to improve

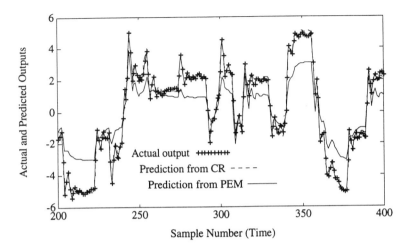

Figure 8.25 Actual and predicted process output for test data prduced by the process shown in Figure 8.23. Note that the prediction from CR (dashed line) is so close to the actual output that it is difficult to see on the graph.

the predictions from on-line process instruments through the use of a Kalman filter. There are many texts that describe the Kalman filter and its properties in detail [9, 10, 32]. We will briefly consider the general form of the Kalman filter and its use in process monitoring.

Consider the one-step-ahead prediction problem with a process described by the state-space form of equations (8.7) and (8.8). The Kalman estimator has the form:

$$\hat{\mathbf{x}}(k+1|k) = \mathbf{\Phi}\hat{\mathbf{x}}(k|k-1) + \mathbf{\Gamma}\mathbf{u}(k) + \mathbf{K}(k)[\mathbf{y}(k) - \mathbf{C}\hat{\mathbf{x}}(k|k-1)] \quad (8.72)$$

where $\hat{\mathbf{x}}(k+1|k)$ is the estimate of the state vector $\mathbf{x}$ at time $k+1$ using information available up to time k, $\hat{\mathbf{x}}(k|k-1)$ is the estimate of the state vector from the previous time step, $\mathbf{\Phi}$, $\mathbf{\Gamma}$, $\mathbf{C}$ and $\mathbf{y}$ have the meaning as defined earlier, $\mathbf{K}$ is the Kalman gain. From the equation it can be seen that the process inputs, the process model and the measurements are used to estimate the value of the process states. The critical parameter, of course, is the Kalman gain, $\mathbf{K}(k)$, which must be chosen so as to minimize the variance between the actual and predicted values of the state vector $\mathbf{x}$. Note that the Kalman gain is assumed to be time varying, as the formulation allows for time varying $\mathbf{\Phi}$, $\mathbf{\Gamma}$ and $\mathbf{C}$.

It can be shown that optimal Kalman filter gain $\mathbf{K}(k)$ is given by

$$\mathbf{K}(k) = \mathbf{\Phi}\mathbf{P}(k)\mathbf{C}^{\mathrm{T}}(\mathbf{R}_2 + \mathbf{C}\mathbf{P}(k)\mathbf{C}^{\mathrm{T}})^{-1} \quad (8.73)$$

where $\mathbf{R}_2$ is the measurement noise covariance matrix and $\mathbf{P}(k)$ is defined by

$$\mathbf{P}(k + 1) = \mathbf{\Phi}\mathbf{P}(k)\mathbf{\Phi}^{\mathrm{T}} + \mathbf{R}_1 - \mathbf{\Phi}\mathbf{P}(k)\mathbf{C}^{\mathrm{T}}(\mathbf{R}_2 + \mathbf{C}\mathbf{P}(k)\mathbf{C}^{\mathrm{T}})^{-1}\mathbf{C}\mathbf{P}(k)\mathbf{\Phi}^{\mathrm{T}} \quad (8.74)$$

where $\mathbf{R}_1$ is the state noise covariance matrix. Note that equation (8.74) gives an estimate of the variance of the estimation error and therefore is known as the variance equation. It is assumed that the covariance between the state noise and measurement noise $\mathbf{R}_{12}$ is zero. If it is not, the gain and the variance equations (8.73) and (8.74) become

$$\mathbf{K}(k) = (\mathbf{\Phi}\mathbf{P}(k)\mathbf{C}^{\mathrm{T}} = + \mathbf{R}_{12})(\mathbf{R}_2 + \mathbf{C}\mathbf{P}(k)\mathbf{C}^{\mathrm{T}})^{-1} \quad (8.75)$$

$$\mathbf{P}(k + 1) = \mathbf{\Phi}\mathbf{P}(k)\mathbf{\Phi}^{\mathrm{T}} + \mathbf{R}_1 - \mathbf{K}(k)(\mathbf{C}\mathbf{P}(k)\mathbf{C}^{\mathrm{T}} + \mathbf{R}_2)\mathbf{K}(k) \quad (8.76)$$

It is clear from equation (8.73) that as the measurement noise increases, i.e.. $\mathbf{R}_2$ becomes large, the Kalman gain $\mathbf{K}$ will become small. In equation (8.72), it is apparent that a small $\mathbf{K}$ will discount the measurements in favor of the predictions of the model. Thus, the Kalman filter achieves the optimum balance between the predictions of the dynamic model and the process measurements when estimating the state of the process.

The Kalman filter described by equation (8.72) gives optimal one-step-ahead prediction of the state vector. It is also possible to formulate the problem as a true filter, i.e. predict the value of the states at time k including all information available up to and including time k. In this case the Kalman filter is defined by

$$\hat{\mathbf{x}}(k + 1|k + 1) = \mathbf{\Phi}\hat{\mathbf{x}}(k|k) + \mathbf{\Gamma}\mathbf{u}(k) + \mathbf{K}(k + 1)[\mathbf{y}(k + 1)$$
$$- \mathbf{C}(\mathbf{\Phi}\hat{\mathbf{x}}(k|k) + \mathbf{\Gamma}\mathbf{u}(k))] \quad (8.77)$$

where

$$\mathbf{K}(k) = \mathbf{P}(k|k - 1)\mathbf{C}^{\mathrm{T}}(\mathbf{R}_2 + \mathbf{C}\mathbf{P}(k|_k - 1)\mathbf{C}^{\mathrm{T}})^{-1} \quad (8.78)$$

$$\mathbf{P}(k|k - 1) = \mathbf{\Phi}\mathbf{P}(k - 1|k - 1)\mathbf{\Phi}^{\mathrm{T}} + \mathbf{R}_1 \quad (8.79)$$

$$\mathbf{P}(k|k) = \mathbf{P}(k|k - 1) - \mathbf{K}(k)\mathbf{C}\mathbf{P}(k|k - 1) \quad (8.80)$$

$$\mathbf{P}(0|0) = \mathbf{R}_0 \quad (8.81)$$

where $\mathbf{R}_0$ is the state covariance at time 0.

8.7.5 Kalman filter example

We will now consider an example of Kalman filtering that will also help illustrate the connection between the ARX and state-space model forms. Consider the first-order system described by

$$y(k) + ay(k - 1) = bu(k - 1) + e(k) + ce(k - 1) \quad (8.82)$$

where the process measurement noise e has standard deviation σ. Note that the measurement noise includes an autoregressive term. In order to assure that the noise does not rise exponentially, assume that $|c| < 1$. A state-space representation of the process is given by

$$x(k + 1) = -ax(k) + (bu(k) + c(k) \qquad (8.83)$$
$$y(k) = (c - a)x(k) + e(k) \qquad (8.84)$$

It can be verified that if the process is at steady-state then $\mathbf{P} = 0$ and $\mathbf{K} = 1$. The one-step-ahead predictor of the process state is then given by

$$\hat{x}(k + 1|k) = -e\hat{x}(k|k - 1) + bu(k) + y(k) \qquad (8.85)$$

and at steady-state the one-step-ahead prediction of the output is given by

$$\hat{y}(k + 1|k) = (c - a)\hat{x}(k + 1|k) \qquad (8.86)$$

Note that since the variance in the estimation of the state is zero, so must be the variance in the estimate of the true output, since the relationship between the output and the state is known exactly. Thus, we have effectively removed all of the noise from the measurement of the process output. Of course, in order to be effective, the process model must be accurate, however, the example illustrates the value in having a model of the system under observation.

8.7.6 Evolving factor analysis

Dynamic modeling is intended to describe time behavior of the system and can be useful for directly improving process measurements and process control. Other techniques, which will be described in the following sections, are useful for extracting information about the operation of a process that may lead to a better understanding of the system. This information may then be used to improve the process operation.

Evolving factor analysis (EFA) is one such method. EFA is a general technique for the analysis of multivariate data that has an intrinsic order [33]. Process data, of course, is typically multivariate and is logically ordered by time. Like PCA, EFA can be used to determine how many 'factors' are present in a data set, i.e. how many independent sources of variation exist. In addition, EFA can be used to determine where in a data set the factors first appear and where they disappear.

The fundamental idea of EFA is to follow the singular values of a data matrix as the rows (samples) are added. Given a data matrix $\mathbf{X}$ (usually not mean-centered when performing EFA) with m samples and n variables, one starts by determining the singular value of the first row (sample), then the first two rows, then the first three and so on until all m samples have been considered. Since the number of samples considered, and thus, number of factors evolves as the method proceeds, the technique is known as EFA. Of course, the number of non-zero singular values can never be greater than the lesser of the number of rows and columns in the submatrix considered. It is also customary to do EFA in the reverse direction, i.e. determine the singular values starting from the last row in the matrix and working upwards until all rows have been considered.

Once the singular values have been determined, they (or their logarithms, since they usually span many orders of magnitude) can be plotted versus the ordered variable, e.g. time. As new factors enter the data new significant singular values will rise from the baseline of small singular values associated with noise in the data. Working in the reverse direction, significant singular values will rise from the baseline when factors disappear from the data matrix.

As an example of EFA, consider a batch chemical process that is being monitored by an NIR spectrometer. Suppose that there are three analytes of interest in the system, and that their concentration profiles are as shown in upper half of Figure 8.26. Furthermore, suppose that the pure component spectra for each analyte is as shown in the lower half of Figure 8.26. Now imagine that neither the concentration profiles or the pure component spectra are known *a priori*. Instead, the available data appear as in Figure 8.27, which shows the measured spectra as a function of time.

Figure 8.28 shows the results of performing EFA on this data. From the forward moving singular values, it is apparent that only one analyte exists in the first several samples. However, a new analyte appears at sample 8, where the singular value goes from the background level to a

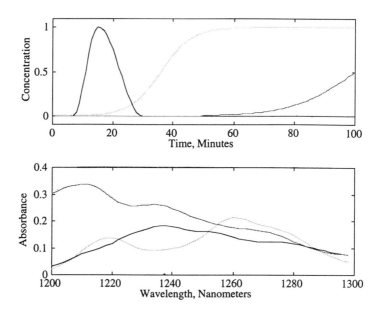

Figure 8.26 Concentration profiles of batch reactor (top) and pure component spectra (bottom).

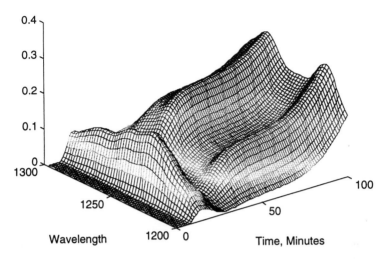

Figure 8.27 NIR spectra as a function of time.

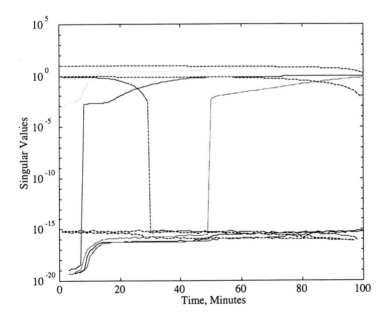

Figure 8.28 Forward (solid lines) and backward (dashed lines) singular values from evolving factor analysis of NIR data from batch process.

significant value. No other analytes appear until sample 50. Working from the backwards singular values, it is apparent that there were two analytes present at the end of the data set since there are two significant singular values. Working from right to left, it is apparent that another singular value becomes non-zero at sample 30. This indicates the addition of analyte when moving backwards in time, which corresponds to the disappearance of an analyte when moving forward in time.

Working from the EFA curves with the assumption that the first analyte to appear was also the first to disappear, we would conclude that the analyte which was present at the beginning of the run disappeared at about sample 30. Additional analytes appeared at sample 8 and sample 48 and remained until the end of the run.

8.7.7 Multivariate curve resolution

EFA gives us important information regarding the system: the total number of factors present and their times of appearance and disappearance. However, it would be impossible to identify the compounds based on this information alone (unless one was very familiar with the system at hand). However, the results of the EFA do provide a starting point for the determination of the concentration profiles and pure component spectra. Techniques for obtaining this additional information are known as self-modelling curve resolution (SMCR) [34] or multivariate curve resolution (MCR) [35].

Simply put, the goal of MCR is to extract the number of analytes, concentration profiles and pure component spectra with as few assumptions about the data as possible. Restated mathematically, given a data matrix $\mathbf{X}$ (not mean centered) that is the product of concentration profiles $\mathbf{C}$ and pure component spectra $\mathbf{A}$,

$$\mathbf{X} = \mathbf{CA} \tag{8.87}$$

we wish to obtain the physically meaningful $\mathbf{C}$ and $\mathbf{A}$. Obviously, this cannot be obtained directly from the principal components decomposition of $\mathbf{X}$ without additional information because the equation:

$$\hat{\mathbf{X}} = \mathbf{TP}^\mathrm{T} = \mathbf{TRR}^{-1}\mathbf{P}^\mathrm{T} = \mathbf{CA} \tag{8.88}$$

where

$$\mathbf{X} = \hat{\mathbf{X}} + \mathbf{E} = \mathbf{TP}^\mathrm{T} + \mathbf{E} \tag{8.89}$$

has an infinite number of solutions for any arbitrary transformation matrix $\mathbf{R}$. This results in a rotational and intensity ambiguity for $\mathbf{C}$ and $\mathbf{A}$ if no other information is known.

Fortunately, more information *is* typically available. In particular, both concentrations and spectra are necessarily non-negative. Using this constraint, it is possible to extract the pure component spectra through a procedure of alternating and constrained least squares optimization. Starting with the results of the EFA, it is possible to obtain an initial estimate of the concentration profiles, or at least, the range of existence in the data set of each component. These concentration profiles are used as the initial estimates in the constrained and alternating least squares optimization. At each iteration of the optimization, a new estimate of the spectra matrix $\mathbf{A}$ and of the concentration profiles $\mathbf{C}$ is obtained. Iteration between the following two equations is performed:

$$\mathbf{A} = \mathbf{C}^+ \hat{\mathbf{X}} \tag{8.90}$$

$$\mathbf{C} = \hat{\mathbf{X}} \mathbf{A}^+ \tag{8.91}$$

where $\mathbf{C}^+$ and $\mathbf{A}^+$ are the pseudoinverses (see equations (8.64) and (8.65) of the matrices $\mathbf{C}$ and $\mathbf{A}$, respectively. At each iteration, the negative elements of $\mathbf{C}$ and $\mathbf{A}$ are reset to zero. Obviously, the selection of the correct number of components in the calculation of $\hat{\mathbf{X}}$ is important. The use of this matrix instead of the original data $\mathbf{X}$ improves the stability of the calculations since $\hat{\mathbf{X}}$ is a noise filtered estimate of $\mathbf{X}$.

It should be noted that there will still be an intensity ambiguity in the results of the MCR, i.e. an ambiguity in the absolute magnitude of the concentration profiles and pure component spectra obtained since, for any single pure component matrix $\mathbf{X}_p$

$$\mathbf{X}_p = \mathbf{ca} = \alpha \mathbf{c} \frac{1}{\alpha} \mathbf{a} \tag{8.92}$$

for any arbitrary constant α. If, however, the (non-zero) concentration of a component is known at any point within the data, it is possible to resolve the ambiguity for that particular component.

If additional information is known about the problem, it is possible to incorporate it into the iteration as a constraint. For instance, it is often assumed that the concentration profiles are unimodal. In other cases, closure or some known stoichiometry may be used. When additional known (correct) constraints are used, the quality of the solution obtained generally improves.

It is also possible to combine data sets from different runs of the process under consideration simply by forming the augmented matrices as follows:

$$\mathbf{X} = \begin{bmatrix} \mathbf{X}_1 \\ \vdots \\ \mathbf{X}_{NP} \end{bmatrix} = \begin{bmatrix} \mathbf{C}_1 \\ \vdots \\ \mathbf{C}_{NP} \end{bmatrix} \mathbf{A} = \mathbf{CA} \tag{8.93}$$

where $\mathbf{X}_1$ to $\mathbf{X}_{NP}$ are the ordered data from the NP individual process runs and $\mathbf{C}_1$ to $\mathbf{C}_{NP}$ are the associated concentration profiles from the runs. Combining data from different runs may also contribute to the stability of the solutions obtained.

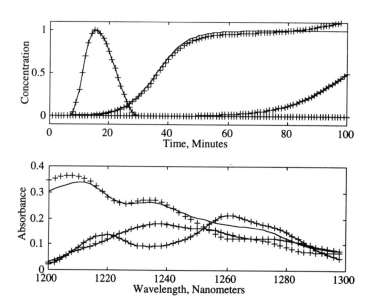

Figure 8.29 Actual (−) and recovered (+) concentration profiles and spectra.

As an example of MCR, we now return to the data used in the EFA example. Recall that we had determined the range of existence of each of the three components present in the data. Thus as the initial concentration profiles, we will use the forward singular values from the EFA with the first singular value re-set to zero after sample 30, where we know (from consideration of the backward EFA) the associated component disappeared.

The resulting spectral estimates and concentration profiles are compared with the known values in Figure 8.29. In this problem the intensity ambiguity exists so the concentration profiles and spectra have been scaled for comparison to the actual values. Note that the spectra and concentration profiles have been recovered almost exactly from the original data.

8.8 Process control

Measurements made by process analytical devices can be used for several purposes. These include verification of the final product quality, as a process diagnostic, as in statistical process control, for learning something fundamental about the process so that it might be improved, or for feedback process control, i.e. real-time adjustment of the process

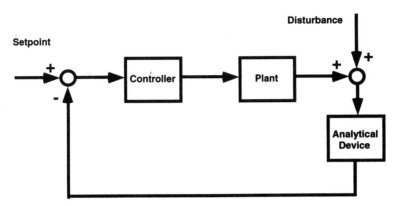

Figure 8.30 Conventional control block diagram.

inputs in order to achieve a particular output. The latter purpose, process control, is perhaps the most important. When measurements are made on the process output without feedback, it is too late to avoid making a product that does not meet specifications. Thus, process efficiency is lost.

Obviously, the quality and timeliness of measurements made by process analytical instruments ultimately impacts the accuracy with which the process can be controlled. Consider the process block diagram shown in Figure 8.30. This block diagram shows the flow of the process materials and analytical signals. On the left, the process setpoint, i.e. the desired output of the process, is entered to the controller, which is the first block on the left. The controller, in turn, acts on the process by manipulating the process inputs through mechanical actuators (such as flow control valves) or electronically (such as power control). The output of the process is a function of the inputs, but there is also the possibility that an unknown disturbance, as shown on the upper right, might effect the process output, on the right. Measurements are made on the process output by the analytical instrument on the right. Note that the analytical instrument can have its own dynamic characteristics. Generally, measurements are not made instantaneously and with absolute accuracy. The measured value is returned to the left-hand side of the figure and compared with the process setpoint. In most simple feedback control schemes, the controller acts on the difference between the measured output and the desired output (the setpoint).

Because process measurements are so critical for control, the selection and placement of analytical instruments should be considered with a control objective in mind explicitly. In industry, it is common for analytical chemists to team up with control engineers when selecting instruments.

8.8.1 PID control

Classical control methods are generally based on the proportional/integral/derivative (PID) controller [36]. The output of the PID controller, is the weighted average of the control error (deviation from setpoint), the integral of the error (receding backwards from the current error to minus infinity) and the current derivative on the error:

$$c = \tau_p e + \tau_i \int_{-\infty}^{\tau} e(t)\mathrm{d}t + \tau_d \frac{\mathrm{d}e(t)}{\mathrm{d}t} \qquad (8.94)$$

The weights, τ_p, τ_i and τ_d, can be determined several ways, many 'tuning rules' having been developed over the years. The first control systems, developed for controlling industrial steam engines in the 1800s, used only proportional control, i.e. τ_i and τ_d were set to zero. It can be shown, however, that proportional control alone cannot drive the process to its setpoint except in the limit where τ_p is infinite. In simple systems, τ_d is often set to zero, resulting in PI control. PI control will drive process errors to zero. This is easy to see, given equation (8.94). The integral of the error term will continue to increase as long as an error exists, thus increasing the controller output (and thus control action) until the error is driven to zero. In more complex systems, particularly those that are underdamped or have time lags, the derivative term is important for stabilizing the process at the setpoint in a timely fashion.

8.8.2 Model-based control

Modern control of complex processes is generally model based, i.e. a dynamic model of the process is an explicit part of the controller design [37]. Such a scheme is shown in Figure 8.31. Note the difference between this and Figure 8.30. Here the model of the chemical process has been added to the flow sheet. To the extent that the model accurately represents the process, the difference between the actual process output and the output of the model is an estimate of the disturbance to the process. Note the position of the process analyzer in the block diagram.

In some sense, model-based control is feedforward, in that the method recognizes the dynamics of the process and can anticipate them effectively. For this reason, model-based control can be more more accurate with complex processes.

8.8.3 Measurement placement

One must also consider the most appropriate place for making a process measurement. Often, the very end of the process is not the best place. Instead, if intermediate measurements can be made that can be related

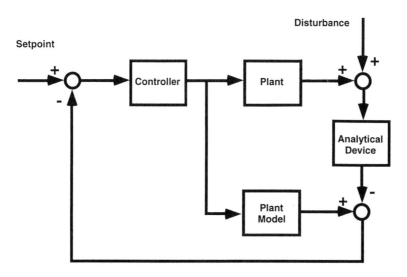

Figure 8.31 Model-based control block diagram.

to final product quality, better control can be achieved. A comon example of this is distillation. In distillation control, it often makes more sense to measure composition on a tray of the column near but not at the top of the column where the product comes out. This allows the system to anticipate the future output of the process. Also, it is often true that in high-purity distillation there is more variation on the product in intermediate locations than at the product end. Measurements taken at a location of higher variability can be less accurate and still useful for control purposes. Thus, it is possible to achieve the same control with a less expensive instrument.

8.8.3.1 An example of the effect of measurement on control. As an example of the effect of measurements on control, consider the chemical process defined by the schematic shown in Figure 8.32. At the left is the process setpoint, which specifies the desired output of the process. The setpoint is compared with the current measurement of the process output, and the difference is fed to the process controller. The controller then acts on the process. The output of the process, however, is affected by an additive disturbance. Thus, the true output of the process is the sum of the expected process output based on the controlled input and the disturbance. The output is measured and returned to be compared with the setpoint, however, the measurement process adds noise, modelled here by a random number generator, and a time delay, since measurements are typically taken at intervals and cannot be made instantaneously.

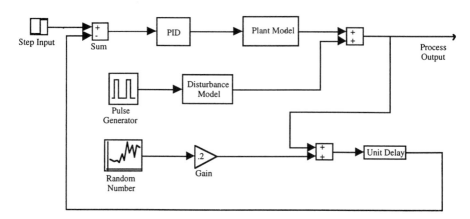

Figure 8.32 Block diagram of system for showing effect of measurement on control.

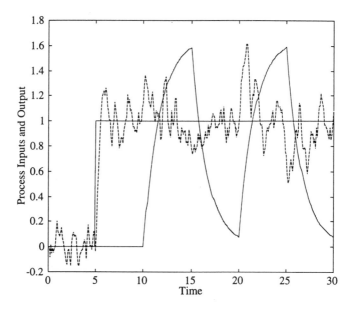

Figure 8.33 Example output from test system of Figure 8.32. Process setpoint shown as solid line changing from zero to one at five time units. Disturbance shown as solid line with sawtooth pattern. Output shown as dashed line.

An example output from the process is shown in Figure 8.33. Here the process is a simple first-order system, one whose output $y(t)$ is modeled by a first-order differential equation. In the case of a linear system, we have

$$a_1 \frac{\mathrm{d}y}{\mathrm{d}t} + a_0 y = bu(t) \qquad (8.95)$$

where $u(t)$ is the process input or forcing function. Assuming $a_0 \neq 0$ and rearranging we obtain

$$\frac{a_1}{a_0}\frac{dy}{dt} + y = \frac{b}{a_0}u(t) \tag{8.96}$$

We can now define

$$\frac{a_1}{a_0} = \tau \quad \text{and} \quad \frac{b}{a_0} = K \tag{8.97}$$

which results in

$$\tau\frac{dy}{dt} + y = Ku(t) \tag{8.98}$$

where τ is the process time constant and K is the process gain.

Step response behaviour (i.e. the response to a step change in the input) is specified by

$$y = u(1 - e^{-t/\tau}) \tag{8.99}$$

where y is the process output, and u is the process input.

The desired process output, or setpoint, is shown as a solid line in Figure 8.33. Note that the setpoint changes from zero to one at five time units. An additional disturbance acts on the process starting at 10 time units. (This disturbance is modelled by a square wave through a simple filter.) The actual output of the process is shown as the dashed line.

The measurement noise gain was varied from 0.05 to 0.3 and delay time was varied from 0.05 to 0.3 time units in a matrix of experiments. In each case, the control parameters were optimized so that the minimum control error was achieved for each pairing of noise gain and delay. The result of this procedure is Figure 8.34, which shows the contours of the control error as a function of measurement delay and noise. Note that the control error gets worse moving from lower left to upper right of the figure.

Imagine now that it is possible to describe the possible trade-off between speed of analysis and noise of the particular analytical instrument. Such a function might have the form

$$n = c_1\sqrt{d} + c_2 \tag{8.100}$$

where n is the measurement noise, d is the delay time in achieving the measurement and c_1 and c_2 are constants. We would suspect that many instruments, such as on-line spectrometers, would have a similar form since the measurement noise declines with the square root of the number of times the measurement is repeated.

Such a line is plotted in Figure 8.34, with $c_1 = 0.08$ and $c_2 = 0.01$. It can be seen from the figure that the optimum control would be achieved

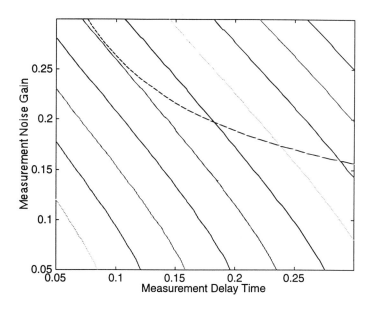

Figure 8.34 Control error surface as a function of delay time and measurement noise showing analytical instrument trade-off.

when the device was operated with a measurement delay time of about 0.1 units and a resulting noise gain of 0.25. Thus, we can see, that through the use of process models and measurement models, it is possible to select the optimum measurement trade-off for a given control application.

8.9 The future

If the current trend continues, the future will bring increasing demands on process efficiency, more environmental constraints and tighter customer specifications. Processes will be more complex as we move into the production of new materials. More information will be needed to handle higher levels of process complexity and tighter control. New process analyzers are already showing that they have the power to produce the required data; however, it is a long way between data and information. This is where the chemometrics will be critical: combining measurements from disparate process sources, compressing the data, extracting information and presenting it in a manner that can be used by both the control systems and the process operators.

A recent workshop, 'Fieldbus for Analyzers,' serves as a paradigm for the future integration of chemometrics with process monitoring and control. The workshop dealt with systematic methods for data acquisition and output for process analyzers, including fault detection, and calibration. Peter Van Vuuren of Exxon Chemicals, one of the workshop organizers summarized, "We are trying to see that chemometrics is supported at the process analyzer level. This is a new development for process analyzers, which have only recently acquired this much capability." Ultimately, these 'data handling' functions will be as much a part of the operation of chemical processes as PID controllers. This will require that chemists and chemical engineers be trained in the fundamentals of chemometrics, and it will be clear that chemometrics is indeed a part of chemical process engineering.

Acknowledgment

This work was supported by the Molecular Science Research Center of Pacific Northwest Laboratory. Pacific Northwest Laboratory is operated by Battelle Memorial Institute for the US Department of Energy under Contract DE-AC06-76RI.O 1830.

References

1. S. D. Brown, Questions and comments. *NAmICS Newsletter* **1**(1) (1991) 3.
2. B. M. Wise, A broad interpretation of chemometrics. *NAmICS Newsletter* **1**(2) (1991) 10.
3. J. V. Kresta, J. F. MacGregor and T. E. Marlin, Multivariate statistical monitoring of process operating performance. *Can. J. Chem. Engng* **69** (1991) 35–47.
4. B. M. Wise, N. L. Ricker, D. J. Veltkamp and B. R. Kowalski, A theoretical basis for the use of principal components models for monitoring multivariate processes. *Process Control Quality,* **1**(1) (1990) 41–51.
5. J. E. Jackson, Principal components and factor analysis: Part 1—Principal components. *J. Qual. Tech.* **13**(2) (1981) 201–213.
6. B. M. Wise, D. J. Veltkamp, N. L. Ricker, B. R. Kowalski, S. M. Barnes and V. Arakali, Application of multivariate statistical process control (MSPC) to the west valley slurry-fed ceramic melter process. *Waste Management '91 Proceedings,* Tucson, AZ, USA, 1991.
7. J. E. Jackson and G. S. Mudholkar, Control procedures for residuals associated with principal component analysis. *Technometrics,* **21**(3) (1979) 341–349.
8. J. E. Freund, *Statistics A First Course* (3rd edn). Printice Hall, NJ, USA, 1981, p. 236.
9. K. J. Åstrom and B. Wittenmark, *Computer Controlled Systems.* Prentice-Hall, Englewood Cliffs, NJ, USA, 1984.
10. H. Kwakernaak and R. Sivan, *Linear Optimal Control Systems.* Wiley, New York, USA, 1972.
11. C. L. Erickson, M. J. Lysaght and J. B. Callis, Relationship between digital filtering and multivariate regression in quantitative analysis. *Anal. Chem.* **64**(24) (1992) 1155A–1163A.
12. T. Næs and H. Martens, Principal components regression in NIR analysis: Viewpoints, background details and selection of components. *J. Chemometrics,* **2** (1988) 155–167.
13. P. Geladi and B. R. Kowalski, PLS Tutorial. *Anal. Chim. Acta.* **185**(1) (1986) 1–17.
14. A. Lorber, L. E. Wangen and B. R. Kowalski, A theoretical foundation for the PLS algorithm. *J. Chemometrics* **1**(19) (1987) 19–28.

15. S. de Jong, *Chemometrics Intelligent Lab. Sys.* **18** (1993) 251–263.
16. H. R. Draper and H. Smith, *Applied Regression Analysis* (2nd edn). John Wiley and Sons, New York, USA, 1981.
17. A. E. Hoerl, R. W. Kennard and K. F. Baldwin, Ridge regression: some simulations. *Commun. Statistics* **4** (1975) 105–123.
18. M. Stone and R. J. Brooks, Continuum regression: cross-validated sequentially-constructed prediction embracing ordinary least squares, partial least squares, and principal components regression. *J. R. Statist. Soc. B*, **52** (1990) 337–369.
19. B. M. Wise and N. I. Ricker, Identification of finite impulse response models using continuum regression. *J. Chemometrics* **7**(1) (1993) 1–14.
20. Y. Wang. D. J. Veltkamp and B. R. Kowalski, Multivariate instrument standardization. *Anal. Chem.* **63**(23) (1991) 2750–2756.
21. T. Næs, T. Isaksson and B. Kowalski, Locally weighted regression and scatter correction for near-infrared reflectance data. *Anal. Chem.* **62**(7) (1990) 668–680.
22. D. E. Rummelhart and J. L. McClelland, *Parallel Distributed Processing, Part 1.* MIT Press, Cambridge, MA, USA, 1986.
23. J. R. M. Smits, W. J. Melssen, L. M. C. Buydens and G. Katemen, Using artificial neural networks for solving chemical problems. Part I and II. *Chemolab.* **22** (1994) 165–189, **23** (1994) 267–291.
24. J. R. Long, H. T. Mayfield, M. V. Henley and P. R. Iromann, *Anal. Chem.* **63**(13) (1991) 1256–1261.
25. S. Sekulic, M. B. Seesholtz, Z. Wang, B. R. Kowalski, S. E. Lee and B. R. Holt, *Anal. Chem.* **65** (1993) 835A.
26. S. J. Qin and T. J. McAvoy, Nonlinear PLS modeling using neural networks. *Computers Chem. Engng* **16** (1992) 379–391.
27. J. E. Frank, NNPPSS: Neural networks based on Per and Pls components nonlinearized by Smoothers and Splines. *First International Chemometrics InterNet Conference. InCINC'94*, 1994. To be published in *Chemometrics and Intelligent Laboratory System.*
28. Y. Wang, M. J. Lysaght and B. R. Kowalski, Improvement of Multivariate calibration through instrument standardization. *Anal. Chem.* **64**(5) (1992) 562–565.
29. Z. Wang, T. Dean and B. R. Kowalski, Additive background correction in multivariate instrument standardization. *Anal. Chem.*, in press.
30. L. Ljung, *System Identification Toolbox User's Guide.* The Math Works, Inc., Natick, MA, USA, 1991.
31. L. Ljung, *System Identification: Theory for the User.* Prentice-Hall, Englewood Cliffs, NJ, USA, 1987.
32. A. P. Sage and C. C. Whitle III, *Optimum Systems Control* (2nd edn.) Prentice Hall, Inc, Englewood Cliffs, NJ, USA, 1977.
33. H. R. Keller and D. L. Massart, Evolving factor analysis. *Chemometrics Intelligent Lab. Sys.* **12** (1992) 209–224.
34. R. Tauler, A. Izquierdo-Ridorsa and E. Casassas, Simultaneous analysis of sevral spectroscopic titrations with self-modelling curve resolution. *Chemometrics Intelligent Lab. Sys.* **18** (1993) 293–300.
35. R. Tauler, B. R. Kowalski and S. Fleming, Multivariate curve resolution applied to spectral data from multiple runs of an industrial process. *Anal. Chem.* **65** (1993) 2040–2047.
36. G. Stephanopolous, *Chemical Process Control.* Prentice Hall, Englewood Cliffs, NJ, USA, 1984.
37. D. Seborg, T. Edgar and D. Mellichump, *Process Dynamics and Control.* Wiley & Sons Inc., New York, USA, 1989.

9 Environmental monitoring for a sustainable future
From pollution control to pollution prevention
E. A. McGRATH K. S. BOOKSH and J. J. BREEN

9.1 Introduction

A sustainable future, based on technologies that meet the needs of the present without compromising the ability of future generations to meet their own needs, will be achieved in large part with the successful implementation of an aggressive strategy for environmental technology development. Environmental technologies that advance sustainable development by reducing risk, enhancing cost effectiveness, improving process efficiency, and creating products and processes that are environmentally beneficial or benign will be the hallmark of the 21st century.

Environmental technologies are divided into four major categories: avoidance or *pollution prevention*, monitoring and assessment, control, and remediation and restoration. Pollution prevention involves activities that avoid the production of environmentally hazardous substances or alter human activities in ways to minimize damage to the environment. Pollution prevention encompasses product substitution or the redesign of entire production processes, rather than simply the use of new pieces of equipment. *Monitoring and assessment* technologies are used to establish and monitor the condition of the environment, including releases of chemicals harmful to the environment. *Control* technologies render hazardous substances harmless before they enter the environment. *Remediation* technologies render harmful or hazardous substances harmless after they enter the environment, while *restoration* technologies are designed to improve ecosystems that have declined due to naturally induced or anthropogenic effects.

It is clear that *environmental monitoring* will play an increasingly important role in achieving the goals of sustainable development. Measurement capabilities are critical to all four environmental technology categories whether to avoid pollution, to assess current conditions, to control processes, or to document remediation and restoration efforts. With the advent of *industrial ecology* as the new paradigm to implement sustainable development in the industrial sector, environmental and industrial monitoring become critical to the process.

Industrial ecology employs a direct comparison of industrial systems to natural ecosystems in which industrial plants consume resources and produce wastes. When industrial systems are small, resources and waste sinks are considered infinite and free. As their relative size and number increase however, resources become more limited (and therefore valuable). Industrial ecology recognizes this shift and accordingly adopts a no-waste or low-waste approach to production. In-stream recovery and reuse of materials are crucial tenants of industrial ecology. In this new paradigm, environmental considerations are incorporated into all aspects of product and process design, and technology plays a more active and positive role in achieving sustainable development.

The city of Kalundborg, Denmark serves as a real world, 30 year, industrial ecology laboratory. It involves the reuse of energy and materials by five partners. A refinery provides gas to a power plant and plasterboard company for their energy needs, while steam from the power plant is passed to a biotechnology company and into a district heating system of the city. Lower temperature energy goes to an experimental fish farm, and the power plant's desulfurization unit produces gypsum that is used by the plasterboard company. By reusing energy and materials, this 'industrial ecosystem' saves 19 000 tons of oil, 30 000 tons of coal, and 600 000 m^3 of water each year. Yearly savings are estimated at US \$12–15 million. The ability of Kalundborg to develop their industrial ecosystem stems in large part on their creative insights and their commitment to implement monitoring regimes that go beyond traditional compliance monitoring to process monitoring to avoid pollution. They have figuratively and literally gone from 'end of pipe' pollution control to 'front end' pollution prevention by designing their community for the environment.

9.2 Environmental monitoring

In this chapter, environmental monitoring is defined as the measurement, often at extremely sensitive levels, of naturally occurring or contaminate analytes in air, soil, water or some combination of these. Until 1990, monitoring for the environment could be categorized into two main areas: monitoring of natural systems to increase our understanding of them (e.g. volcanic eruptions and deep ocean CO_3 levels) and monitoring for regulatory compliance (e.g. lead in tap water). Each area of monitoring has developed specific (and, mandated, in the case of regulatory compliance) technologies and methods necessary to obtain the desired measurements. Unfortunately, as the sensitivity of analytical technology improves it has revealed the startling fact that few, if any, environments exist that are unaffected by man. Air-borne pollution has contaminated even the remotest areas of the earth while surface water throughout the world is threatened by human activities. The challenge of the future will be to monitor and protect the global environment from further deterioration while improving the lives of all the world's inhabitants. Wise use of dwindling resources and protection of the natural environment requires a new approach to manufacturing that is both efficient and clean. In this chapter, we will discuss the shift from pollution control to pollution prevention, what is driving this shift, and what technologies show promise for meeting these goals. Today's students in science and engineering must help develop the measurement technologies necessary to provide the robust monitoring capabilities that are essential for sustainable development and a safe, sustainable environment in the 21st century.

The US, together with the rest of the world, must deal with the complex environmental problems resulting from unprecedented population growth and worldwide industrial expansion. Past practices of controlling pollutants at the 'end-of-the-pipe' (where they enter the environment) are no longer effective in dealing with today's discharges. While the prescriptive command-control approach to controlling pollution has been successful to a point in easing egregious pollution, it has a major drawback of shifting pollution from air to land to water and back again. Today, 'end-of-pipe' treatment as a means of 'pollution control' cannot keep pace with increases in pollution from nonpoint sources and the expanding scale of worldwide economic activity. For the first time in history, the burgeoning numbers and activities of mankind are capable of drastically affecting the natural systems of the planet.

In addition, the task of remediating the damage to the environment 'after-the-fact' has become too difficult and expensive for governments and industries involved. Past activities have resulted in environmental insults which cross political boundaries and encompass entire geographic areas (e.g. Eastern Europe, acid rain, ozone depletion, rainforest destruction, etc.). Once generated, the cost of clean-up and disposal can be enormous, as exemplified by the Superfund program (estimated at US$ 10 million/acre), clean-up of former US and Soviet nuclear production sites, and treatment and disposal of hazardous chemicals already in use. To deal with this escalating and potentially lethal environmental degradation, it is imperative that new approaches to resource use be considered. Past industrial practices which relied on seemingly unlimited supplies of raw materials and easy and unconstrained access to disposal of wastes are no longer acceptable or competitive. The concept of 'sustainable development' was first voiced at the 1992 United Nations' Earth Summit in Rio de Janerio. Advocating that society make the transition for the 21st century from a 'limitless nature' industrial era to a 'finite planet' spaceship earth era, the sustainable development movement realizes that even as the world's raw materials continue to diminish the waste produced by their production and disposal remains with us. For 'sustainability' to occur, technological advances in monitoring and process control are essential for efficient resource use in production while simultaneously reducing pollution. The ultimate goal of this effort strives for zero emissions from any industrial process.

In the US, two programs arising from this new strategy – *Pollution Prevention* (P2) and *Design for the Environment* (DfE) – serve as the keystone of federal, state, and local environmental policy. Similar programs for pollution prevention are being adopted by the European Union and have attracted attention throughout the developing nations of Asia. Pollution prevention is a proactive multimedia management approach to pollution aimed at 'front-end' reduction of pollutants in the waste stream. P2 advocates the tightening of industrial processes to prevent process upsets and unwanted emissions, improve product quality, reduce loss of raw materials to the waste stream and recycle coproducts. P2 eliminates the transfer of pollutants from one media to another because pollutants are not generated in the first place. Industry saves not only by using its raw materials more efficiently but also by avoiding the costs and liabilities of waste disposal and clean-up. The success of the P2 effort will require a multidisciplinary approach. Process analytical chemists must design the robust measurement tools that provide real-time chemical information about the process, which in turn allows the chemical engineer better process control and optimization.

Design for the Environment is the next logical step in advancing the pollution prevention model. In addition to tightening the process under which products are manufactured it is also necessary to consider how the products are made including a consideration of the individual synthetic and overall synthetic sequence in the production of a chemical substance. DfE asks scientists to consider the substitution or elimination of hazardous chemicals wherever possible. Historically, the most important criterion for selection of one synthetic step in the formation of a chemical product over an alternative has been which one had the better 'yield'. The use of yield as a criterion was kinetically and thermodynamically sound and, in most instances, favorable from an economic standpoint as well. However, in view of the new emphasis on pollution prevention by both industry and the regulatory agencies, in addition to the skyrocketing costs of disposal, treatment, and compliance a selection scheme based solely on yield is no longer valid. DfE advocates the concept of environmentally benign synthesis which incorporates all the impacts: scientific, economic, and environmental, into the selection of how to make a chemical product. By putting forethought into the synthesis of a chemical product such that all the impacts of a particular process are considered, the synthetic chemist will also become a major force in achieving pollution prevention.

Today, throughout the industrialized world end-of-pipe control and waste disposal is being superseded by a policy shift to front-end control exemplified by pollution prevention; a shift to less hazardous chemicals as advocated by the design-for-environment program; and increased environmental stewardship promoting industrial ecology and good operating practices. Industrial ecology promotes a systems view of industrial products and processes that considers the total materials and energy cycle to minimize adverse environmental effects. Through total quality management which advocates updating processes based on state-of-the-art technology and providing reduced raw materials losses, reduced end-of-pipe treatment costs and reduced yield losses, industry can avoid future Superfund or other corrective action liabilities, reduce liability from future legislations, reduce waste disposal costs and liabilities while conserving H_2O, raw materials and energy. This paradigm shift for the future accents pollution prevention by ensuring that industrial processes run under tight control, substitute less hazardous chemicals whenever possible, and recycle by-products or redesign production processes.

This reshaping of industrial systems for both economic and environmental success has broad based support from industrialists, environmentalists, law-makers, academicians, government regulators and policymakers, and the general public. However, the challenge will be to switch from two decades of environmental policy based on pollution controls and government mandated regulations to a future environmental policy based on pollution prevention, source reduction, recycling, and waste minimization. The switch will require a new social compact amongst environmental, industrial, and regulatory interests. The roles and contributions of the chemical engineer, synthetic organic and inorganic chemist, and the process analytical chemist will be integral to the full articulation and implementation of this new vision.

9.3 Regulatory concerns

During the past 20 years the USA has responded in an evolutionary fashion to its pollution problems from pollution control to waste management to waste minimization to pollution prevention. Since the 1970s the US has passed a lengthy list of environmental legislation relevant to users and producers of chemicals and aimed primarily at pollution control (Table 9.1). The focus of the majority of this legislation is self-evident from the titles and deals with industrial facilities and other sources of pollution. In addition to industrial facilities, the Clean Air Act also covers other stationary sources of air pollution, motor vehicles and consumer products, while the Clean Water Act also includes publicly owned and other point sources of water pollution and certain nonpoint sources. The Resource Conservation & Recovery Act (RCRA) deals with industrial facilities and other entities that generate, treat, transport, store, or dispose of hazardous waste. Comprehensive Environmental Response, Compensation & Liability Act (CERCLA or Superfund) covers all entities that generate, treat, transport, store, or dispose of hazardous substances, or did so in the past. The Emergency Planning & Community Right-to-Know Act (section 313 of Superfund Amendments & Reauthorization Act (SARA)) arose as a response to heightened public concern after the catastrophic accidents in Bhopal, India, the town of Institute, West Virginia and other sites about the possibility of such an event occurring in the USA. Highlights of this law include sections on Emergency Planning especially in regards to extremely hazardous substances; Emergency-Release Notification pertaining to accidental release of hazardous substances; Hazardous Chemical Reporting through the use of such vehicles as material safety data sheets (MSDSs), and Toxic Chemical Reporting, which requires that covered facilities submit annual reports, on yearly toxic chemical releases to states and EPA. From this chemical information, the EPA has established a national toxic chemical release inventory (Toxic Release Inventory) which identifies the pounds of listed hazardous chemicals released to air, water, and soil by industry.

Table 9.1 Recent federal regulations mandating environmental monitoring

Toxic Substances Control Act (TSCA)
Clean Air Act
Clean Water Act
Resource Conservation & Recovery Act
Safe Drinking Water Act
Occupational Safety & Health Act
Hazardous Materials Transportation Act
Comprehensive Environmental Response, Compensation & Liability
  Act (CERCLA or Superfund),
Federal Hazardous Substance Act
Emergency Planning & Community Right-to-Know Act

In 1990, passage of the Pollution Prevention Act and Amendments to the Clean Air Act emphasized the shift of federal policy from control to prevention as a more effective strategy for dealing with pollution and as a potential market for developing pollution prevention technology. These two Acts directly impact all manufacturing in the USA. Much of this legislation specifically requires the use and development of monitoring technology not only for its preventative value but also for potential economic value both in savings of resources and energy and as a marketable technology.

The Pollution Prevention Act of 1990 states: " . . . There are significant opportunities for industry to reduce or prevent pollution at the source through cost-effective changes in production, operation, and raw materials use . . ." With this Act, Congress declared pollution prevention to be the national policy of the US and established a hierarchy for pollution, declaring that:

- pollution should be prevented or reduced at the source wherever feasible;
- pollution that cannot be prevented should be recycled in an environmentally safe manner whenever feasible;
- pollution that cannot be prevented or recycled should be treated in an environmentally safe manner whenever feasible;
- disposal or other release into the environment should be employed only as a last resort and should be conducted in an environmentally safe manner.

Pollution prevention is intended to include source reduction and is not limited to hazardous waste or chemicals subject to toxic release inventory reporting, but to encompass any hazardous substance, pollutant or contaminant. Source reduction is defined to mean any practice which reduces the amount of any hazardous substance, pollutant or contaminant entering any waste stream or otherwise released into the environment (including fugitive emissions) prior to recycling, treatment or disposal; and reduces the hazards to public health and the environment associated with the release of such substances. The term includes equipment or technology modifications, process or procedure modifications, reformation or redesign of products, substitution of raw materials and improvements in housekeeping, maintenance, training, or inventory control.

Passage of the Clean Air Act amendments of 1990 (the Act) has required industry to become accountable for air toxic emissions as part of a source's compliance strategy. One of the biggest impacts comes from Title III of the Act, Hazardous Air Pollutants (HAPs), and mandates that EPA address emissions of HAPs from both area and point sources. Title III also requires EPA to categorize sources, regulate emissions of 189 HAPs, determine applicability of minimum achievable control technology (MACT), evaluate residual risk after MACT and address sudden accidental releases. The amendments provide for state permitting programs which require monitoring, calculating and recording emissions *from every process*, down to *every valve and connection* (emphasis mine). With passage of the Act, industry must now be prepared to know exactly what is being emitted from its facilities.

Under the Act, existing sources must comply within three years of established MACT standards, while new sources must meet more stringent standards than existing sources. The sudden accidental release section of the Act deals with operator safety and requires analysis of potential public health hazards. In summary, all of the above regulatory initiatives will require sources to quantitate their emissions in order to demonstrate compliance.

Compliance with the 1990 Clean Air Act Amendments is anticipated to cost US industry US $25–40 million per year by the time the law is fully implemented in 2005. A trade publication, *Chemical Week* declares that the rule will cut to the heart of the chemical industry. Although the chemical industry will employ hardware solutions (incineration, activated charcoal, etc.) to meet some of these reduction requirements, the most innovative change in emission reduction technology will come from pollution prevention by engineered process changes. Many industries are using the air quality audit as not just a means of demonstrating compliance but also an opportunity to redesign a facility to minimize waste generation at the source. Currently, the voluntary Industrial Toxics Project, better known as the 33/50 Program, has given industry a head start on reducing emissions of 17 of the most hazardous chemicals (Table 9.2) listed under the EPAs Toxic Release Inventory and potential credit for amendment required reductions. Many of the targeted chemicals are chlorinated organics or metals. With 1988 as the baseline year, the program's goal is a 33% reduction by 1992 and a 50% reduction by 1995 and is designed to get an early start on emissions reductions required by the Clean Air Act as well as other legislation. For 1990, 134 million pounds of the 17 listed chemicals were released or transferred by chemical manufacturers who reported figures, a 27% decrease from 1988

Table 9.2 Emission of certain chemicals in 1988

Target chemicals	Millions pounds released in 1988
Benzene	33.1
Cadmium	2.0
Carbon tetrachloride	5.0
Chloroform	26.9
Chromium	56.9
Cyanide	13.8
Dichloromethane	153.4
Lead	58.7
Mercury	0.3
Methyl ethyl ketone	159.1
Methyl isobutyl ketone	43.7
Nickel	19.4
Tetrachloroethylene	37.5
Toluene	344.6
Trichloroethane	190.5
Trichloroethylene	55.4
Xylene	201.6

levels. The extent of continuous emission monitoring (CEM) of vents and stacks required by the Act has not been formalized but calls for 'enhanced monitoring' may signal an increase in continuous monitoring. A list of expenditures resulting from the Act fix industries bill for source testing, testing equipment, ambient and source continuous monitoring, reporting, data logging and network designs at close to US$ 1 billion.

The Clean Water Act was due for reauthorization in 1994 and did not pass. Its effects would have the same implications to industry in the area of surface water as the Clean Air Act had to industries producing airborne emissions.

9.4 Environmental monitoring technology

The ideal instrumentation for environmental monitoring has many desirable features. It must be rugged to withstand the elements for weeks or months without the need for repair. It must be stable such that frequent recalibration is unnecessary and then the intrinsic ability for self-recalibration should be included. It should perform analysis on a small scale to reduce reagent use and lessen further impact on the environment. It should perform analysis *in situ* or remotely to eliminate errors associated with sample transportation. It should perform analysis in a short time-frame compared to the effect being monitored. It must be free of chemical or environmental interferences that degrade analytical ability. It should have a large dynamic range. It must be smart to diagnose unreliable information resulting from sampling problems or instrumental errors. It should be simple enough for an unskilled technician to install, maintain and service. And above all, it should it should be very inexpensive with negligible maintenance costs. Needless to say, this piece of instrumentation has not yet been built.

Fortunately, however, environmental monitoring devices that include one or more of the above features are currently being perfected. It is these devices, and their ilk, that constitute the future of environmental monitoring. Presented below is a small subset of such environmental probes. Examples were chosen from the fields of water, air, and soil analysis.

9.4.1 Example 1

A self-contained, flow injection based analyzer for *in situ* determination of nitrate has been developed [1]. Instrumentally, this device epitomizes many of the ideals listed at the beginning of the chapter. It is relatively inexpensive and environmentally rugged, performs automatic recalibration, and uses less than 10 ml of reagent per month. On the negative side, it has a linear dynamic range of only one order of magnitude, although the non-linear dynamic range extends at least to a second order of magnitude. Furthermore, its analytical and diagnostic abilities are limited to that of a univariate probe [2].

Capable of enduring months of continuos analysis without servicing, this colorimetric analyzer relies on a miniature osmotic pump to propel the reagents and sample through the reaction chamber. Operating at flow rates of 1 and 12 $\mu l/h$, the analyzer consumes between 0.7 and 8 ml of reagent per month. The nitrate from the standards and sample are reduced to NO_2 by a Cd/Cu surface. The NO_2 then reacts with p-aminobenzenesulfonamide and N-(1-napthyl) ethylenediamine dihydrochloride to form an azo dye. The dye concentration subsequently attenuates the transmittance from a green LED (565 nm max emission) to a single photodiode. Analysis is completed in approximately 30 min. The entire apparatus is powered by a datalogger that requires less that 35 mW of battery power.

Two-point calibration with blank and 20 μM NO_3 standards is automatically performed at predetermined intervals to correct for instrumental drift. Under controlled situations, this results in a limit of determination of 0.11 μM NO_3 when recalibration is performed every second day. On the short term, the analyzer demonstrates a determination limit of 0.05 μM NO_3 with a signal to noise of 5.

Potential environmental interferences result from the temperature dependence on the performance of the osmotic pump that controls reagent flow through the reaction chamber. The pump demonstrates a factor of five increase in the flow rate between 5 and 37°C. Rapid temperature changes manifest fluctuations in the signal as the thermal equilibrium time for the large reagent pumps is greater than that of the small sample pump. Very rapid temperature changes can ultimately hinder performance by backwashing reagent and desorbing the Cu from the Cd/Cu surface. Ill effects from environmental temperature fluctuations can be minimized by insulating the analyzer or performing analysis at faster flow rates. It is important to note, that while the temperature fluctuations are important in surface analysis of nitrate, this would not be an issue when the analyzer is employed in applications where the temperature does not change between recalibrations (e.g. deep inland lakes or well water monitoring).

9.4.2 Example 2

The above example relies on the existence of a highly selective chemical reaction to insure accurate analysis. Furthermore, due to the univariate nature of this approach, it is impossible to determine when the analysis is biased by the presence of a spectroscopically active species. An alternative approach to fully selective chemistry is to employ instrument-ation amicable to multivariate analysis (e.g. diode array detectors). These types of instrumentation allow for simultaneous analysis of multiple chemical species and accurate quantitation of any species included during instrumental calibration. Furthermore, multivariate data contains enough information to permit diagnostic tools that reject spurious data stemming from instrumental or sampling problems.

The Fujiwara reaction, with single wavelength detection, has been employed in many efforts to quantitate halogenated hydrocarbons in water and air samples: In the Fujiwara reaction, the halogenated hydro-carbon reacts with a pyridine and tetrabutyl amonium hydroxide solu-tion. The reaction rate and intermediate products formed vary between hydrocarbons. Unfortunately, as 22 distinct halogenated hydrocarbons form light absorbing species when reacted with the Fujiwara reagent, the lack of selectivity of this method has hamstrung efforts to achieve reliable quantitative results.

However, by developing a sensor that incorporates both temporal and spectroscopic resolution, more reliable quantitative results can be achieved [3]. The temporal response is dependent on both the permeation of the chlorinated hydrocarbons through a membrane and the analyte-specific reaction kinetics. Table 9.3 shows preliminary analytical results for the multivariate determination of 1,1,1-trichloroethane, trichlo-roethylene and chloroform at levels expected for certain hazardous waste sites.

Table 9.3 Prediction errors for determination of TCA, TCE, and $CHCL_3$

Actual concentration (ppm)			Error of estimation (ppm)		
TCA	TCE	$CHCL_3$	TCA[*]	TCE[†]	$CHCL^3$
66.9	0	0	21.6	0.00	0.05
133.8	0	0	3.2	0.00	0.05
0	1.46	0	4.4	0.00	0.04
0	29.3	0	14.5	1.1	0.01
0	0	4.45	1.7	0.04	1.1
33.5	7.32	0	21.6	0.30	0.03
133.8	2.93	0	42.0	0.00	0.04
33.5	2.93	1.48	16.1	0.05	0.00

[*]60 min reaction time.
[†]5 min reaction time.
[‡]5 min reaction time.

Further investigation of the applicability of this reaction kinetics–spec-troscopic method for quantitation simultaneously incorporates the tem-poral and spectroscopic profiles in the analysis [4–6]. Ideally, the advanced mathematical algorithms employed permit accurate, unbiased quantita-tion with only one standard of known analyte concentration. In this instance, the best results were obtained by a standard addition method. Here prediction errors of 2.6% and 5.5% were obtained for 0.48 and 0.12 μM TCE in the presence of chloroform.

It is conceived that this chemistry will eventually be encorporated in a fiber-optic based probe. However implementation will be limited to thermally stable areas as the reaction rate is heavily influenced by temperature. Another possible impediment to field applications stems from the fact that the pyridine reagent used has a greater toxicity that the contaminants monitored. Hence, a mishap with the probe could lead, in itself, to regulatory non-compliance.

9.4.3 Example 3

Another example of a so-called 'hyphenated' probe employs chemically facilitated Donnan dialysis followed by reaction with a semiselective colorimetric reagent for determination of free Pb(II) [7,8]. This *in situ* sensor has the advantages of being small, having a flexible dynamic range, and being immune to chemical interferents. On the negative side, it is instrumentally more complex than the first example and hence would be much more expensive to built and service.

The initial step employed in this probe is the preconcentration of cations in a tubular cation exchange membrane. By judiciously selecting the preconcentration time the dynamic range has be adjusted from 10^{-1} to 10^{-5} M Pb(II). The membrane is then removed from the sample and surrounded by distilled water. The diffusion rate of the cations into the center of the membrane is enhanced by a flowing sodium thiosulfate receiving solution. The sodium thiosulfate temporally enhances the diffusion of the Pb(II) from that of other diffusing metal cations. The metal cations are then transported to a flow cell and react with 4-(2-pyridylazo)resorcinol to form orange–red complexes with each cation. Spectra are acquired on a multichannel diode array detector at 30 s intervals while the Pb(II) is leaching from the membrane.

Table 9.4 Comparison of estimated Pb(II) molarity ($\times 10^9$) and standard deviation ($\times 10^9$) between the probe and GFAAS

	One standard		Four standard		GFAAS	
Sample	Mean	SD	Mean	SD	Mean	SD
Tap water	31	7.7	30	4.3	29	2.4
Lake water	28	11	27	5.8	48	1.9

Table 9.4 compares the performance of this probe to that of graphite furnace atomic absorption spectroscopy (GFAAS) for analysis of Pb(II) in tap and lake waters. Analysis with the probe was performed with both 1 (5×0^{-8} M Pb(II)) and 4 ($4–7 \times 10^{-8}$ M Pb(II)) standards. The probe

results for the tap water agree closely with the predicted concentration by GFAAS. And the precision of analysis is close to the 3×10^{-9}M detection limit realized under optimal conditions. The differences in estimated Pb(II) concentrations between the probe and GFAAS in lake water are attributed to complexation of the Pb(II) with large organic compounds. *Ergo*, the probe is capable of only monitoring free Pb(II).

It is important to note that the probe accurately, with only a single standard, estimates the Pb(II) concentration in tap water despite the presence of 4.49×10^{-7} M Cu(II) and 4.5×10^{-8} M Zn(II). This highlights the power of multivariate analysis of hyphenated instruments. Reliable quantitation is accomplished despite the fact that the maximum ratio of Pb(II) signal to interferent signal is only 12.2 for Cu(II) and 23.3 for Zn(II). In this instance, at the *most* selective channel, the interferent signal is *greater* than that of the Pb(II).

9.4.4 Example 4

In situ monitors, like those presented in the three previous examples, are not applicable to all environmental monitoring situations. When high resolution spacial characterization of a large area is desired, it could easily be cost prohibitive to blanket the whole area with small probes. Likewise, for harsh environments such as inside of large radioactive waste storage tanks, *in situ* probe degradation becomes a serious problem. One solution to both problems is construction of a remote system capable of monitoring multiple spacial locations. One fine example of such technology is a portable Raman system designed for remote characterization of liquid, solid, and gas phase hazardous compounds before and during remediation of mixed waste storage tanks [9].

The spectrometer consists of a 0.35 m f/4 image-corrected spectrograph with a liquid nitrogen cooled CCD detector. The spectrometer could be used with either a 488 nm line argon laser or a 809 nm GaAlAs diode laser. Appropriate filters and collection optics are employed to remove the laser line, maximize rejection of Rayleigh scattered light, and focus the returned Raman scattered light onto the spectrograph slit. The typical instrumental configuration results in approximately 10 cm^{-1} spectral resolution in the visible and 7 cm^{-1} resolution in the near-infrared (NIR).

Flat mirrors steer the laser beam and the field of view of the spectrograph. A 50 mm aluminum-coated steering mirror and a 100 mm gold-coated mirror are used in conjunction with the visible and NIR lasers, respectively. The argon laser is focused to a 5 mm diameter spot at the sample. This delivers approximately 100 mW of power to the sample. The diode laser is focused to a 2 mm diameter spot at the sample. This delivers approximately 20 mW of power.

The Raman system is capable of producing Raman spectra with excellent signal to noise ratios from samples between 10 and 100 m distant from the spectrograph with short integration times. Pure $K_4[Fe(II)CN_6]$ was analyzed from 16.7 m with 1 s integration resulting in a S/N of 250/1 at the 2055 cm^{-1} ferrocyanide peak. Based on this experiment, the Raman system could produce in 1 s the spectrum of pure ferrocyanide 355 m distant with a S/N of 4/1. Equivalently, it should be possible to detect 0.1% ferrocyanide concentrations from 16.7 m with a 4/1 S/N in only a few minutes. A mixture of 10% ferrocyanide and 10% $NaNO_3$ in a NaCl matrix produces an excellent Raman spectra from 16.7 m distance with only a 60 s integration. ferrocyanide, $NaNO_3$, and atomaspheric O_2 and N_2 peaks are all clearly distinguishable. The average S/N is approximately 40/1.

The Raman system is also capable of remotely analyzing liquid samples. Samples of 3 M $NaNO_3$ and 3 M $NaNO_2$ in 1 cm cuvettes were analyzed from 16.7 m with 60 s integration. This achieves S/N of approximately 150/1 for the NO_3^- and 50/1 for the NO_2^- with no spectral overlap of the most intense peaks of the two compounds. Similarly, neat CCl_4 measured under identical conditions produces a Raman spectrum with a S/N of approximately 150/1.

9.4.5 Example 5

What the previous examples all have in common is that they are high-tech solutions to environmental monitoring problems. Therefore, they are all doomed to violate the final tenant of an ideal environmental monitoring device – low cost. In general, there is a strong positive correlation between technologic level and cost of a device. Concurrently, this strong positive correlation extends to complexity of the device, and with complexity comes the need for highly trained users which, again, leads to greater cost of operation. *Ergo*, inexpensive, low-tech solutions to environmental monitoring problems must not be overlooked. Especially when these low-tech solutions perform comparably to more expensive options.

One case in point is a long-term soil gas flux sampler for measurement of volatile hydrocarbons [10]. This passive device consists solely of an insulated stainless-steel housing 150 mm in diameter and 170 mm tall. The inside top of the chamber is a mounted packet of activated carbon. This maintains the soil–air concentration gradient inside of the sampler. An absorbent sampling element is suspended below the activated carbon. Different commercially available sorbents could be employed if the analyte of interest has little affinity for carbon.

The open-ended housing is pressed in the soil and left unattended for a number of hours or days. The sampling element is removed from the housing. The absorbed organics are extracted and the extract is analyzed. The total soil gas flux is calculated based on the exposed areas of the activated carbon and sampling element.

Tables 9.5a and 9.5b compare the performance of the passive sampler to that of a pump based sampler. In limited laboratory tests, the performance quality is equivalent. The pump based sampler, naturally, has a shorter sampling time. Samples were collected with a 10–30 min exposure time for the active sampler compared to a 88–1367 min exposure time for the passive sampler. But, the measured toluene fluxes for the two systems are not statistically different. The passive sampler performed comparably to the active sampler in field tests. Benzene, toluene, and ethylbenzene soil gas fluxes were measured over a 4-day period. The average measured flux for benzene and toluene are nearly identical for the two samplers. However, the active sampler did estimate a greater ethylbenzene flux than the less expensive sampler.

Table 9.5a Determined fluxes from active and passive samplers using a soil gas simulator

Toluene (ppm)	Sampler type	Exposure time (min)	Flux rate ($ng/cm^2 s$)
115	Passive	120–1367	0.70 ± 0.06
	Active	10	0.77 ± 0.06
220	Passive	120	1.65 ± 0.04
	Active	10–30	1.60 ± 0.16
532	Passive	88–856	2.70 ± 0.32
	Active	10–12	3.20 ± 0.20

Table 9.5b Measured soil gas flux over 4 days at a land treatment unit

Sampler	Exposure time (min)	Flux ($pg/cm^2 s$)		
		Benzene	Toluene	Ethylbenzene
Passive	16–14440	23	55	22
Active	13–43	23	50	38

9.5 The future

In 1991 the USA spent US$115 billion/year (2.1% of GNP) on pollution control. By 2000 the EPA estimates the cost of such protection could rise to as high as US$185 billion (2.8% GNP). In addition to the economic benefits gained by wise use of raw materials, industry has an opportunity to develop and market new pollution prevention technologies. Much more needs to be done to capitalize on this potential. Current best practice must be diffused more rapidly. Environmentally superior products and processes need to be brought to market more quickly. Research and development should be encouraged and channeled in environmentally sound directions.

Unfortunately, technology transfer faces a number of impediments many of which are the result of Federal and corporate policy. Members of the World Resource Institute have addressed a number of these issues in the publication, *Transforming Technology: An Agenda For Environmentally Sustainable Growth in the 21st Century*. The issues discussed in this publication are relevant and timely.

The first issue describes the need for reformation of environmental regulations to encourage technological change. To quote WRI, "Relying on best available technology standards tends to entrench existing control technologies at the expense of long-term innovation". Regulations "provided no incentive for doing better than standards dictate and do not recognize activities which exceed standard requirements. Cumbersome administrative procedures also impede innovation". Administrator Browner of the EPA has vigorously advocated the adoption of the pollution prevention paradigm. It will require the EPA to change the focus of its activities to "promote the development and deployment of new technology".

Thus, if the ultimate in environmental protection is pollution prevention—eliminating problems before they occur, then the ultimate environmental analytical chemistry may be process analytical chemistry.

References

1. H. W. Jannasch, K. S. Johnson and C. M. Sakamoto, *Anal. Chem.* **66** (1994) 3352.
2. K. S. Booksh and B. R. Kowalski, *Anal. Chem.* **66** (1994) 782A.
3. J. M. Henshaw, L. W. Burgess, K. S. Booksh and B. R. Kowalski, *Anal. Chem.* **66** (1994) 3328.
4. R. Tauler, A. K. Smilde, J. M. Henshaw, L. W. Burgess, and B. R. Kowalski, *Anal. Chem.* **66** (1994) 3337.
5. A. K. Smilde, R. Tauler, J. M. Henshaw, L. W. Burgess, and B. R. Kowalski, *Anal. Chem.* **66** (1994) 3345.
6. K. S. Booksh, J. M. Henshaw and B. R. Kowalski, *J. Chemometrics*, in press.
7. Z. Lin and L. W. Burgess, *Anal. Chem.* **66** (1994) 2544.
8. Z. Lin, K. S. Booksh, L. W. Burgess and B. R. Kowalski, *Anal. Chem.* **66** (1994) 2552.
9. S. M. Angel, T. J. Kulp and T. M. Vess, *Appl. Spectrosc.* **46** (1992) 1085.
10. S. A. Batterman, B. C. McQuown, P. N. Murthy and A. R. McFarland, *Environ. Sci. Technol.* **26** (1992) 709.

10 Non-invasive techniques

R. M. BELCHAMBER

10.1 Introduction

There are numerous advantages in making on-line measurements non-invasively. This has lead to considerable investment by manufacturers in developing a range of non-invasive techniques. These have almost entirely focused on measurement of physical parameters. At present there are only a small number of chemical measurements that can be made this way. The purposes of this chapter are to discuss the merits of the non-invasive approach, to describe some of the existing and developing technologies for physical and chemical measurement and, perhaps, to stimulate further development.

Before embarking on a discussion of the various techniques it is necessary to define a few terms. A transducer, in this context, is a device for converting a chemical or physical parameter into an electrical signal. The terms non-invasive and non-intrusive are widely used in the realm of process analysis [1]. Figure 10.1 illustrates these approaches. Non-invasive means that in all cases the transducer does not come into contact directly with the process medium. A familiar example of this is the thermowell which is simply a piece of tubing welded into the wall of the vessel and closed at the innermost end. To measure the process temperature a thermocouple or platinum resistance thermometer is inserted into the tube. To be able to make a reasonably representative measurement of the process fluid temperature this tube needs to extend some distance into the fluid. It is therefore intrusive. An obvious disadvantage of an intrusive technique is that the probe may disturb the flow of process fluid or may be eroded by the passage of entrained abrasive material.

An example of an invasive intrusive analyser is a pH electrode. For this to function the pH sensitive element needs to be in intimate contact with the process fluid, preferably where there is some kind of flow. This suffers the disadvantages of an intrusive monitor. Additionally, there are further disadvantages, namely corrosion, or possibly, poisoning of the sensing element. However, in many cases the only possible means of making a measurement is to use an invasive approach because of the lack of the equivalent non-invasive technology.

All non-invasive measurements rely on the transmission of information through the walls of the vessel. This must involve, with a couple of minor

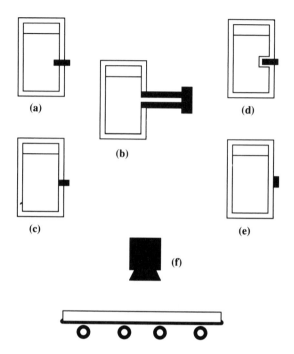

Figure 10.1 Schemes for on-line analysis: (a) invasive, intrusive; (b) invasive, intrusive (extractive); (c) invasive, non-intrusive; (d) non-invasive, intrusive; (e) non-invasive, non-intrusive; and (f) non-enclosed, non-invasive.

special exceptions, the use of a penetrating radiation field; i.e. heat, γ-rays, ultrasound, microwaves, etc. Measurements can either be passive or active. In a passive monitor the radiation field is generated directly by the process under study, e.g. heat, infrared emission, acoustic emission. In an active technique the process modulates a radiation field imposed by the analyser. The analyser then detects and interprets this modulation. Active techniques include X-ray fluorescence, ultrasonics and nucleonics. One example where this does not apply is the measurement of process pressure by measuring the expansion of the vessel using an optical technique.

When contemplating the use of a non-invasive technique, one needs to consider not only the process material, but also the composition and structure of the containment.

To perform a non-invasive measurement a radiation field must be capable of penetrating the vessel wall and of being modulated by the process parameters of interest. This rules out the adaptation of many laboratory techniques. Techniques that would never be considered for a laboratory analysis may have to be applied such as acoustic emission monitoring or γ-ray attenuation. Appropriate windows can be installed in the process plant to act as entry and exit points for the probing radiation. For instance plastics and ceramics are used with microwaves while sapphire and silica are used with ultraviolet, visible or near infrared radiation. It may be argued that this is not a truly non-invasive measurement. Certainly the installation of windows, negates many of the benefits provided by truly non-invasive techniques. There are many possible technologies that sit in this rather grey area.

Cases where the material is not enclosed at all are especially amenable to non-invasive monitoring. These include solid material on a conveyor belt, extruded polymer film, sheet steel, or paper in a paper mill. Because there is now no necessity to penetrate the containment a much larger range of techniques is applicable such as X-ray fluorescence, diffuse reflectance spectroscopy, Raman spectroscopy and infrared emission spectroscopy.

10.2 Benefits and drawbacks

The main arguments for non-invasive measurement are the economic and environmental benefits achievable. A non-invasive monitor may cost as much, or more, than an invasive device. However, as these devices gain market penetration their price is likely to fall. The process development team must, however, also consider the cost of ownership of the device. This not only includes the cost of purchase but also the cost of installation and ongoing maintenance. The installation cost of a non-invasive monitor will always be lower than an equivalent invasive or extractive analyser. In the case of invasive measurement, in a plant handling corrosive or toxic liquids at high temperatures and pressures, the cost of multiple isolation, purge and drainage valves and associated manifolds, often made from fairly exotic materials; e.g. hastelloy or zirconium alloy, can be very large. In certain extreme cases, this can result in millions of dollars additional expenditure. More typically, the installation cost of a process analyser such as a gas chromatograph, near-infrared analyser or mass spectrometer will be more than double the original cost of the hardware. For cheaper techniques this cost differential is much greater. On a new plant this results in a large capital outlay. On an existing plant, retro-fitting invasive devices results in loss of production while the plant is shut down, cleaned, the analyser is installed, the vessel is pressure tested and recertified, and the process restarted.

During the lifetime of the analyser further economic benefits realise themselves. Because the transducer does not encounter the process fluid, problems of erosion and corrosion are eliminated and transducer lifetime is likely to be very long. When a transducer has to be repaired or replaced this is an easy operation which does not disrupt the process. Non-invasive monitors are more flexible in their usage. It is relatively easy to reposition these devices during the lifetime of the process.

Not only is the transducer completely protected from the process fluid, there is no risk of the chemical or microbiological contamination of the process fluid during maintenance. This is of particular importance in the pharmaceutical and food industries.

Apart from cost there are significant environmental and safety benefits. By reducing the number of valves, flanges, and seals required plant design is simplified. This decreases the number of places where leaks could develop during process operation, and eliminates the possibilities of accidental spillages during maintenance. In plant handling hazardous materials, there is no need to decontaminate the transducer before maintenance, storage or disposal.

In the process environment there is a drive for reducing the exposure of plant personnel to all process materials. In many cases where high toxicity materials are being handled the only sure way of achieving the necessary isolation is to use non-invasive techniques.

Because the transducer does not breach the vessel containment, additional safety benefits are derived from the increased vessel strength.

It may prove that the overriding benefits are environmental. As green and health and safety issues become even more prevalent non-invasive monitoring may be the only acceptable means, at least in some areas, of operating a commercial chemical plant.

With all the advantages provided by a non-invasive approach to process analysis the obvious question is why most measurements are not made this way. The main reason for this is the limitation of available technology. Many measurements particularly of chemical parameters can only be made by invasive techniques.

Nucleonic measurements have their own problems. γ-rays are an obvious choice of penetrating radiation for steel vessels. Throughout industry γ-rays are used for level gauging and density measurement. However, the use of radioactive sources is currently unpopular because of the potential safety hazard. This argument is not entirely emotive. While offering many of the economic benefits of low installation cost, the provision of the necessary radiological safeguards may outweigh many of the savings provided by a non-invasive measurement.

Calibration of non-invasive analysers can be difficult. A simple device such as a thermocouple can be removed from the plant and calibrated independently. Ultrasonic time-of-flight flowmeters provide an absolute measurement of velocity and do not require calibration. Other techniques involving ultrasonic attenuation and acoustic emission are much harder to calibrate because of their high dependence on the thickness, properties and condition of the containment. These need to be calibrated *in situ* often by an invasive measurement or sampling system.

There will be a need for recalibration of most non-invasive analysers on a periodic or as required basis. With all these analysers, build up of material inside the vessel, near the transducer, is likely to cause calibration and performance problems irrespective of the complexity of the measurement. Deposition of layers of rubbery polymer around thermowells increases their response time and leads to erroneous measurement. In at least one incident, failure to monitor temperature precisely due to this cause, has lead to catastrophic damage occurring to process plant. With a non-invasive ultrasonic level switch this type of material adhering to the wall of the containment could cause it to indicate the presence of liquid, when in reality it ought to be indicating a gas. Erosion or corrosion of the inner walls will also affect the response of most non-invasive monitors. These problems can be dealt with by internal inspection but this is obviously costly and would eliminate many of the benefits so-far discussed.

A more attractive approach to validating the transducer output is to use intelligent non-invasive sensors. These devices are equipped with a local microprocessor which continually monitors the analyser performance. Parameters such as long-term drift, changes in response time and changes in the noise present on the output are logged and analysed. When these changes become significant the process operator is alerted. A particularly dangerous mode of failure for a measurement system is the covert failure. Many analysers and sensors are installed to warn of hazardous conditions. The sensor does not usually experience the conditions that it is designed to warn against. This means that it may have become deactivated, or even disconnected, during normal process operation without being noticed. Intelligent sensors offer a means of detecting this type of failure. Further security may be achieved by the installation of multiply redundant sensors. If a particular sensor is identified as performing badly, its output can be ignored, until a suitable time has been found to replace or repair it. Thus, disruption to the process is minimised. For safety critical applications some redundancy in the monitoring system is likely to be mandatory. For a summary of the advantages and disadvantages of non-invasive techniques see Table 10.1.

10.3 Techniques

It is not proposed to discuss exhaustively all the analytical techniques that could, given the right set of circumstances, be used for making non-invasive measurements. Instead, a number of techniques which are

Table 10.1 Summary of the advantages and disadvantages of non-invasive measurement techniques

Advantages	Disadvantages
Low installation costs	Limited measurements currently available
Reliability (erosion of probes eliminated, blockages in sampling systems eliminated)	Calibration of non-invasive technique can present problems
Environmentally sound—possibility of leaks and spillages eliminated at the analyser	Build up of materials inside the vessel may influence performance and remain unnoticed
	Erosion/corrosion of containment walls may also influence performance and remain unnoticed
Safety – pressure vessel containment is not breached	
Easily maintainable	Difficult to calibrate
	Nucleonic techniques may pose environmental hazard and require extra expenditure
Easily relocatable	
Provides real-time information	May provide real-time information

currently being applied to monitor real processes or are active research areas will be treated in more detail (see Table 10.2). The techniques themselves share little in common apart from their ability to make non-invasive measurement. The order that these techniques will be dealt with in this chapter does not necessarily relate to their importance. An attempt has been made, however, to group together similar technologies.

Table 10.2 Non-invasive techniques application and status

Technique	Radiation	Applications	Status
Ultrasound	Compressional elastic waves in the range 50 kHz–10 MHz	Level measurement, density, concentration, flowrate	Available from a large number of suppliers
Acoustic emission	Compressional elastic waves in the range 50 kHz–600 kHz	Process characterisation, fluid beds, flow monitoring	In various states of development/ research area
Infrared emission spectroscopy	Electromagnetic radiation $4000\ cm^{-1}$ $600\ cm^{-1}$	Composition analysis of films coatings polymers, etc.	Research area
Infrared thermography	Similar to infrared emission spectroscopy	Plant problem identification	Available from a limited number of suppliers
X-ray fluorescence (XRF)	Electromagnetic radiation 10–100 keV	Quantitative elemental analysis	Suited to non-contained systems, e.g. material on conveyer belts
X-ray imaging	Electromagnetic radiation 40–160 keV	Pilot plant studies of fluid-bed processes	Research area—facilities available

Technique	Radiation	Applications	Status
γ-rays	Electromagnetic radiation 0.04–1.3 MeV	Density measurement, level determination	Available from a large number of suppliers, still developing
Neutrons	Elementary particles	Sensitive to elements rich in hydrogen, moisture measurement	Available
Microwaves	Electromagnetic radiation λ = 0.1–30 cm	Dielectric measurement, water content of oils	Available from a large number of suppliers

10.3.1 Ultrasound

Ultrasound has many applications for process measurement. One of its most attractive features is that it offers the possibility of non-invasive measurement. Mostly, it is used to measure physical parameters rather than chemical composition, although in the case of binary mixtures and solutions, chemical concentration can be arrived at via ultrasonic density measurement. The main applications found on process plant include level measurement, flow measurement and density determination.

Ultrasound is the term given to pressure (elastic) waves set up in gases, liquids or solids at frequencies above the range of human hearing (this is usually taken to be above 20 kHz). Very high frequency acoustic waves are severely attenuated in most materials. For most applications 10 MHz is the upper practical limit. At more modest ultrasonic frequencies attenuation is still significant and interference from noisy process equipment, such as pumps and compressors, is negligible.

There are a number of modes of ultrasonic propagation. The two main modes are compressional waves (p-waves) and transverse or shear waves (s-waves). In compressional waves, the motion of the molecules is in the same direction as the wave propagation. In transverse waves, the motion of the molecules is at 90° to the direction of propagation. There are also other modes such as surface waves, plate waves and many special modes associated with regular structures such as tubes. Compressional waves are the only mode propagated any distance in liquids or gases. Very viscous liquids (e.g. tars, polymers and waxes) will propagate transverse waves very short distances. Almost all reported process measurements are concerned with the propagation of compressional waves through the material of interest. Some of the other modes of propagation are useful in directing the ultrasound to the appropriate part of the structure [2].

The source of most ultrasound is a piezoelectric transducer. Ultrasonic devices that use continuous wave transmission require a separate receiving transducer, while many pulsed devices use a receiver and transmitter combined in a single element. Measurements generally rely on measuring the acoustic velocity (transit time) of a pulse of ultrasound travelling between two points or by the attenuation of the signal.

For single-phase liquids ultrasonic velocity measurements are used. For suspensions, slurries, etc. attenuation measurements provide a means of obtaining the concentration of the solid or second liquid phase.

There are a number of non-invasive ultrasonic devices for measuring fluid flowrates. Mostly, these rely on measuring the time a pulse takes to transit between two transducers mounted axially along a pipe. The pulse is sent in both directions; the difference of the two transit times is obviously related to the velocity of the material in the pipe. An alternative to this is the cross-correlation flow meter which uses two sets of two transducers. Pulses or a continuous wave signal are sent between the first pair which are mounted radially on the pipe. The received signal is cross-correlated with signal received by the second pair mounted further along the pipe. Anomalous conditions such as temperature, density and composition pass by both sets of transducers with a slight time delay. The time delay in the cross-correlation gives the transit time, which is used to calculate the average flow velocity. This type of arrangement is better suited to multiphase flow applications than the simpler time-of-flight device.

10.3.2 Ultrasonic density determination

The velocity of an acoustic wave is related to the density of the transmission medium by the equation:

$$V = \sqrt{\rho \times K}$$

where V is the velocity, ρ is the density and K is the bulk modulus. As density varies with concentration, determination of acoustic velocity can provide the composition of binary mixtures. Density also varies with temperature and this must be corrected for.

To determine density time-of-flight measurements are made. Two transducers are mounted on opposite walls of the vessel and the transit time, of a pulse of ultrasound, between them is measured. Alternatively a single transducer can be used to transmit a pulse and detect the echo from the far wall of the vessel.

Transducers are available from a number of manufacturers, and in a range of formats. Several versions are certified to the highest standards of intrinsic safety making them suited for measurement in the hazardous environments associated with many chemical plant. Some may be used in atmospheres containing potentially explosive mixtures such as hydrogen or acetylene! This would be considered a very demanding environment for most analytical equipment.

The transducers are simple to deploy. For temporary mounting, magnetic holders may be used. For permanent installation the face of the transducer may be bonded with epoxy resin or strapped to the vessel. The only requirement is that a good acoustic contact is made between the transducer and the vessel.

10.3.3 Ultrasonic near-wall attenuation

Near-wall attenuation of ultrasound is used to obtain information on process fluids at a single point on a vessel. The effect depends on the acoustic impedance of the material in contact with the inner wall of a vessel. A pulse of ultrasound is launched into the outer wall of the vessel. When this pulse encounters the inner wall some of the power is reflected back towards the transducer and some passes into the contained material (Figure 10.2). The ratio of acoustic power reflected at the inner interface (a_r), to the power transmitted into the vessel (a_t), is governed by the acoustic impedances (Z) of the vessel's wall and the material it contains:

$$a_r = (Z_1 - Z_2)^2/(Z_1 + Z_2)^2$$
$$a_t = 4Z_1 Z_2/(Z_1 + Z_2)^2$$

The acoustic impedance of a material is dependent on its density and its compressibility or bulk modulus.

For steel, water and air the characteristic impedances are 46.5×10^6, 1.5×10^6 and $4.0 \times 10^2 \ \mathrm{kg \ m^{-2} \ s^{-1}}$ respectively. If these values are put into the above equations it can be seen that at a steel–water interface 88% of the acoustic power is reflected. At a steel–air interface the reflected power is almost 100%. Thus, the reflected signal power is 12% lower when the vessel contains water than when it contains air. If the pulse is allowed to reverberate within the vessel wall, encountering the inner

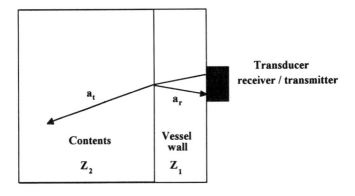

Figure 10.2 Near-wall acoustic attenuation.

interface a number of times, the effect is considerably heightened and the sensitivity increased.

Point level switches utilising near-wall attenuation are commercially available. These devices have two output states (i.e. liquid or no-liquid) and are used to monitor liquid levels in tanks, pressure vessels, oil–gas separators, boilers, etc.

Near-wall attenuation methods work well with suspensions and slurries which would ordinarily attenuate ultrasound too strongly for a transmission method to be used. Because acoustic impedance is mainly a function of density, other information could in principle be derived such as the density of foams, position of liquid–liquid interfaces and concentration of binary mixtures.

10.3.4 Acoustic emission

Acoustic emission, as a technique, is well established for monitoring the condition of rotating machinery, detecting cavitation in pumps, detecting defects in boilers during pressure testing, detecting gas leaking through relief valves, etc. However, there are growing numbers of applications where acoustic emission technology is used for process monitoring applications. While direct chemical information is not provided by acoustic emission, it provides a sensitive means of inferring the condition of many processes. Acoustic emissions are rich in real-time information which can be extracted by signal processing. In contrast to other non-invasive techniques, such as nucleonics, it responds to dynamic events, has very high sensitivity and has no special safety restrictions.

Acoustic emissions are elastic waves generated by a rapid release of energy at a localised source. They are produced by events such as particle impact, gas evolution, boiling, phase transitions, precipitation, etc. Some processes produce emissions that can be heard. A lot more emit either outside the audible frequency range or at too low an intensity to be heard [3]. The very high sensitivity of acoustic emission transducers enables many of these apparently silent processes to be monitored effectively.

Acoustic emission sensors are piezoelectric sensors similar to those used for other ultrasound applications. There are usually designed to resonate at a frequency in the range 70–600 kHz. Resonant transducers have great sensitivity. High operational frequency transducers have limited spatial range. For these reasons high operational frequency transducers are virtually immune to background noise and can be used to make sensitive measurements in very noisy industrial environments.

The impact of a single particle on the wall of a vessel will produce a detectable transient acoustic emission signal. In some cases it is possible to resolve individual acoustic signals, but in the majority of applications the individual events are far too numerous. The resultant appears as an amplitude modulated continuous signal. The carrier is the resonant wave from the transducer, and the modulation is characteristic of the process.

The usual approach to acquire these signals is to use an RMS-DC converter which removes the carrier wave and produces an output which is proportional to the average acoustic emission signal level. This reduces the measurement bandwidth requirements thus simplifying and reducing the cost of the data acquisition system. The resulting waveform is digitally sampled at a rate appropriate for the process. For many processes, sampling rates in the order of 50 Hz are adequate, allowing many channels to be multiplexed to a single data acquisition system.

Fluid-bed processes are ideal candidates for acoustic emission monitoring [4]. The main solid transportation mechanism in a fluid-bed is the upward motion of low density regions or gas bubbles. These move, through the denser regions of the bed, drawing streams of particles in their wakes. The larger bubbles move at a higher velocities and transport correspondingly greater numbers of particles. An acoustic emission transducer will detect an increase in signal as a bubble passes due to the surge of accompanying particles. The regular flow of bubbles appears as a series of rhythmic pulses in the average intensity of the acoustic emission signal. Changes in the process fluid dynamics is often indicative of process problems such as particle agglomeration. These may be observed as changes in the signal. Pattern recognition software can automatically detect these and assign signals to known classes of process condition.

Other changes in the physical properties of the particles, such as shape, cohesiveness and hardness also produce significant changes in the acoustic emission signals. These can often be attributed to temperature or compositional changes of the particles and can warn of impending problems such as agglomeration.

Acoustic emission coupled with *chemometric* techniques can reveal surprisingly detailed information about process conditions. In the minerals industry the wet grinding of ore, in ball mills, is an important process. Particle size control of the material issuing from the mill is important. Without this the next stage of the process, usually froth flotation, will not operate efficiently. Sampling of mill streams is difficult. Off-line particle size analysis is extremely time consuming and therefore not applicable to real-time control. Conventional on-line particle size analysers are known to be unreliable in this environment.

Audio frequency acoustic emission has been successfully applied in this application [5]. Analysis of the mill's acoustic emission frequency spectrum with fast Fourier transformation and partial least squares (PLS) modelling allows particle size distribution to be predicted. This is a two-stage procedure. The first is to develop a calibration database. Particle size distributions are obtained using conventional off-line techniques while simultaneous acoustic emission spectra are recorded. After a representative data set is obtained, which covers a broad range of operating conditions, PLS is used to develop a model linking the acoustic emission and particle size data. The second stage is prediction. The process acoustic emissions are monitored and the model used to predict the particle size distribution. The result is real-time particle size analysis. The model needs to be periodically updated as the mineral composition changes and mill liners wear out.

Chemical reactions that involve gas evolution, precipitation and other phase changes tend to produce acoustic emissions. Even if the reaction does not directly produce acoustic emission, it is often possible to sense secondary effects. For instance, success has been reported in monitoring precipitation in stirred vessels. In this application acoustic emission detects the impact of small crystals either on a submerged acoustic probe or the walls of the vessel [6].

There are many potential process applications of acoustic emission monitoring. As experience grows with the technique, the range of applications will continue to expand. Developments in the areas of smart sensors and neural networks for signal processing will no doubt have their part to play.

10.3.5 X-ray fluorescence spectroscopy

X-ray fluorescence (XRF) spectroscopy is a routine laboratory technique for quantitative elemental analysis which can also be used for on-line applications. XRF can measure all elements from ^{12}Mg to ^{92}U. In some cases these on-line measurements can be made non-invasively.

The technique consists of bombarding the target material with X-rays from some primary source. These stimulate atoms in the material to emit secondary or fluorescent X-rays which are characteristic of the elemental composition.

Primary X-rays are generated by either a high-vacuum X-ray tube or a radioactive source. These excite inner shell (K, L and M) electron transitions in the target species. As the atom returns to the ground state a characteristic secondary or fluorescent X-ray is emitted. As in all fluorescent techniques, the primary radiation must be of a higher energy (shorter wavelength) than the secondary or fluorescent emission. Quantum mechanics permits only a limited number of electronic transitions which result in quite simple X-ray emission spectra. The X-ray spectral lines are designated by the shell from which the electron was initially removed K, L or M, by the subshell α, β, γ, etc., and a subscript indicating the relative intensity of the emission. X-ray emission lines for a particular element are thus identified as Mn $K\alpha_1$ (5.90 keV), S $K\alpha_2$ (2.31 keV), Au $L\beta_4$ (80.39 keV), etc.

There are two basic ways of performing XRF analysis. Wavelength dispersive XRF provides the highest sensitivity and resolution combined with the shortest analysis time. The source of the primary X-rays is usually a vacuum tube. Commonly used X-ray tube targets are rhodium and titanium. The fluorescent X-rays are then dispersed by a crystal, such as lithium fluoride which acts as a diffraction grating, and detected using a scintillation counter (elements of atomic number 24 and below) or flow proportional counter (elements of atomic number 25 and above). The other approach is energy dispersive XRF. Usually the primary source of exciting radiation is a radioactive isotope such as ^{241}Am, ^{244}Cm or ^{55}Fe. The fluorescent X-rays are separated into a number of energy groups by means of semiconductor (Si(Li)) or proportional detector linked to a multichannel analyser. This arrangement allows a compact analyser to be constructed. It provides greater flexibility in operation than a wavelength dispersive instrument as the elements to be measured are easily reconfigurable.

XRF is applicable to quantitative analysis of virtually all types of solid and liquid material. There are very many reported applications, these include: mineral composition, plating solutions, metals such as vanadium and nickel in oils, lead and sulphur in gasoline, cement composition and paper coatings to name only a few.

Because of the limited penetration of the longer wavelength X-rays excitation occurs mainly within the surface layer. In the laboratory surface preparation and particle size control of samples for analysis is crucial to obtain a satisfactory level of reproducibility. This is not such a severe constraint for on-line processes. In most cases the material is either a flowing liquid or material moving on a conveyor belt. In these cases time averaging of the signal from the material as it flows past the XRF detector will remove many effects that are noted in the laboratory. Particle size distribution must however be controlled. Failure to do this will have a very adverse affect on the analytical results. When monitoring light element compositions it is particularly important to ensure cleanliness of the windows on the XRF measurement head. X-rays associated with the lighter elements (longer wavelengths) can be totally absorbed by a few tens of micrometres of material. This may involve routine cleaning of the windows. The longer wavelength X-rays are attenuated strongly in air and care must be taken to minimise the distance between the measurement head and the material. With these on-line analysers it is not usual practice to use any form of internal standardisation. Calibration is achieved using material of known concentration.

There is no single XRF analyser that will suit all applications. X-ray source, configuration, analyser crystal and detector all need to be selected. The choice will be influenced by the elements to be determined, the sensitivity required, the sample form and the elemental matrix.

The major application of non-invasive XRF is the analysis of bulk materials on conveyor belts. This has been particularly taken up by the minerals processing industry. Another commerically available non-invasive system is designed to measure the thickness of silicone coatings on paper. This uses XRF to monitor the silicon content of the coating. This device enables savings to be made by preventing overcoating. Other applications of on-line XRF usually involve a sampling system and are therefore invasive in their usage. Because of the high X-ray attenuating properties of steel it is unlikely that XRF measurements can be made non-invasively through most normal process containments.

10.3.6 X-ray imaging

X-ray imaging is mainly applicable to process development. Because of the thick walls and attenuating materials used in the construction of commercial plant, X-ray imaging cannot usually be used. However, during process development X-ray imaging provides quantitative real-time information about the internal structure of rapidly changing three-dimensional processes. It has been mainly used in the area of solids handling and multiphase flow. Examples of this include the study of defluidisation and agglomerate formation in fluid-beds, catalyst injection profiles, dispersion characteristics of gases and catalyst behaviour in fluidised catalytic cracker units (Figure 10.3).

X-rays are generated at between 40 and 160 keV depending on the absorptivity and path length of the material. By pulsing the X-rays internal movement can be frozen allowing a clear image to be obtained. Pulsing also enables high peak power to be achieved from a comparatively modest source, allowing vessels up to a metre in diameter to be penetrated. Processes then can be simulated, or run, under industrially relevant conditions. The X-rays pass through the vessel and are attenuated depending on the amount and distribution of solids and liquids. They are detected by a CsI image intensifier and converted into a visual image that can be recorded in conventional video format [7].

10.3.7 Nucleonic methods

Methods employing γ-ray sources are used to measure density, determine levels and to detect the passage of interfaces in process plant. γ-rays are electromagnetic radiation emitted from the atomic nucleus, whereas X-rays result from electronic transitions in the inner orbitals. Industrial γ-ray sources are generally either ^{137}Cs (0.66MeV), ^{60}Co (1.33MeV) or ^{241}Am (0.06MeV). These are usually encapsulated in glass and contained within a stainless steel tube. Radiological protection is provided by a substantial thickness of lead.

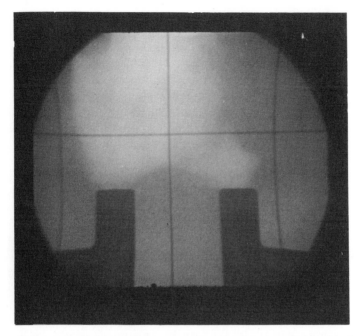

Figure 10.3 X-ray image of gas flow from two feed nozzles in a fluidised bed of FFC catalyst. This shows that the nozzles are too close together causing coalescence of the gas jets. (Reproduced by permission of BP Chemicals Ltd, Sunbury, UK.)

The choice of source depends largely on the density of the material, the path length and the composition and thickness of the vessel walls. Typical source strengths range from 1 to 200 mCi. Anything above about 100 mCi would be considered as an exceptionally strong and its use would be discouraged because of the associated hazards. The choice of isotope is largely a matter of penetration and half-life consideration. ^{137}Cs has a half-life of 30.2 years and ^{60}Co 5.3 years. This means that the lifetime of a ^{137}Cs source in a particular application is essentially indefinite whereas ^{60}Co may have to be replaced when its count rate has decreased to a level where measurement inaccuracies become unacceptable. Included in most instrumentation software is a correction for the gradually decreasing count rate due to radioactive decay. It is preferred to use ^{137}Cs wherever possible as containment is easier than with the more penetrating higher energy ^{60}Co source. ^{241}Am is only used for special applications which will be discussed later.

The interaction of γ-rays with material is fairly complicated. There are three modes of interaction with matter which are responsible for γ-ray absorption. There are dependent on the energy of the γ-ray and the nature of the material. Below 1 MeV the major interaction with materials of high atomic number is the photoelectron effect. The important outcome of this is that almost all the γ-ray energy is given up in a single interaction. The second interaction is Compton scattering which is the interaction of the γ-ray with bound or un-bound electrons. In this process the γ-ray only loses part of its energy. Above 1.02 MeV electron–positron pair production is possible. Here the γ-ray loses its energy as the kinetic energy of the two particles. γ-ray absorption is directly related to the density of the material, the linear absorption factor and the path length. As γ-ray beams are not collimated, the radiation intensity decreases with the square of the distance between the source and the detector.

The detectors are either Geiger–Müller tubes or scintillators. Plastic and NaI(Tl) scintillators are used in preference to Geiger–Müller tubes as they offer better sensitivity. Currently, plastic scintillation rods are an active area of development. These consist of a synthetic rod scintillator, a photomultiplier tube and a control unit (Figure 10.4). On entering the scintillator each γ-quanta is attenuated and generates a minute flash of light. These are detected by the photomultiplier. All the signals which lie above a pre-set threshold are counted in a given time. Over the last 10 years these devices have become extremely sensitive. The important consequence is that this has allowed the γ-ray measurements to be made with correspondingly lower source strengths. Accordingly, many of the safety constraints have lessened leading to a greater utilisation of the technology.

Process applications include continuous level measurement, measurement of separation layers (e.g. oil and water in a separator), mass of material on conveyor belts and density measurements. The usual mode of measurement is transmission with the source and detector on opposite sides of the material to be monitored. An alternative to this is to use back-scattering where the source and detector are mounted adjacent to one another. This is most useful in situations when plant geometry does not facilitate easy installation of a transmission arrangement.

An extension to normal γ-ray absorption methods is double energy absorption. The absorption of the low energy γ-rays from ^{241}Am increases as a function of the atomic number of the material it passes through. Whereas the attenuation of the higher energy γ-rays from ^{137}Cs depends only on the mass of material it passes through (Figure 10.5). By measuring the absorption of these two γ-rays it is possible to derive more detailed compositional information. A commercial system exists for determining the ash content of coal as it passes on a conveyor belt. The ash components in coal have higher atomic numbers than the coal. Another system uses ^{137}Cs and ^{244}Cm (0.04MeV) sources. The low energy ^{244}Cm source has a high sensitivity to sulphur. This is used for measuring the sulphur content in finished combustion oil products, oil blending operations, etc. However, this system is invasive and requires a sampling system for the process liquids.

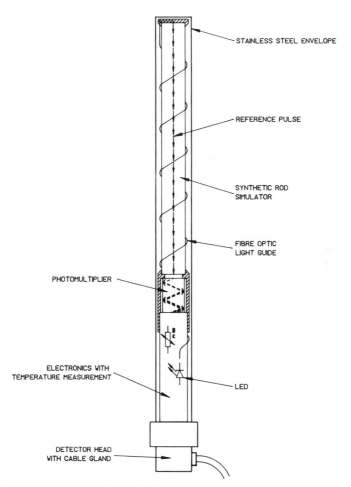

STAINLESS STEEL ENVELOPE

REFERENCE PULSE

SYNTHETIC ROD SIMULATOR

FIBRE OPTIC LIGHT GUIDE

PHOTOMULTIPLIER

ELECTRONICS WITH TEMPERATURE MEASUREMENT

LED

DETECTOR HEAD WITH CABLE GLAND

Figure 10.4 A plastic rod scintillator. Before the start of the measurement cycle a light pulse is generated by the LED and used as a reference measurement. (Redrawn by permission of Endress and Hauser, Weil am Rhein, Germany.)

Neutrons are also used to make non-invasive measurements. This technique is known as neutron scattering. Fast neutrons from a suitable source are scattered strongly by nuclei of low mass, particularly hydrogen. A considerable fraction of the energy of the incident neutron is lost this way. Thus, a fast neutron source will produce a cloud of slow neutrons in a hydrogen-rich material. These can be selectively detected using an appropriate slow neutron detector such as a helium detector (^{3}He). The most commonly used sources of neutrons are ^{241}Am, a γ-ray source, mixed with beryllium and ^{252}Cf a radionuclide which produces neutrons by spontaneous fission.

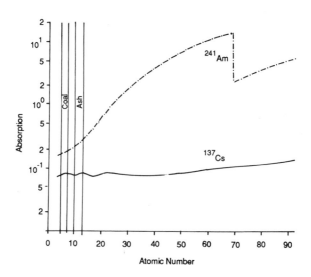

Figure 10.5 γ-ray absorption as a function of atomic number.

Applications of this technique include level measurement in refinery plant, such as coke drums, and moisture of bulk goods in tanks and bunkers. There are number of possible source/detector arrangements: (i) the source is mounted inside the vessel and the detector on the outside wall; (ii) the source and detector are mounted within separate wells within the vessel; and (iii) totally non-invasively, with both source and detector on the outside of the vessel. The tank wall will also attenuate the neutron signal by about 5% per millimetre of steel. The totally non-invasive measurement therefore requires a considerably stronger neutron source which may, for certain vessels, be in excess of the limits imposed by many national authorities.

Nucleonic methods will increase in application particularly as new technology allows measurements to be made with far weaker sources. An example of this, currently under development by a number of instrument manufacturers, utilises the natural radioactivity of certain elements, e.g. ^{40}K. The technique is obviously without hazard. However, this work is at an early stage and little information is currently available.

10.3.8 Infrared emission spectroscopy

Infrared emission spectroscopy (IRES) is the study of radiation in the range from 4000 to 600 cm^{-1}, emitted from warm or hot bodies. It is equivalent to the more familiar mid-infrared absorption measurements routinely made in laboratories (Figure 10.6). In this spectral region

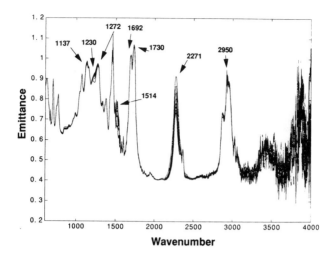

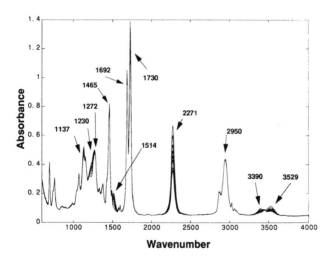

Figure 10.6 Comparison of infrared absorbance and emission spectra of a two-component paint mixture at 100°C. (Reproduced with permission from Ref. 8.)

absorption bands correspond to vibrational transitions in molecules. These are well understood and tabulated. Like infrared absorption the technique gives a powerful insight into the structure of organic molecules. It can be used for qualitative identification of compounds and quantitative analysis of mixtures.

IRES measurements are easily made. However, due to the non-linear behaviour of the emission signal, it is more difficult to extract quantitative information from IRES than from a similar mid-infrared absorption spectrum. Recent developments in data analysis techniques and software now enable quantitative information to be obtained [8]. Most mid-infrared spectrometers may be adapted to make emission measurements. The prerequisites for this technique are: the sample must be at a different temperature to its surroundings, the sample must be at a different temperature to the detector, and the sample must not be a uniformly heated optically opaque substance. It is not a problem to insure that the detector temperature is lower than that of the sample because most infrared spectrometers already use cooled detectors. If the sample is uniformly heated and opaque, the IRES will be obscured by the background 'black body' radiation. However, it has been demonstrated that reasonable spectra can still be obtained by controlled cooling or heating of the surface layer of the material [9].

IRES is well suited to non-invasive monitoring of processes that allow open access to the material of interest. Applications include extruded products such as polymer films and coatings.

10.3.9 Infrared thermography

Infrared thermography is capable of remotely measuring the true temperature distribution of a 'scene' by detecting and mapping the amount of thermal (black body) radiation being emitted.

The detectors are usually liquid nitrogen, 'Stirling cycle' or thermoelectrically cooled HgCdTe or InSb. They respond in the wavelength range 3–12 µm, roughly equivalent to the range used in IRES. They allow temperatures to be mapped anywhere between −40 and 400°C. By the use of neutral density filters this upper limit can be extended. There are systems commercially available which will operate up to 1500°C. Most thermography systems are capable of resolving approximately 0.1°C at 30°C. The detector is mounted in a camera and the scene is raster scanned by means of oscillating mirrors or rotating optical polygons. The resultant image is displayed on a conventional video monitor, usually in false colour. The colours represent the approximate surface temperature of the object.

In the process environment, thermography is used to identify: hot spots occurring within reactors, detecting defective insulation on steam pipes, detection of wet insulation, detecting liquid levels in tanks, condition monitoring of refractory linings, detecting build up of scale and other deposits in pipelines, etc. A potential application of infrared thermography is to visualise gas leaking from flanges and valves. If the gas is infrared absorbing, such as a hydrocarbon, in daylight it will be seen as a cool plume against a hotter background. Because many process problems manifest themselves by generating heat or modifying heat flow locally the number of potential applications for this technique are huge.

The pictorial output makes this technique very amenable to interpretation. Complex process problems can often be identified simply by qualitative inspection of the images.

10.3.10 Microwaves

The microwave region of the electromagnetic spectrum extends from below 1 m to several tens of centimetres. Microwaves interact with matter by coupling energy from the electromagnetic field by electric and magnetic dipole interactions. They penetrate materials with a low dielectric constant such as: plastics, glass, ceramics and composite materials, even when several centimetres thick but are reflected by metal surfaces. This limits their usefulness as a non-invasive monitor. For metal vessels microwave window materials such as plastics (e.g. PTFE) and ceramics must be used. The microwave sources used in process applications are almost always solid-state Gunn or tunnel diodes.

Traditional microwave spectroscopy, which provides molecular structure information, is only applicable to molecules in the gas phase at low pressures. It is therefore not particularly useful for non-invasive analysis. Gases at high pressure resemble the condensed phase in their interaction with microwaves. As a process monitoring tool microwave absorption has been mostly applied to the measurement of moisture content. Due to its high dielectric constant the microwave absorption of water is several orders of magnitude greater than most dry materials. Reported applications include: profiling water content of sheet materials, monitoring water content of oil shale, water in cotton, etc. [10]. Some work has shown that microwaves may be suitable for determining the water content of crude oil. Accurate determination of the water content of oil is important at the 'custody transfer stage' and for other fiscal measurements. The form of the water is important. Dissolved water, well separated aqueous and oil phases, water droplet size distribution and salt content will affect the microwave absorption characteristics.

Microwaves are widely used for liquid and solids level detection in tanks and hoppers. The majority of applications are invasive with microwave antennae installed within the vessel. There are two modes of operation; either a radar type method (pulse echo) that measures the time-of-flight of a very short pulse (< 1 ns) of microwaves between an antenna and the surface of the material, or an attenuation method utilising a separate transmitter and receiver. Time-of-flight microwave measurements have the advantage, over similar ultrasonic measurements, that density changes in the vapour space do not greatly affect the transit time of the pulse.

The attenuation method may be applied non-invasively only if the material of the containment is sufficiently transparent to microwaves (see above). One possible application of this is to detect the passage of oil–water interfaces in separators, as low dielectric materials such as oil, are far less attenuating than water.

10.3.11 Electrodeless conductivity

Electrodeless conductivity measurement is a non-invasive but intrusive technique requiring a probe to be immersed in the process fluid. Its main area of application is with aqueous liquids, the most important of which is water quality monitoring, e.g. boiler feed water quality assurance, detecting acid leaks into cooling water streams, etc. Other applications include the control of acid dilution systems and blending systems. In principle, conductivity measurements may be of use in organic process streams but there is little evidence of its widespread application.

Conventional conductivity measurements rely on two electrodes immersed in the process fluid. A low frequency (in the order of 1 kHz) AC voltage is applied across them. The current passing is proportional to the solution conductivity. The conductivity of a solution has a contribution from all the ions present. In aqueous solutions H_3O^+ and OH^- ions have the greatest influence on conductivity because of their very high mobility. Contamination of the electrode surfaces may result in a low reading.

Electrodeless conductivity measurement uses an electromagnetic-induction principle that avoids the necessity of having electrodes in contact with the solution. Two toroidally wound coils, on a common, axis, are encapsulated to form the sensor. This is immersed in the process liquid. An AC current is applied to the transmitter coil which induces a current in the receiver coil. The current induced in the receiver coil is dependent on the current flowing through the solution, which is proportional to the conductivity of the solution (see Figure 10.7).

Electrodeless conductivity probes have a wide measurement dynamic range (1 µS cm^{-1} to 2 S cm^{-1}) enabling a single probe to cover a wide range of conditions. Electrodeless conductivity is not sufficiently sensitive for monitoring the purity of water used in semiconductor manufacture (< 1 µS cm^{-1}), conventional conductivity techniques must be used. In commercial probes the encapsulation is usually plastic either PEEK (polyetheretherketone) or a fluorocarbon. Because conductivity is highly temperature dependent sensors invariably incorporate an integral temperature sensor.

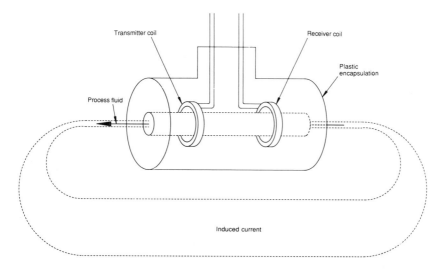

Figure 10.7 An electrodeless conductivity probe.

10.3.12 Other techniques

There are many other analytical techniques that with further development could be used in certain circumstances for non-invasive monitoring.
One such technique is nuclear magnetic resonance (NMR). NMR has
been reviewed as a potential non-invasive technique [1]. NMR is a
powerful analytical technique that is sensitive to the environment of
hydrogen atoms in organic compounds. It can be used for structural
determination, qualitative and quantitative analysis. It has not been
widely used for process analysis due to certain practical difficulties.
These include not being applicable to material contained within metal
tubes, the NMR tubes need to be small bore and the technique is costly
and difficult to implement in a process environment.

10.4 Conclusions

The field of non-invasive measurement is an active one. At present there
are only a limited number of chemical determinations that may be
performed non-invasively. There are, however, a much larger number of
physical measurements that can be made non-invasively.

Non-invasive density measurement can be made via γ-ray absorption,
ultrasonic velocity and ultrasonic near-wall attenuation. Water content
can be measured using microwave attenuation and neutron scattering.
Elemental concentration may measured using X-ray fluorescence and

limited number of elements measured directly by γ-ray absorption. Direct non-invasive organic compositional analysis is currently not possible in most process containment. In the case of binary mixtures, organic composition may be arrived at by density determination using a number of non-invasive methods. However, a much wider range of techniques such infrared emission spectroscopy are applicable to organic materials that are being extruded, rolled, etc.

Non-invasive measurement alone, generally provides relatively little information compared to an invasive measurement. It is likely that in the future that combination measurements will be common place as they have the potential of providing much more information. Already microwave moisture measurements are made in conjunction with γ-ray measurements to compensate for density changes, multienergy γ-ray absorption provides better compositional analysis, etc. Developments in intelligent sensors and improved process communication technology, such as 'Fieldbus', will further enhance the capabilities of non-invasive techniques.

Due to the increasing demand placed on process operation by environmental issues and the economic benefits achievable it is inevitable that the total number of applications of non-invasive monitoring will increase over the next decade and well into the foreseeable future.

References

1. I. M. Clegg, Non-invasive techniques for on-line measurement. *Measure. Control* (1989) **22** 102–104.
2. R. C. Asher, Ultrasonic sensors in the chemical and process industries. *J. Phys. E. Sci. Instrument.* **16** (1983) 959–963.
3. D. Betteridge, M. T. Joslin and T. Lilley, Acoustic emission from chemical reactions. *Analy. Chem.* **53** (1981) 1064–1072.
4. R. Belchamber and M. Collins, Sound processing for your plant? *Control Instrument.* **October** (1993) 41–42.
5. European Patent Application 88303551.6, Method for Determining Physical Properties, 1988.
6. J. G. Bouchard, M. J. Beesley and J. A. Salkeld, Acoustic diagnosis in process plant: crystallisation monitoring. *Process Control Quality*, **4** (1993) 261–277.
7. D. Newton, C. S. Grant and B. Gamblin, X-ray imaging to reveal FCC catalyst and gas interactions. *Hydrocarbon Technol. Internat. Quart.* **Winter** (1994–1995) 41–45.
8. R. J. Pell, J. B. Callis and B. R. Kowalski, Non-invasive polymer reaction monitoring by infrared emission spectroscopy with multivariate statistical modelling. *Appl. Spectrosc.* **45** (5) (1991) 808–818.
9. R. J. Pell, C. E. Miller, B. R. Kowalski and J. B. Callis, Infrared emission spectroscopy with transient cooling. *Appl. Spectrosc.* **47** (12) (1993) 2064–2071.
10. J. F. Alder, M. F. Brennan, I. M. Clegg, P. K. P. Drew and G. Thirup. The application of microwave-frequency spectrometry, permittivity and loss measurements to chemical analysis. *Trans. Inst. Measurement Control* **5** (2) (1983) 99–111.

11 Future developments of chemical sensor and analyzer systems

W. P. CAREY and L. W. BURGESS

11.1 Introduction

The benefit of employing continuous chemical analysis during manufacturing processes is widely recognized as providing very high rates of return. Improvements in process control, which affect process efficiency and product quality, can be critically dependent upon on-line chemical analysis. This is evident by the current rapid growth in the development of on-line analytical instrumentation designed specifically for process control applications. Current process automation demands distributed feedback closer to events that require monitoring. As markets become increasingly more competitive, feedstocks more costly, and environmental issues escalate, the need will continue to grow for robust, stable, and affordable on-line chemical analysis systems. Distributed chemical sensors or sensor-based chemical analyzer systems provide an attractive approach to meet this demand. However, the development of sensors for on-line chemical monitoring presents some interesting challenges [1].

Clearly, the development of chemical sensor technology has not kept pace with that of physical sensors. Even within the subset of chemical sensors, the driving forces for development of process chemical sensors has been restricted due to the specialized nature of each application. For example, a pH electrode may work well in a laboratory setting, but fail for monitoring pH in a particular fermentation process because the process matrix fouls the sensor. Thus, several approaches may be required to develop devices for the same analyte depending on the process matrix and measurement conditions. More restrictive however, are the requirements for long-term reliability, specificity, and *in situ* function verification. Sensitivity and device dynamic range can also prove to be problematic in matrices where one wishes to monitor relatively small changes in a major constituent. Physical size and even cost are much less of an issue for process than in other sensor applications, if the above requirements can be met.

So how does one approach developing devices for process sensing applications? The challenge is to incorporate what is classically thought of as sensor technology into a system that will result in the *in situ* chemical analyzer. This will require that the system be an integrated package that can perform the functions normally associated with classical laboratory analysis; sampling, sample conditioning, analysis, data reduction, and reporting. Ideally, the system will be self-calibrating, and be capable of detecting and correcting (or at least signaling) problems when they occur. This chapter focuses on sensor technology utilizing multivariate analysis techniques. This is the first step toward sensors capable of operation in complex samples with more than one analyte. It provides the basis for multianalyte analysis, background detection and data reduction.

It is obvious that a high degree of functionality needs to be economically combined into each system. Currently we are seeing rapid technological advances in the areas of micromachining and micro-optics. This technology allows the construction of sensor systems into viable analytical instruments [2]. Micromachining permits precise and reproducible alignment of optical elements (fibers), better delineation of flow paths, and more functionality to be constructed into a given device. Coupling micromachining with current trends in silicon-based sensor technology will lead to sensor systems with a great deal of versatility. Many of these concepts are described in this chapter.

11.2 The higher order approach to array sensing

As with most of the sensor development currently being pursued, the individual sensor (zero-order sensor) is the most common researched and developed sensor type. Implementation of single sensors in the laboratory are the basis of many routine chemical analyses. Sample preparation is usually a required step in order to isolate the analyte of interest since most sensors have interfering responses to other compounds. The basis of using higher orders of sensing is to implement sensor technology without complex sample preparation [3]. Sensor analyzer systems will focus on field measurements and must address the multicomponent sample problem.

The advantage of the single sensor is the simplicity of its implementation. Whether it is a linear or nonlinear responding sensor, calibration using least-squares or non-linear regression techniques are statistically sound and understood. The sensor itself, regardless of its type, is always easier to implement and characterize by itself than working with more than one sensor simultaneously as a group. However, to attempt multicomponent samples, multiple sensors in the array form must be used.

11.2.1 Arrays of sensors or first-order sensors

Sensor arrays have several advantages over the single sensor due to the type of information carried in its signal. First of all, arrays can perform multicomponent qualitative studies using pattern recognition techniques [4]. They also can perform multicomponent quantitative analysis using multivariate regression models [5]. Lastly, they can compensate for interfering species which are known by calibrating them into the model, and unknown species by detecting them as outliers from the calibration models [6,7].

The definition of the sensor array is that the instrument response gives a vector of data for each sample. This vector of data is a grouping of responses from several sensors. The only requirement is that each sensor has some differential selectivity for each analyte in the sample. A common example of a sensor array is a light spectrometer where a group of wavelengths are used to probe a sample. The resulting spectrum gives a response that is unique for each component in the sample. The response pattern for an analyte must be unique to that analyte in order to resolve it from other components.

The mathematical term for lack of uniqueness in the array response pattern for each analyte is collinearity. The degree of collinearity is directly correlated with how well the array can resolve a mixture of two or more components for both identification and quantitation. The parameters of an array that collinearity influence are the figures of merit (array sensitivity, selectivity and limit of determination) and the propagation of error in calibration [8,9]. As one might imagine, the use of perfectly selective sensors in the array would be ideal; the degree of collinearity would be extremely small. Therefore, the sensor array advantage is a compromise in allowing less selective sensors perform the measurement and designing the array to optimize the above figures of merit.

11.2.2 Figures of merit

The figures of merit for an array is specific to a set of analytes in the measurement sample. For two components in a mixture, each pure component has a unique response pattern. This response pattern an be represented as a vector in space with dimensions equal to the number of sensors in the array. Figure 11.1 shows two analyte response vectors, **A** and **B**, measured with two sensors. The angle, θ, between the two vectors is a measure of collinearity. In regression, each analyte response vector is projected onto a new vector which is orthogonal to all of the other analyte response vectors. This projection is the unique part of each analyte's response pattern and is defined as the net analyte signal. **B*** is the net analyte signal for **B**.

To calculate the figures of merit, comparisons of **B*** and **B** are made. The array selectivity for **B** is the ratio of the length (two norm of a vector) of **B*** to **B** as follows:

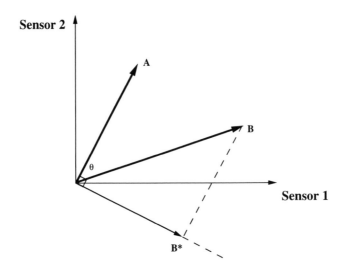

Figure 11.1 Geometric representation of the net analyte signal used in multivariate calibration and calculating sensor array figures of merit.

$$\text{Selectivity} = \frac{\|B^*\|}{\|B\|} \qquad (11.1)$$

By examining Figure 11.1, the selectivity is a function of collinearity or θ. As θ nears zero, the selectivity would approach zero. As θ increases to $90°$, **B** and **B*** would be the same vector and selectivity would approach one.

The calculation of sensitivity of the array to analyte B is the length of its net analyte signal divided by the concentration that gave vector **B**.

$$\text{Sensitivity} = \frac{\|B^*\|}{C_B} \qquad (11.2)$$

The sensitivity of an array to an analyte is similar to the sensitivity of an individual sensor. When a single sensor is calibrated, the sensitivity is the slope (a scalar value) which describes the change in response for a unit change in concentration. When arrays are used, the calibration line (regression vector) is used as the sensitivity and is identical to the net analyte signal represented by equation (11.2).

The limit of determination (LOD) is the lowest concentration that the array can detect of one analyte with the other analytes present in the sample. The same expression as for limit of detection of a single sensor is used, except with the array figure of merits substituted. The LOD for analyte B is the concentration which gives an array response of three standard deviations above the background noise. The expression for this is as follows:

$$LOD_B = \frac{3\varepsilon_d}{\text{sensitivity}_B} \qquad (11.3)$$

where ε_d is the error of the array response. The form of this equation is dependent upon the type error of the array signal (absolute or relative) and its magnitude in both calibration and test sample.

An interesting point to investigate is the comparison of zero-order and first-order sensitivity calculations when designing an array. To set up an array, sensors must be selected from a group of promising devices. But how to select the right sensors for array is a problem. If equations (11.1) and (11.2) and Figure 11.1 are studied, it becomes apparent that the sensors selected should have low collinearity. The angles between response vectors, θ, must be maximized. However, when looking at individual sensor responses, the sensitivities to each analyte may be drastically different. This magnitude can have an negative influence on subjective selection of sensors for an array. One way to objectively select sensors for an array is to test a group of candidate sensors to the analytes of interest and build a matrix of responses with dimensions of analytes by sensors. If principal components analysis (PCA) is applied to the matrix, we can observe which sensors correlates best with each eigenvector [10]. Since the eigenvectors are orthogonal, the sensors selected will minimize collinearity.

To prove this point, a test was made using polymer coated quartz crystal microbalance (QCM) sensors. To build an array of seven polymer coated QCMs, a larger group of 27 sensors were tested individually with each analyte of interest. The PCA method was used to build an array for three analytes in the application. The figures of merit for this PCA selected array are then compared to the figures of merit for a similar array built by taking the most sensitive sensors of the 27 to the three analytes of interest. Table 11.1 compares the figures of merit for the three

Table 11.1 Figures of merit comparison for seven-element QCM sensor array

Analytes	Sensitivity (Hz/ppm)	Selectivity	LOD (ppm)
A. Sensor array designed from maximum sensitivity			
Benzene	1.41	0.76	3.26
Dodecane	0.73	0.67	6.29
DMMP*	4.98	0.73	0.93
Mean	2.37	0.72	3.49
B. Sensor array designed by PCA for maximum selectivity			
Benzene	1.45	0.90	2.79
Dodecane	2.90	0.89	1.40
DMMP*	4.94	0.82	0.82
Mean	3.10	0.87	1.67

*DMMP—dimethylmethylphosphonate.

analyte application. The most striking feature is that the PCA selected array had on average a greater sensitivity for the three analytes than the array designed from the most sensitive sensors. The PCA method selects sensors that maximize selectivity of the array to the analytes. As the selectivity increases, the sensitivity increases too. Additionally, the limit of determination is significantly lowered using well designed arrays. This example shows the connection between collinearity, figures of merit and array performance.

11.2.3 Quantitation with sensor arrays

Quantitative calibration for sensor arrays involves the same type of regression approaches as found in the single sensor case extended into multivariate statistics [11]. Linear algebra is used where the concentration and sensor data are assembled into vector and matrix forms. The regression coefficients for a sensor array is a matrix of coefficients, relationships between each sensor and each analyte. The fundamentals of array calibration are presented in chapter 9.

11.2.4 Quantitation with non-linear sensor arrays

One of the challenges of multivariate statistics is the ability to handle non-linear data. Non-linear responses are a bit more difficult to handle since iterative techniques are often needed to fit individual portions of the response curve. In the multivariate case, with non-linear sensor arrays, calibration models depend upon iterative algorithms to optimize several functions simultaneously.

Calibration of non-linear sensor arrays has taken several approaches using methods based on non-parametric regression. The techniques used most often for sensor array calibration are projection pursuit regression (PPR) [12,13], multivariate adaptive regression splines (MARS) [14], and artificial neural networks (ANN) [15–17]. PPR projects the response vectors onto new vectors which are found through an optimization algorithm. These projections are then transformed into linear functions much like a log transform of an exponential function. MARS is a multivariate spline fitting routine which operates in a similar fashion to one-dimensional splines. A response curve is divided up into sections and linear regressions are fit into the intervals. A final third-order polynomial smoothes the entire curve to make a continuous fit.

11.2.5 Two-dimensional arrays or second-order sensors

As outlined previously in this chapter, definite advantages arise when progressing from the single sensor to the sensor array. By examining the mathematics for the next level of complexity, the second order, even more advantages are evident. These advantages include the ability to quantitate a specific analyte without knowing any information on the rest of the sample. In addition, if an unknown interferent entered into a later prediction sample, it would not disrupt the quantitation of the analyte of interest. From these advantages, the implications for sensor applications is enormous. If a second-order sensor which monitored groundwater or industrial processes was available, it could easily be implemented into complex systems without the effort of knowing all of the interferences that may possibly come into contact with it.

The evolution of second-order sensor calibration has lead to investigations to find equivalent instrumentation that gives rise to second-order data. The definition of second-order instruments is simply two sensor arrays which are independent of each other (Figure 11.2). However, in order for the arrays to be independent, one of the arrays must modulate the sample's analyte concentrations. The best known instrument currently is gas or liquid chromatography with an array detector such as mass spectrometry or light spectroscopy. The response for a single sample is a matrix of data where the columns may indicate the response of the array detector and the rows are the retention times of the chromatograph. The chromatograph modulates the analyte concentrations while the sensors monitor the sample at given time intervals.

11.2.6 Second-order quantitation

Second-order problems were first investigated by rank annihilation as a technique to subtract a know response matrix of a pure component from a matrix of responses due to the component of interest and other interfering compounds [18]. The rank of the complex matrix decreases as the amount (concentration) of the pure matrix is subtracted. When the rank drops one value, i.e. from five to four, the amount of the pure component has been completely removed and the concentration of that component is the amount subtracted. This is an iterative process which consumes a lot of time.

A new technique called rank annihilation factor analysis (RAFA and GRAM) was developed which did this process non-iteratively [19,20]. Now an exact solution can be estimated directly on the sample matrix if the pure component response matrix of the analyte of interest is known.

The basis of RAFA is the direct eigenvalue solution of the mixture matrix of responses based on a singular value decomposition. Let N_k represent a data matrix of the pure component response of analyte k from the two-dimensional sensor array and M represent the sample response of the same instrument to the analyte of interest and all

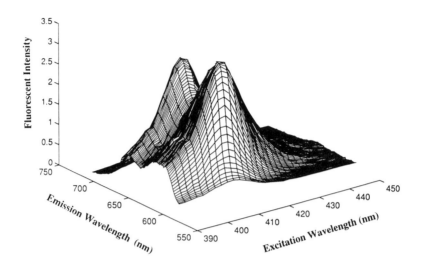

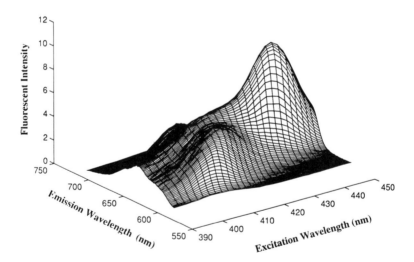

Figure 11.2 Examples of data from a second-order sensor array. Data are excitation-emission fluorescence spectra from mixtures of porphyrins. In the bottom plot, a new component is added to the original two-component mixture (top). To quantitate this new porphyrin, only its pure component spectra needs to be known and subtracted.

interfering components, the solution of the rank annihilation problem can be stated as follows:

$$\mathbf{M} = \sum_k c_k \mathbf{N}_k \qquad (11.4)$$

where

$$\mathbf{N}_k = \mathbf{x}_k \mathbf{y}_k^T \qquad (11.5)$$

The vectors $\mathbf{x}$ and $\mathbf{y}$ are response patterns of the two separate arrays used. If the array is truly second order, these two vectors are orthogonal. c_k is the concentration of analyte k as $\mathbf{N}_k$ is defined as the bilinear response pattern for a unitary concentration level. If the matrix $\mathbf{M}$ has a rank of p, then subtracting $c_k \mathbf{N}_k$ from $\mathbf{M}$ will drop one unit of rank $(p - 1)$ at the correct concentration level of c_k. Therefore, the determinate of $\mathbf{M}$ will approach zero at the correct concentration.

$$\det(\mathbf{M} - c_k \mathbf{N}_k) = 0 \qquad (11.6)$$

To solve for c_k, the generalized eigenvalue problem is used with the singular value decomposition technique. The results of the problem indicate both the pure component response patterns $\mathbf{x}$ and $\mathbf{y}$ and the ratio of concentrations of the pure components to the standard response concentration.

11.3 Optical sensing

Optical sensing is the most utilized first-order sensor array. As an array of sensors, multiple wavelengths are ideal for most colorimetric measurements. All of the advantages of first-order sensor arrays are inherent in spectroscopic analyses. The spectra obtained are linear responses due to Beer's Law, and spectral overlap of two or more components can be resolved by using linear first-order calibration routines such as PLS or PCR. In the case of process sensors, spectroscopic analyses can be utilized by the use of fiber-optic technology as both a light transfer medium to a remote location and as part of the sensor itself. In this section we will discuss two application areas of fiber-optic sensors, the first is a first-order array with the fiber as part of the sensing unit and the second is a second-order sensor array.

11.3.1 High acidity sensor

At Los Alamos National Laboratory, chemical processes to purify metals use a solvent of 7M nitric acid. The acidity of these processes is critical to precipitation and ion exchange reactions. To monitor these processes, a fiber-optic sensor approached based on evanescent field spectroscopy was developed [21]. Evanescent field spectroscopy or internal reflection spectroscopy is based on the light reflection inside an optical element such as a fiber or prism. At the site of reflection, a fraction of the light wave penetrates into the lower refractive index media on the outside of the element. This field penetration can be used for optical spectroscopy purposes. The advantage of this technique is that absorption measurements in highly adsorbing or opaque media can be performed. Metal ion solutions are usually highly colored due to the metal complexations that occur. The disadvantage is the signal strength of the evanescent measurement. Usually many bounces or reflections are needed to build up the equivalent pathlength of the absorption cell.

To detect acidity at high concentrations of acids, Hammett indicators are used. Hammett indicators are organic molecules which are weak bases and are protonated at low pH. Upon protonation, the indicator shifts its visible light absorption characteristics. In the case of the range of 2–12M nitric acid, the indicator chromazurol-S was used. The indicator was immobilized onto the surface of the optical element by physical entrapment inside a polyimide polymer matrix. The sensor is constructed by silane coupling the polymer/indicator blend onto the cylindrical waveguide or the stripped bare fiber-optic core. The thickness of the polymer layer is kept thin to insure fast response rates. In the case of the glass core fiber-optic, the refractive index of the polymer is higher than the core. Therefore, the tendency to couple the light out of the core and guide inside the polymer is favorable. To stop this coupling, the film thickness of the polymer must be less than the wavelength of light, 500 nm. The sapphire rod has a higher refractive index and was not affected by film thickness.

Spectra of the indicator, and both the fiber-optic and sapphire rod sensors is shown in Figure 11.3 and 11.4. By using multiple wavelengths, it is clear that both the acid and base forms of the indicator can be monitored. In this example, there are no interfering compounds, but it is possible to separate out effects due to other sources than the indicator itself. One case of multiparameter monitoring would be in the fiber-optic example. The indicator absorbs at 550 nm but has no absorbance above 600 nm. The absorbance offsets between 650 and 700 nm are direct measurements of the refractive index of the solution. Both the indicator absorbance and refractive index signal intensity are codependent. Using first-order calibration techniques, the two effects can be resolved and predicted. Another example of how the array approach would benefit this example would be to immobilize a metal ion chelating indicator in the polymer. As long as there is some spectral resolution, both acidity and metal ion concentration could be determined simultaneously.

11.3.2 Metal ion probe using second-order techniques

Fabricating sensors with second order properties require two stable sensing dimensions. The most logical source for the first dimension is an optical array using multiple wavelengths and absorption spectroscopy

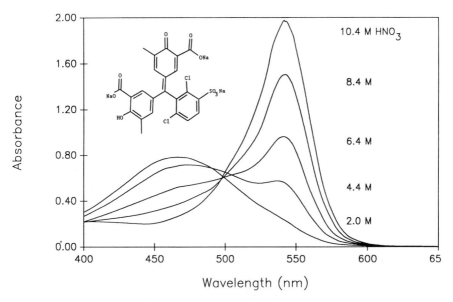

Figure 11.3 Absorbance spectrum and structure of chromazurol-S acidity indicator.

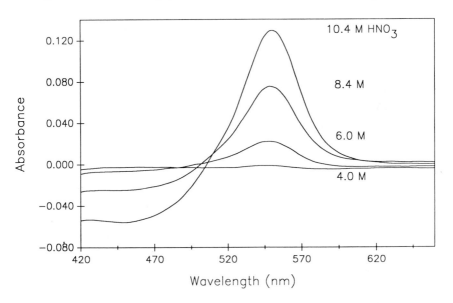

Figure 11.4 Evanescent field spectra of chromazurol-S immobilized with polybenzimidazole onto a sapphire rod.

because of its linearity. The second sensing dimension must be an analyte concentration modulator. As an example, the development of renewable

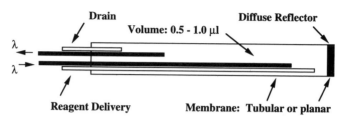

Figure 11.5 Schematic of a renewable reagent–fiber optical probe.

reagent optical fiber based sensor (flow probe) is a first step in developing this analyzer concept [22,23]. In one possible geometry for this
sensor, shown in Figure 11.5, a reagent chemistry is delivered under
direct control to the sensor tip which consists of a tubular semipermeable
membrane. The analyte is passively sampled by diffusion across the
membrane into the reagent stream, forms a detectable product, is
spectroscopically analyzed via fiber-optics, and is cleared from the sensor
by fresh reagent. The dimensions of the probe tip are such that the
diffusion distances are kept short while maintaining reasonably long
paths for optical detection. The selectivity of the sensor may be obtained
by selecting the semipermeable membrane and analyte-specific reagent
chemistry, monitoring differential reaction and diffusion kinetics of the
permeating materials, and spectral analysis of the reagent–analyte product. It is possible to alter the selectivity, sensitivity, and even the
operating range of the device by changing reagent composition or
delivery rate while the probe remains *in situ*.

Determination of multiple analyte concentrations is addressed by
temporal modulation, i.e. flushing the reagent analyte product from the
probe head and collecting time dependent responses. The rate of color
formation in the sensor for a given analyte is a function of both the
permeation rate through the membrane and the rate of reaction with the
reagent. Each of these variables is affected by several physical and
chemical parameters and within these, the variability may be used to help
discriminate between analytes in multicomponent systems. Permeation
through the membrane is affected by the membrane composition, thickness, and porosity. Factors affecting the formation of absorbing species
in the sensor include reagent composition, reaction kinetics, temperature,
secondary reaction pathways and reaction product stability.

The system produced by Lin and Burgess was specific for the divalent
ions of lead, cobalt and zinc. Lead determination in water samples (lake
and tap water) in the 6 µg/l level had standard deviations of 1 µg/l using
second-order methods. Most of the error of second-order methods is in
the reproducibility of the temporal dimension. As the membrane fouls or
ages, ions transport rates are directly affected.

11.4 Silicon processing for sensor development

Some of the standard processing techniques used in VLSI design are directly applicable to construction of sensor structures on silicon. The chemistry of silicon processing involves a number of standard steps to manufacture a layered electronic device based on CMOS technology. In the final chip the CMOS structure has functions of transistors, capacitors, resistors, and other basis electronic modules which comprise the integrated circuit [24]. Of course, VLSI has to do with packing as much electronics into a square micrometer of silicon as possible, and therefore, submicrometer resolution is of great interest.

In the sensor world, however, the needs are different, and densely packed electronics are not necessary. Common dimensions for electrochemical sensors are electrodes of 50–100 µm widths and spacing. Typically sensors are only a couple of layers thick of different CMOS structures which reduces the number of masks and processing steps. Another major difference with integrated sensors is the deviation from standard materials. VLSI processing uses only a limited set of processes and materials, whereas, sensors may need a material such as platinum or palladium as an electrode. Since, this cannot be performed by standard fabrication plants, a specialized sensor fabrication laboratory is required.

The basic chemical reactions of silicon needed for CMOS development are oxidation, etching, and diffusion. Oxidation of silicon to form silicon dioxide results in extremely good adhesion and stability of the two materials. The oxide is an excellent electronic insulator which can withstand electric fields approaching 10^7 V/cm. Manipulation of the conductive characteristics of silicon is performed by diffusion of atoms into silicon to make the material either electron rich or electron poor. Typically boron or phosphorus is diffused into silicon by high temperature processes. The ion source is either a wafer of pure dopant or a paste painted onto the silicon wafer, which is then placed into a diffusion oven at 800–1000°C. Etching either by dry or wet chemical methods of all materials (silicon, oxides, metals) is used to transform the basic silicon wafer into the complex structure of a sensor. Wet chemical etching consists of isotropic methods for silicon and oxides. Anisotropic etching of silicon is oriented along the crystal dimensions of the wafer and is commonly used in micromachining channels and reservoirs.

Other steps that are typically necessary to construct a design are developing masks with the sensor pattern and placing and developing photoresist on the silicon wafer. These steps are usually the limiting factor in how high a resolution one can achieve in the non-standard laboratory (usually 10–20 µm). Deposition techniques are needed to place both silicon and non-silicon materials onto the surface of the silicon wafer. Metals are either evaporated or sputtered in a vacuum chamber. Insulators and silicon are deposited on top of metals by vapor deposition reactions from gas reactants such as silane and water vapor for silicon dioxide, silane and heat for polysilicon, and silane and ammonia for silicon nitride.

Many of the devices mentioned in this chapter are a result of silicon processing and micromachining. The idea of a miniaturized sensor system may depend upon the combination of technologies with silicon as its basis. Pumps, fluid reservoirs, valves, sensors, signal processing, and communications are all structures that can be incorporated onto silicon.

11.5 Acoustic sensing

The piezoelectric effect occurs in certain crystal lattices when a potential field is applied to a specific orientation and the crystal lattice physically shifts to form a dipole along this field. If a radio frequency potential is applied, the crystal lattice will begin to oscillate in various patterns. The frequency of oscillation is precise due to the crystal type and its geometry, and it is used in most timing circuits such as watches in the form of a quartz oscillator. In the early 1960s, quartz crystals were investigated as gas and liquid sensors by monitoring oscillation frequency shifts, ΔF, due to mass loading, Δm, on the surface just as the effect of mass loading on any harmonic oscillator (spring or pendulum)

$$\Delta F = (-2.3 \times 10^{-6})F^2 \frac{\Delta m}{A}$$

where F is the fundamental frequency of the oscillator and A is the area of the sensitive region of the crystal. Since the 1960s, many different type of sensors have been made by applying a sorptive layer on top of the crystal to dissolve to bind to analyte molecules in vapor or liquid media [25,26].

The problem with this type of sensor is the selectivity of the sorptive phase. It is nearly impossible to construct a phase that is purely selective for only one analyte. To address this problem, arrays of piezoelectric sensors have been tested to perform multicomponent analysis. By using multiple sensors, each with a different sorptive phase, a response pattern similar to the absorption spectrum of a molecule can be obtained. This response pattern is used to both identify and quantitate a specific analyte in the presence of other responding molecules. As mentioned in section 11.2.2, the most important design factor is that the different sorption phases must have enough uniqueness to discriminate between the analytes present in a mixture.

Using these sensor array design techniques coupled with calibration techniques that decrease collinearity, arrays of piezoelectric sensors can perform favorably. For a group of closely related chlorinated hydrocarbons, an array of nine quartz crystals was designed based on gas chromatographic stationary phases. This array worked well using PLS as the calibration algorithm when compared to multiple linear regression. However, collinearity effects due to a lack of resolution of the array to the analytes increase error of prediction significantly. In this example, the three analytes were dichlorobenzene, dichloropropane, and trichloroethane. Because the analytes are all chlorinated compounds, their interactions with the sorptive layers are similar. Figure 11.6 shows the response patterns of the three compounds to the array. With calibration on two-component mixtures the error of prediction is approximately 5% relative. When all three compounds are present, the error increases to approximately 9% relative.

The figures of merits for this array to three different analyte mixture samples are in Table 11.2. The array has much better selectivity with water and 2-methyl-2-pentanol than it does to the chlorinated hydrocarbon samples. The vapor–solid interaction is much different for these two analytes. The chlorinated hydrocarbons have a high degree of similarity in adsorption which reduces the selectivity of the array to below 0.2 for each analyte. In going from the two-component to the three-component sample set, the selectivity and sensitivity of m-dichlorobenzene and trichloroethane decreases. These collinearity effects directly impact the calibration and prediction performance of the array as outlined in the above paragraph. Therefore array design is critical to ensure success of the array performance as well as using stable noise reducing calibration techniques.

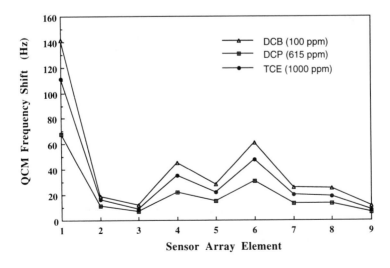

Figure 11.6 Sensor array response pattern of a nine-element QCM array to three chlorinated hydrocarbons: dichlorobenzene (DCB), dichloropropane (DCP) and trichloroethane (TCE).

Table 11.2 Figures of merit for a nine-element QCM sensor array

Calibration set	Sensitivity (Hz/ppm)	Selectivity
2-Methyl-2-pentanol	0.058	0.62
Water vapor	0.010	0.62
m-Dichlorobenzene	0.141	0.17
Trichloroethane	0.023	0.17
m-Dichlorobenzene	0.093	0.13
1,2-Dichloropropane	0.010	0.19
Trichloroethane	0.018	0.15

Surface acoustic wave devices (SAW) are also piezoelectric mass detectors using a different oscillatory motion [27,28]. In this crystal, waves, much like ocean waves transverse the surface of the crystal. The most common type of SAW device available commercially for sensing is the 250 MHz resonator, Figure 11.7 [29]. The surface wave, a Rayleigh wave, is excited by a pair of interdigitated transducers located in the center of the crystal. The wave travels in both directions away from the center. The wave reflects off of the reflection plates back to center of the crystal. This resonating mode increases the signal-to-noise ratio of the device while allowing higher sensitivity due to the high frequency of the harmonic wave.

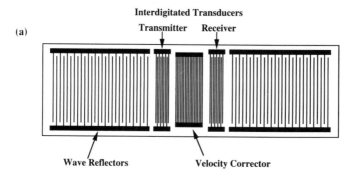

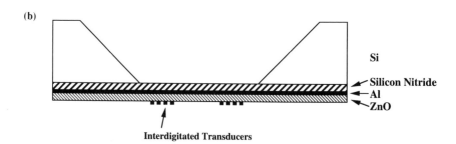

Figure 11.7 (a) Top view of a SAW resonator sensor. (b) Cross-sectional view of a Lamb-wave oscillator.

The response function between frequency shift and applied mass or tension is as follows,

$$\Delta f = (k_1 + k_2)F^2hp - k_2F^2h\left(\frac{4\mu}{V_R^2}\left(\frac{\lambda + \mu}{\lambda + 2\mu}\right)\right) \qquad (11.8)$$

where F is the resonant frequency of the SAW oscillator and coating without sorption, k_1 and k_2 are materials constants for the crystal substrate and the coating film respectively, h is the coating thickness, p is the coating density, λ is the Lame' constant, μ is the shear modulus of the coating and V_R is the Rayleigh wave velocity in the piezoelectric substrate. The first term of this expression represents the effects of mass loading on frequency shift, while the second term represents the effect of elasticity of the coating. The coating elasticity effect can usually be ignored with thin films and older, lower frequency delay lines. However, with the newer, higher frequency resonant devices, this term can be more significant due to the higher sensitivity of the device.

Sensor sensitivity of both QCMs and SAWs are based on their noise characteristics in each application. QCMs at 10 MHz can have noise levels of approximately 0.1–0.5 Hz with detection limits on the order of 10^{-8}–10^{-9} grams or low (10–20) ppm levels of gases. Lower levels in the ppb range can be reached with more rigorous instrumental approaches with QCMs. SAW devices based on the resonator configuration have been a breakthrough in technology for chemical sensing. The sensitivity increases with the square of the frequency of the device. In the past, 50 and 150 MHz SAW delay lines were used with noise levels of 15 Hz or higher. Now, 250 MHz SAW resonators are commercially available with noise levels in the 1–5 Hz range. The detection limits can be routinely extended down to 1 ppm of gas concentration and below that with good instrumental practices. Mass detection of 10^{-10} grams or lower can be sensed.

The use of piezoelectric devices in first-order arrays has been investigated extensively in the past decade. Primarily in the detection of vapors mixtures, sensors arrays of QCMs and SAWs have been studied and applied using pattern recognition and multivariate quantitation. Areas of interest include odor sensing, hazardous vapor monitoring, chemical agent detection for the military and process analytical measurements in the chemical industry. The linear response function is one advantage of this sensor when used in the array format. Linear models such as PCA, PCR and PLS can be used directly.

11.5.1 Lamb-wave devices

A new development in the past few years involving acoustic sensors is the Lamb-wave oscillator [30,31]. It is similar to a SAW device except that the thickness of the piezoelectric substrate is small compared to the ultrasonic wavelength. Wenzel and White micromachined a 3.5 μm thick piezoelectric substrate on silicon. The Lamb-wave oscillator at 5.5 MHz has an acoustic wavelength of 100 μm. The physical action is similar to a flag flying in a breeze. Figure 11.7(b) shows an example of how this sensor is machined out of silicon. Since the entire membrane oscillates, both sides are equally sensitive to mass changes. Therefore, sensing membranes on one side can be isolated from the electronics used to induce oscillation from the other. Another significant feature is the ability to use the devices in liquid media. Sensing is the obvious function of the liquid applications, but due to the flexural properties, the Lamb-wave can also be used as a liquid pump. These devices will undoubtedly play a role in future systems based on SAW technology.

11.6 Metal oxide semiconductor sensors

Electrochemical gas sensors based on metal oxide semiconductors have been in use since the early 1960s. Taguchi first explored the principal using a SnO_2 substrate heated to approximately 400°C. The principal of this sensor is much like a catalytic converter, by adsorbing oxygen and oxidizable gases, reduction and oxidation reactions occur resulting in the exchange of electrons between the surface and the adsorbed molecules. The relationship between semiconductor resistance, R, and analyte (oxidizable gas) partial pressure is as follows:

$$\frac{R}{R_0} = P_{O_2}^{-\beta}(1 + P_i^{n_i})^{\beta} \tag{11.9}$$

where P_{O_2} is the partial pressure of oxygen, P_i is the partial pressure of analyte i, n is a function of analyte–surface interaction kinetics and β describes a surface energy term that is temperature dependent. By measuring the change in conductivity with respect to vapor phase concentration of a gas or vapor, an analytical sensor is produced. To aid in selectivity a variety of metal oxides and dopant materials are used. The sensor response is non-linear with analyte concentration. Various methods have been used to calibrate individual metal oxide sensors. Using polynomial regression only about 80% of the response curve can be modelled. Using non-linear regression with equation (11.9) is possible, but is difficult due to all of the parameters needed to be estimated.

One of the major problems with metal oxide gas sensors is the selectivity issue. Above, we discussed the selectivity issue with piezoelectric crystal sensor based on films of organic polymers or macromolecules. Selectivity was achieved by different functional groups. In metal oxide gas sensors, selectivity is a function of the adsorption or chemisorption of a molecule onto the semiconductor. Therefore, semiconductor stoichiometry and surface temperature play a major role in selective adsorption. To help selectivity of the semiconductor, dopants are added which change the energy required for chemisorption. Dopants range from 0.01 to 5% by weight of metals or oxides such as Pt, Pd, Ag, and V_2O_5.

To look at multicomponent mixtures, arrays of metal oxide gas sensors have been investigated. Two problems exist that are critical to their success. The first is the degree of differential selectivity between sensors. The resolving power of the array is limited by the lack of selectivity between different metal oxide sensors. This results in having to implement more sensors in the array to resolve few compounds. The second issue is the nonlinearity of the sensor response. The use of non-linear first-order arrays requires multivariate non-parametric calibration models. Models such as these require more calibration samples and are more sensitive to noise and sensor error.

Early experiments with first-order arrays of metal oxide semiconductors were based on commerically available Taguchi sensors [32–34]. These sensors are SnO_2 semiconductors with varying dopants for selectivity. Primary investigations were focused on mixtures of carbon monoxide, methane, propane, butane and alcohol vapors. Schierbaum showed an example of the three-dimensional non-linear mapping of responses from an array using two components and varying the concentration from 100% of carbon monoxide to 100% methane. Currently studies investigating the proper use of calibration models are underway. Two techniques for quantitative calibration being used are PPR and MARS. These techniques have proven quite useful in non-linear array calibration by reducing the prediction error by over 50% when compared to standard linear calibration techniques. Neural networks are also promising as methods for calibration and quantitation using arrays of non-linear sensors [35]. These techniques are relatively new and are still being explored.

The sensor analyzer system using first-order arrays based on thin film metal oxide semiconductors is an example of how solid-state sensors can be integrated onto silicon. Using microlithography techniques (Table 11.3), multiple sensors can be constructed in a small amount of space. Wang has shown that integrating eight thin film metal oxide sensors onto a

Table 11.3 Process steps in fabricating integrated sensor array

12 photolithographic processes with 10 masks
8 sputtering processes with 7 different target materials
6 wet chemical etching processes
3 thermal evaporations, 1 electron beam evaporation
2 ion beam milling processes
1 oxidation of silicon
1 boron diffusion into silicon
2 cleaning steps with H_2SO_4, HCl, HF
1 anisotropic etching of silicon (backside process)
wafer dicing and wire-bonding to LCC carrier

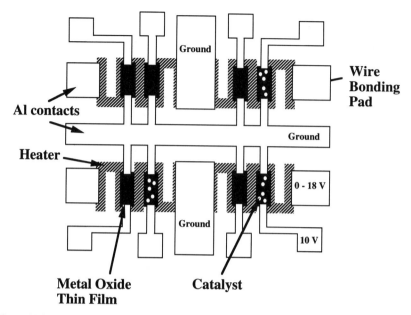

Figure 11.8 Top view of an eight-element integrated metal oxide semiconductor gas sensor array. Sensor elements are approximately 30 μm × 60 μm. The overall chip is 3 mm².

3 × 3 mm silicon chip results in a first-order array in a small implementable package. Figure 11.8 shows the top view of the layout of the sensor array. It has four heating elements with surface temperature ranging from room temperature to 350°C. The base metal oxides are composed of SnO_2, ZnO, and WO_3. Dopant materials are Pd, Ag, and Cr. Results with two-component mixtures of benzene and toluene vapors showed that prediction errors on the order of 20 ppm in the range of 50–500 ppm were achievable with the analogous structure of the two compounds.

The integrated sensor array concept has some interesting challenges that may prove to increase its resolving power. The temperature of the device could be scanned between two temperatures to get a second array order. It is possible that this could become a second-order sensor technique. Presently, the device is not design to scan temperature effectively due to the thermal mass loading of the silicon substrate. Current designs will incorporate a raised bridge on which the sensor will sit without the thermal heat sink of the silicon. This will allow scanning of temperature very rapidly.

11.7 Ion selective electrodes

Ion selective electrodes (ISE) are probably the most abundant sensor found in analytical laboratories. Their response functions follow the

Nerstian relationship between potential and the logarithm of the activities of ions of interest. Again, a non-linear relationship between sensor response and analyte concentration, however, there are some linear portions of the curve in which to perform calibration. Otto and Beebe first showed the application of arrays of ISE using multivariate calibration techniques [36–37]. Otto used linear techniques based on partial least squares regression with both a five and eight element array. The sensors were commercial glass and membrane ISEs. Composition of the four-component mixtures and their prediction error using PLS were calcium (4.54%), magnesium (6.84%), sodium (2.28%) and potassium (1.44%) ions.

Beebe provided the first example of using PPR for non-linear calibration with a five-element ISE array and a two-component mixture of sodium and potassium. Relative prediction errors were 0.4% and 5.3%, respectively, for the two analytes. Based on the results of this array, PPR is now being used for other non-linear array applications such as the metal oxide arrays mentioned above.

11.8 Conclusion

The use of higher order sensor arrays are making an impact on the role of sensors in industrial process control and in stand-alone analyzer systems. The multivariate capability of arrays is necessary to investigate complex samples where many constituents may respond to any one sensor. The more complex the model and system, the more difficulties will arise in analyzer implementation. The technology is still in its infancy and many details need to be addressed. However, the past and present work have proved that the goal is achievable.

References

1. J. Janata, M. Josowicz and D. M. De Vaney, Chemical sensors. *Anal. Chem.* **66** (1994), 207R–228R.
2. J. Janata, *Principles of Chemical Sensors*. Plenum Press, New York, USA, 1989.
3. W. P. Carey, K. R. Beebe, E. Sanchez, *et al.*, Chemometric analysis of multisensor arrays. *Sensors Actuators* **9** (1986), 223–234.
4. J. R. Stetter, P. C. Jurs, and S. L. Rose, Detection of hazardous gases and vapors: pattern recognition analysis of data from an electrochemical sensor array. *Anal. Chem.* **58** (1986), 860–866.
5. W. P. Carey, K. R. Beebe, and B. R. Kowalski, Multicomponent analysis using an array of piezoelectric crystal sensors. *Anal. Chem.* **59** (1987), 1529–1534.
6. D. W. Osten and B. R. Kowalski, Background detection and correction in multicomponent analysis. *Anal. Chem.* **57** (1985), 908–917.
7. J. H. Kalivas and B. R. Kowalski, Compensation for drift and interferences in multicomponent analysis. *Anal. Chem.*, **54** (1982), 560–565.
8. A. Lorber, Error propagation and figures of merit for quantification by solving matrix equations. *Anal. Chem.* **58** (1986), 1167–1172.

9. W. P. Carey, and B. R. Kowalski, Chemical piezoelectric sensor and sensor array characterization. *Anal. Chem.* **58** (1986), 3077–3084.

10. W. P. Carey, K. R. Beebe, B. R. Kowalski, *et al.* Selection of adsorbates for chemical sensor arrays by pattern recognition. *Anal. Chem.* **58** (1986), 149–153.

11. D. M. Haaland and E. V. Thomas, Partial least-squares methods for spectral analyses. 1. Relation to other quantitative calibration methods and the extraction of qualitative information. *Anal. Chem.* **60** (1988), 1193–1202.

12. J. H. Friedman and W. Stuetzle, Projection pursuit regression. *J. Am. Stat. Assoc.* **76** (1981), 817–823.

13. K. R. Beebe and B. R. Kowalski, Nonlinear calibration using projection pursuit regression: Application to an array of ion-selective electrodes. *Anal. Chem.* **60** (1988), 2273–2278.

14. J. H. Friedman, Multivariate adaptive regression splines. *The Annuals of Statistics* **19** (1991), 1–141.

15. T. Nakamoto, K. Fukunishi and T. Moriizumi, Identification capability of odor sensor using quartz-resonator array and neural-network pattern recognition. *Sensors Actuators B* **1** (1990), 473–476.

16. H. Sundgren, I. Winquist, I. Lukkari and I. Lundstrom, Artificial neural networks and gas sensor arrays: quantification of individual components in a gas mixture. *Measurement Sci. Technol.* **2** (1991), 464–469.

17. X. Wang, J. Fang and W. P. Carey, Mixture analysis of organic solvents using nonselective and nonlinear Taguchi gas sensors with artificial neural networks. *Sensors Actuators B* **13** (1993), 455–457.

18. C. Ho, G. D. Christian and E. R. Davidson, Application of the method of rank annihilation to quantitative analysis of multicomponent fluorescence data from the video flurometer. *Anal. Chem.* **50** (1978), 1108–1113.

19. A. Lorber, Quantifying chemical compositions from two-dimensional data arrays. *Analytica Chimica Acta* **164** (1984), 293–297.

20. E. Sanchez and B. R. Kowalski, Generalized rank annihilation factor analysis. *Anal. Chem.* **58** (1986), 496–499.

21. W. P. Carey, M. D. DeGrandpre and B. S. Smith, Polymer-coated cylindrical waveguide absorption sensor for high acidities. *Anal. Chem.* **61** (1989), 1674–1678.

22. R. J. Berman, G. D. Christian and L. W. Burgess, Measurement of sodium hydroxide concentration with a renewable reagent-based fiber-optic sensor. *Anal. Chem.* **62** (1990), 2066.

23. Z. L. Lin, K. S. Booksh, L. W. Burgess and B. R. Kowalski, Second-order fiber optic heavy metal sensor employing second-order tensorial calibration. *Anal. Chem.* **66** (1994), 2552–2560.

24. W. R. Runyan and K. E. Bean, *Semiconductor Integrated Circuit Processing Technology.* Addison-Wesley Publ. Co., New York, USA, 1990.

25. W. J. King, Piezoelectric sorption detector. *Anal. Chem.* **36** (1964), 1735.

26. J. Hlavay and G. G. Guilbault, Applications of the piezoelectric crystal detector in analytical chemistry. *Anal. Chem.* **49** (1977), 1890.

27. J. W. Grate, S. L. Rose-Pehrsson, D. L. Venezky, *et al.*, Smart sensor system for trace organophosphorus and organosulfur vapor detection employing a temperature-controlled array of surface acoustic wave sensors, automated sample preconcentration, and pattern recognition. *Anal. Chem.* **65** (1993), 1868–1881.

28. S. L. Rose-Pehrsson, J. W. Grate, D. S. Ballantine and P. C. Jurs, Detection of hazardous vapors including mixtures using pattern recognition analysis of responses from surface acoustic wave devices. *Anal. Chem.* **60** (1988), 2801–2811.

29. W. D. Bowers, R. L. Chuan and T. M. Duong, A 200 MHz surface acoustic wave resonator mass microbalance. *Rev. Sci. Instruments* **62** (6) (1991), 1624.

30. S. W. Wenzel and R. M. White, A multisensor employing an ultrasonic Lamb-wave oscillator. *IEEE Trans. Electron Devices* **35** (6) (1988), 735.

31. J. W. Grate, S. W. Wenzel and R. M. White, Flexural plate wave devices for chemical analysis. *Anal. Chem.* **63** (1991), 1552–1561.

32. W. P. Carey and S. Yee, Calibration of nonlinear solid-state sensor arrays using multivariate regression techniques. *Sensors Actuators B* **9** (1992), 113–122.

33. C. Hierold and R. Muller, Quantitative analysis of gas mixtures with non-selective gas sensors. *Sensors Actuators* **17** (1989), 587–592.
34. K. D. Schierbaum, U. Weimar and W. Gopel, Multicomponent gas analysis: an analytical chemistry approach applied to modified SnO_2 sensors. *Sensors Actuators B* **2** (1990), 71–78.
35. X. Wang, J. Fang, P. Carey and S. Yee, Mixture analysis of organic solvents using nonselective and nonlinear Taguchi gas sensors with artificial neural networks. *Sensors Actuators B* **13–14** (1993), 455–457.
36. M. Otto and J. D. R. Thomas, Model studies on multiple channel analysis of free magnesium, calcium, sodium, and potassium at physiological concentration levels with ion-selective electrodes. *Anal. Chem.* **57** (1985), 2647–2651.
37. K. R. Beebe, D. Uerz, J. Sandifer and B. R. Kowalski, Sparingly selective ion-selective electrode arrays for multicomponent analysis. *Anal. Chem.* **60** (1988), 66–71.

Index